BOOK TWO

Exploring Modern Mathematics

BOOK TWO

Exploring Modern Mathematics

MERVIN L. KEEDY

PATRICIA L. JOHNSON

STANLEY A. SMITH

RICHARD E. JAMESON

Holt, Rinehart and Winston, Inc.

New York, Toronto, London, Sydney

ABOUT THE AUTHORS

Mervin L. Keedy

formerly Associate Director of the University of Maryland Mathematics Project (UMMaP); subsequently served as Supervisor of Junior High School Mathematics and Science in Baltimore County Public Schools, Maryland; is now Professor of Mathematics at Purdue University, Lafayette, Indiana.

Patricia L. Johnson

formerly a contributing teacher in the University of Maryland Mathematics Project, and Supervisor of Mathematics, Montgomery County Public Schools, Maryland.

Stanley A. Smith

formerly a teacher of junior high school mathematics and science and senior high school mathematics; is now Supervisor of High School Mathematics in the Baltimore County Public Schools, Maryland.

Richard E. Jameson

formerly one of the participating teachers in the University of Maryland Mathematics Project; is now a mathematics teacher at Newton South High School, Newton, Massachusetts.

SBN: 03-084094-5 1234567890 19 987654321

Preface

This two-book series is based upon the works of the University of Maryland Mathematics Project (Jr. H. S.) and the School Mathematics Study Group. Mindful of the enormous changes the 1960's wrought in school mathematics curriculums but equally aware of traditional requirements, the authors have framed a completely new program for today's students and teachers. They have created a well-developed "second generation" mathematics program – one which incorporates the best of the old and the best of the new.

The authors' approach in **Exploring Modern Mathematics** is embodied in the *Let's Explore* lessons that introduce new material. More than a unique teaching device, *Let's Explore* is a tool of discovery that involves the student directly. With it, he works from the known to the unknown, looking for similarities and differences, exploring and thinking. Introduced by the authors in the early 1960's, the *Let's Explore* technique has been used successfully by hundreds of thousands of students for nearly ten years.

Provision for individual differences is an integral part of **Exploring Modern Mathematics.** In addition to the numerous practice, review, and testing exercises, the authors include diagnostic exercises regularly throughout the program. *Unexplored Territory* exercises challenge the student and extend the experiences given in the basic development.

Exploring Modern Mathematics differs from other modern junior high school programs in several important respects: the authors omit or de-emphasize those topics introduced in recent years that do not really contribute to the student's understanding or skill. Unnecessary vocabulary and terminology has been stripped away, and words are employed in such a way that the student always concentrates on mathematics rather than on reading and language.

Too many modern junior high programs assume that students are proficient in computing with whole numbers, fractions, decimals, and per cents. This is not so in **Exploring Modern Mathematics.** The authors supply more than ample material for developing and maintaining the student's computational skills. Throughout the program verbal problems are included, often in the regular lessons as well as in full-page sets on *Problem Solving* pages.

The authors wish to thank the countless teachers, supervisors, and administrators who have received the first editions so well, and whose suggestions have so greatly influenced this revision.

Contents

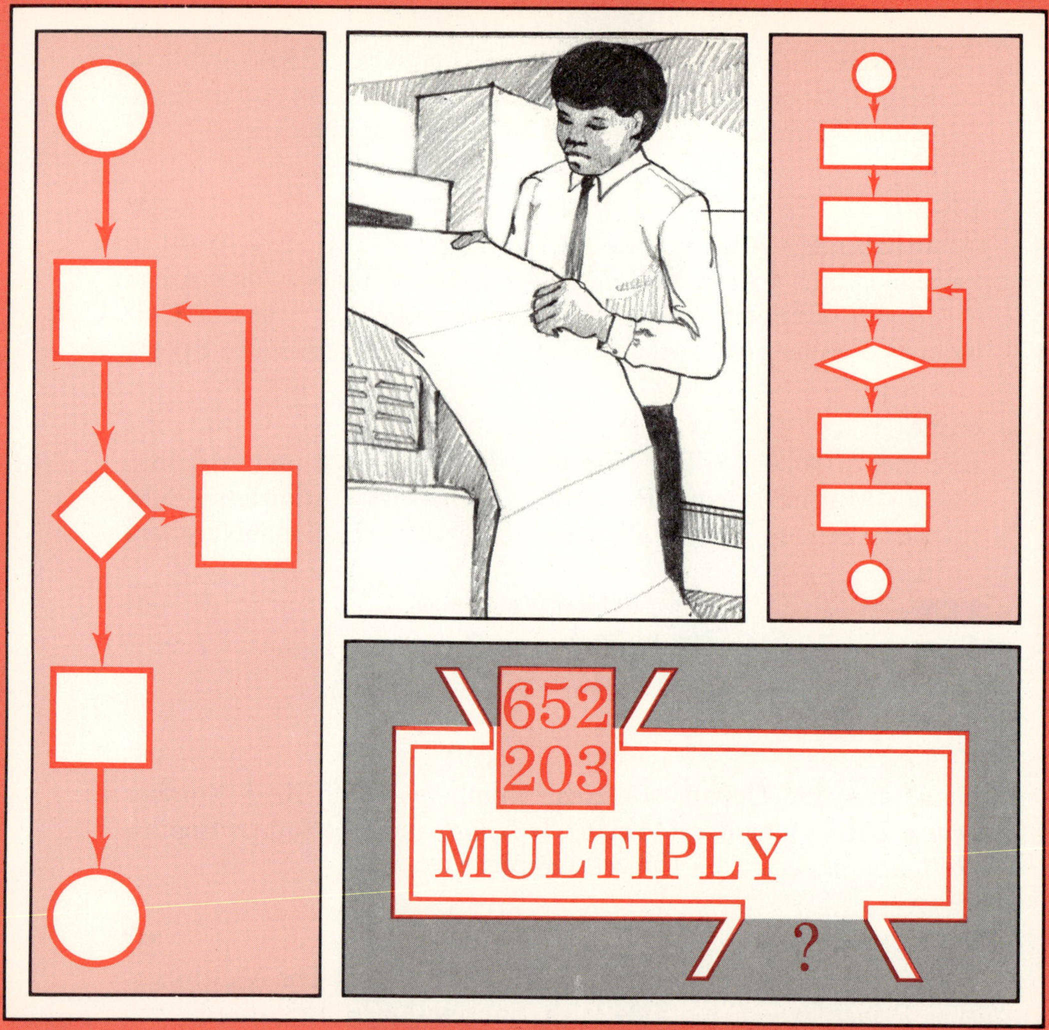

CHAPTER ONE

Computations and Flow Charts

WHOLE NUMBERS AND OPERATIONS

We use natural numbers for counting. The natural numbers together with 0 make up the set of whole numbers.

$\{1, 2, 3, 4, \ldots\}$ ⟵ *natural numbers*

$\{0, 1, 2, 3, 4, \ldots\}$ ⟵ *whole numbers*

1. Which whole number is not a natural number?

2. Are all natural numbers also whole numbers?

3. Addition is an operation on whole numbers. Name some other operations on whole numbers.

In an operation we start with two numbers. We have a rule to find a third number. There cannot be more than one answer. Sometimes there is no answer.

4. Which of the following have no whole number answer?

a. $18 - 9$	**b.** $19 + 4$	**c.** 5×11
d. $20 \div 5$	**e.** $4 \div 7$	**f.** $19 - 74$

5. Which operations on whole numbers always have whole number answers?

EXERCISES

Which of the following have no whole number answer?

1. $15 + 9$	**2.** 6×5	**3.** $3 \div 5$	**4.** $18 - 23$
5. $42 \div 6$	**6.** $19 \div 3$	**7.** 8×7	**8.** $21 - 34$
9. $19 + 0$	**10.** $39 \div 6$	**11.** $21 - 31$	**12.** 4×8
13. $9 \div 11$	**14.** $12 + 15$	**15.** $23 \div 9$	**16.** 2×0

17. Which operations on whole numbers do *not* always have whole number answers?

DO YOU REMEMBER?

1. Race time! Add as fast as you can.

3 + 4	4 + 2	5 + 4	6 + 2
2 + 1	2 + 2	6 + 3	2 + 7
1 + 8	1 + 3	5 + 3	4 + 6
2 + 0	4 + 3	2 + 8	2 + 3
6 + 4	3 + 0	4 + 8	6 + 9
7 + 4	1 + 1	9 + 4	9 + 8
8 + 9	5 + 7	7 + 9	5 + 8
3 + 7	6 + 6	0 + 4	3 + 3
5 + 5	7 + 8	8 + 2	6 + 7
6 + 8	5 + 6	9 + 7	4 + 9
4 + 7	0 + 6	4 + 4	0 + 0
1 + 4	9 + 9	1 + 6	8 + 4
9 + 2	2 + 9	3 + 9	0 + 7
5 + 9	3 + 9	4 + 6	7 + 3
8 + 1	7 + 5	8 + 6	1 + 9
4 + 3	0 + 0	3 + 8	2 + 7

2. Race time! Subtract as fast as you can.

2 − 1	12 − 8	10 − 3	13 − 4
14 − 5	13 − 9	11 − 3	8 − 7
7 − 4	15 − 8	8 − 2	11 − 8
10 − 8	8 − 4	7 − 2	5 − 2
9 − 3	11 − 6	10 − 9	7 − 5
11 − 5	7 − 6	9 − 8	8 − 6
10 − 2	3 − 3	3 − 2	9 − 1
9 − 6	12 − 4	2 − 0	11 − 7
14 − 8	10 − 7	15 − 6	15 − 9
18 − 9	4 − 2	6 − 3	13 − 6
16 − 9	11 − 9	12 − 9	11 − 4
17 − 8	15 − 8	8 − 4	16 − 8
8 − 7	7 − 0	10 − 6	12 − 3
6 − 4	12 − 6	7 − 5	13 − 7
11 − 2	9 − 2	9 − 4	6 − 1
13 − 5	17 − 8	11 − 7	17 − 8

ALGORITHMS

An *algorithm* is a procedure for doing a computation. You already know many algorithms.

1. Here is an addition where we use the ordinary algorithm. Explain the steps.

$$\begin{array}{r} {\scriptstyle 1}\ \ \ \\ 973 \\ 422 \\ 568 \\ \hline 3 \end{array} \qquad \begin{array}{r} {\scriptstyle 1\ 1}\ \ \\ 973 \\ 422 \\ 568 \\ \hline 63 \end{array} \qquad \begin{array}{r} {\scriptstyle 1\ 1}\ \ \\ 973 \\ 422 \\ 568 \\ \hline 1,963 \end{array}$$

2. Complete: In the ordinary addition algorithm we first add ____, then ____, then ____, and so on.

3. Add.

a. $\begin{array}{r} 521 \\ +\,260 \\ \hline \end{array}$ b. $\begin{array}{r} 297 \\ +\,446 \\ \hline \end{array}$ c. $\begin{array}{r} 487 \\ 63 \\ 239 \\ \hline \end{array}$ d. $\begin{array}{r} 21,632 \\ 8,437 \\ 923 \\ 1,230 \\ \hline \end{array}$

4. Here is a subtraction where we use the ordinary algorithm. Explain the steps.

$$\begin{array}{r} 857 \\ -325 \\ \hline 2 \end{array} \qquad \begin{array}{r} 857 \\ -325 \\ \hline 32 \end{array} \qquad \begin{array}{r} 857 \\ -325 \\ \hline 532 \end{array}$$

5. Here is another subtraction. Explain the renaming.

$$\begin{array}{r} {\scriptstyle 5\ 12} \\ 7\not{6}\not{2} \\ -475 \\ \hline 7 \end{array} \qquad \begin{array}{r} {\scriptstyle 15\ \ \ } \\ {\scriptstyle 6\ \not{5}\ 12} \\ 7\not{6}\not{2} \\ -475 \\ \hline 87 \end{array} \qquad \begin{array}{r} {\scriptstyle 15\ \ \ } \\ {\scriptstyle 6\ \not{5}\ 12} \\ \not{7}\not{6}\not{2} \\ -475 \\ \hline 287 \end{array}$$

6. Subtract.

a. $\begin{array}{r} 865 \\ -\,542 \\ \hline \end{array}$ b. $\begin{array}{r} 783 \\ -\,466 \\ \hline \end{array}$ c. $\begin{array}{r} 923 \\ -\,568 \\ \hline \end{array}$ d. $\begin{array}{r} 5,280 \\ -\,3,974 \\ \hline \end{array}$

EXERCISES

Add.

1.	54	2.	86	3.	28	4.	56
	+ 23		+ 13		+ 46		+ 15

5.	235	6.	567	7.	607	8.	817
	+ 128		+ 435		+ 24		+ 435

9.	56	10.	97	11.	742	12.	3,826
	24		30		67		908
	83		62		824		435

13.	6,024	14.	832	15.	18,654	16.	1,654
	271		568		9,673		83
	1,936		395		35		749
			287		2,002		8,237

Subtract.

17.	56	18.	87	19.	943	20.	1,453
	− 34		− 29		− 826		− 862

21.	1,904	22.	2,065	23.	5,682	24.	7,823
	− 1,325		− 1,347		− 2,097		− 5,674

25.	2,943	26.	9,560	27.	5,438	28.	7,943
	− 1,457		− 3,782		− 3,729		− 6,249

Solve these problems.

29. George kept a record of the number of problems missed on his math drills. On the first 5 drills he missed 1, 3, 2, 6, and 4. Each drill had 10 problems on it. How many did he get right in all?

30. Bob has $84. He needs $152 to buy new skis. How much more does he need?

31. Joe sold 289 newspapers on Monday and 305 on Tuesday. Each paper costs 10¢. How much money did he collect?

A NOMOGRAPH

Sums and differences of whole numbers can be found on a simple computer called a *nomograph.*

LET'S EXPLORE

1. Using graph paper, draw three vertical segments. Space them equally and make them the same length. Label them *A, C,* and *B.*

2. Label the scales as shown to form a nomograph.

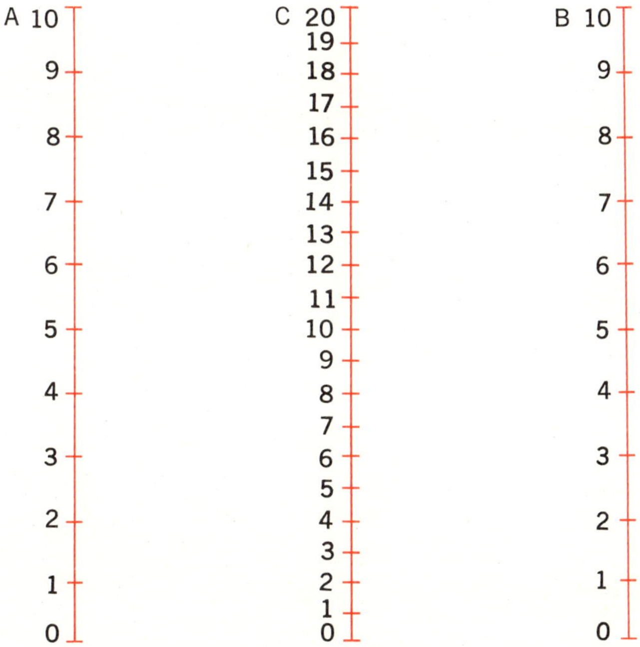

3. a. Choose any two whole numbers from 0 to 10.

   b. Locate one number on the *A* scale and the other number on the *B* scale.

   c. Connect these two points with a segment.

   d. Where does this segment cross the *C* scale?

   e. What is the sum of your two numbers?

4. Let's add 5 and 2 on the nomograph.

   a. Locate 5 on the *A* scale and 2 on the *B* scale.

   b. Connect these two points with a segment.

   c. Where does the segment cross the *C* scale?

   d. What is 5 + 2?

5. Add. Use the nomograph.

   a. 8 + 5 b. 9 + 2 c. 3 + 7 d. 4 + 9

6. Let's subtract 6 from 9 on the nomograph.

   a. Locate 9 on the *C* scale and 6 on the *A* scale.

   b. Connect these two points with a segment.

   c. Where does the segment cross the *B* scale.

   d. What is 9 − 6?

7. Extend the *A* and *B* scales on your nomograph to 20. Do these subtractions on your nomograph.

   a. 8 − 2 b. 9 − 5 c. 17 − 4 d. 18 − 9

EXERCISES

Add. Use the nomograph.

1. 6 + 4	2. 10 + 8	3. 5 + 4	4. 6 + 10
5. 5 + 9	6. 0 + 4	7. 3 + 2	8. 7 + 4
9. 6 + 5	10. 8 + 0	11. 3 + 6	12. 1 + 8

Subtract. Use the nomograph.

13. 10 − 4	14. 6 − 2	15. 5 − 3	16. 18 − 3
17. 20 − 11	18. 19 − 5	19. 17 − 14	20. 16 − 12
21. 15 − 11	22. 10 − 6	23. 16 − 5	24. 9 − 8

DO YOU REMEMBER?

1. Race time! Multiply as fast as you can.

$0 \cdot 3$	$5 \cdot 2$	$4 \cdot 9$	$1 \cdot 5$
$5 \cdot 8$	$6 \cdot 9$	$2 \cdot 8$	$2 \cdot 4$
$7 \cdot 2$	$3 \cdot 1$	$3 \cdot 5$	$5 \cdot 1$
$2 \cdot 3$	$8 \cdot 4$	$4 \cdot 6$	$6 \cdot 7$
$8 \cdot 1$	$7 \cdot 6$	$7 \cdot 3$	$6 \cdot 6$
$5 \cdot 4$	$2 \cdot 5$	$9 \cdot 4$	$9 \cdot 7$
$6 \cdot 0$	$0 \cdot 0$	$9 \cdot 8$	$3 \cdot 9$
$5 \cdot 7$	$5 \cdot 3$	$5 \cdot 6$	$9 \cdot 9$
$8 \cdot 3$	$7 \cdot 5$	$4 \cdot 7$	$6 \cdot 4$
$4 \cdot 4$	$8 \cdot 8$	$0 \cdot 9$	$7 \cdot 7$
$1 \cdot 0$	$6 \cdot 8$	$9 \cdot 5$	$0 \cdot 7$
$6 \cdot 2$	$9 \cdot 6$	$3 \cdot 7$	$4 \cdot 3$
$7 \cdot 8$	$2 \cdot 7$	$5 \cdot 5$	$3 \cdot 3$
$9 \cdot 3$	$6 \cdot 3$	$4 \cdot 2$	$1 \cdot 8$
$7 \cdot 1$	$8 \cdot 9$	$2 \cdot 6$	$7 \cdot 9$
$5 \cdot 5$	$0 \cdot 2$	$7 \cdot 6$	$8 \cdot 5$

2. Race time! Divide as fast as you can.

$8 \div 1$	$36 \div 6$	$54 \div 6$	$4 \div 1$
$0 \div 5$	$2 \div 1$	$12 \div 3$	$45 \div 9$
$6 \div 3$	$25 \div 5$	$56 \div 7$	$18 \div 3$
$12 \div 4$	$64 \div 8$	$63 \div 9$	$24 \div 4$
$15 \div 5$	$81 \div 9$	$14 \div 2$	$40 \div 5$
$8 \div 1$	$5 \div 5$	$0 \div 8$	$15 \div 3$
$35 \div 7$	$9 \div 9$	$48 \div 8$	$0 \div 2$
$9 \div 3$	$0 \div 3$	$27 \div 9$	$30 \div 6$
$7 \div 7$	$24 \div 6$	$32 \div 8$	$42 \div 6$
$16 \div 4$	$35 \div 5$	$28 \div 7$	$5 \div 1$
$48 \div 6$	$7 \div 1$	$0 \div 4$	$8 \div 2$
$3 \div 1$	$0 \div 9$	$9 \div 1$	$18 \div 9$
$18 \div 6$	$72 \div 8$	$10 \div 2$	$21 \div 7$
$21 \div 7$	$12 \div 6$	$18 \div 2$	$24 \div 6$
$63 \div 7$	$56 \div 8$	$42 \div 7$	$49 \div 7$
$42 \div 6$	$81 \div 9$	$16 \div 4$	$21 \div 3$

MULTIPLICATION

Each student in Chuck's science class brought in 12 rocks for the geology unit. There were 32 students in the class.

1. Here is how Chuck found the number of rocks that were brought to class. Explain each step.

$$\begin{array}{r} 32 \\ \times\ 12 \\ \hline 64 \end{array} \qquad \begin{array}{r} 32 \\ \times\ 12 \\ \hline 64 \\ 320 \\ \hline \end{array} \qquad \begin{array}{r} 32 \\ \times\ 12 \\ \hline 64 \\ 320 \\ \hline 384 \end{array}$$

2. The multiplication below shows how the distributive property is used. Copy and complete.

$$\begin{array}{rl} 426 & \\ \times\quad 7 & \\ \hline 42 & 7 \cdot \underline{\qquad} \\ 140 & 7 \cdot \underline{\qquad} \\ 2{,}800 & 7 \cdot \underline{\qquad} \\ \hline 2{,}982 & 7 \cdot (400 + \underline{\qquad} + \underline{\qquad}) \end{array}$$

3. Here is another multiplication. Explain each step.

$$\begin{array}{r} 54 \\ \times\ 37 \\ \hline 378 \end{array} \qquad \begin{array}{r} 54 \\ \times\ 37 \\ \hline 378 \\ 162 \\ \hline \end{array} \qquad \begin{array}{r} 54 \\ \times\ 37 \\ \hline 378 \\ 162 \\ \hline 1{,}998 \end{array}$$

4. Explain each step in these multiplications.

```
  2,158                       5,372
 × 476                     × 2,635
 12948                       26860
15106                       16116
8632                       32232
1027208, or 1,027,208     10744
                          14155220, or 14,155,220
```

5. Here is a multiplication with zeros. Explain each method.

```
   293              293              293
 × 405            × 405            × 405
  1465             1465             1465
  000            11720            1172
1172             118665 or        118665 or
118665 or          118,665          118,665
   118,665
```

EXERCISES

Multiply.

1. 56 × 24	2. 87 × 19	3. 60 × 42	4. 823 × 16
5. 930 × 56	6. 438 × 20	7. 874 × 93	8. 143 × 342
9. 835 × 624	10. 423 × 839	11. 1,624 × 634	12. 9,365 × 403
13. 1,642 × 2,356	14. 2,050 × 1,975	15. 8,354 × 4,883	16. 7,235 × 1,046
17. 9,006 × 7,512	18. 3,692 × 5,930	19. 3,333 × 2,777	20. 2,865 × 5,076
21. 5,069 × 5,069	22. 3,889 × 2,567	23. 1,158 × 8,511	24. 7,676 × 6,767

Solve these problems.

25. Tracey bought 12 cases of soda for the class party. Each case has 24 bottles. How many bottles did she buy?

26. There are 42 medical centers in a large city. They average 321 patients per day. About how many people receive medical help each day?

27. A school photographer can take about 220 pictures per day. About how many pictures will he take during a work year of 260 days?

Let's Practice

Recall that if cross products are the same, then fractional numerals are equivalent.

Example $\frac{9}{12} = \frac{3}{4}$ since $9 \times 4 = 3 \times 12$.

Which pairs are equivalent?

1. $\frac{6}{8}, \frac{9}{12}$
2. $\frac{3}{5}, \frac{6}{9}$
3. $\frac{3}{8}, \frac{6}{16}$
4. $\frac{0}{6}, \frac{0}{10}$
5. $\frac{2}{7}, \frac{6}{21}$
6. $\frac{2}{9}, \frac{14}{63}$
7. $\frac{3}{5}, \frac{5}{10}$
8. $\frac{7}{8}, \frac{14}{16}$

The prime factorization of 15 is $5 \cdot 3$. Find the prime factorization of each.

9. 10
10. 6
11. 8
12. 12
13. 24
14. 33
15. 35
16. 40
17. 18
18. 25
19. 28
20. 50
21. 42
22. 63
23. 51

UNEXPLORED TERRITORY

A secretary said that the digits 0 through 9 would naturally be arranged like this.

8, 5, 4, 9, 1, 7, 6, 3, 2, 0 Why?

COMPUTERS – INPUTS AND OUTPUTS

A nomograph is one kind of simple computer. Modern electronic computers can do calculations at amazing speeds.

The *input* is the information that goes into the computer.

The *program* is the procedure that the computer follows.

The *output* is the answer that the computer gives.

1. Suppose 25 and 68 are *input* and the program is to add. What is the *output*?

2. What is the output?

a.

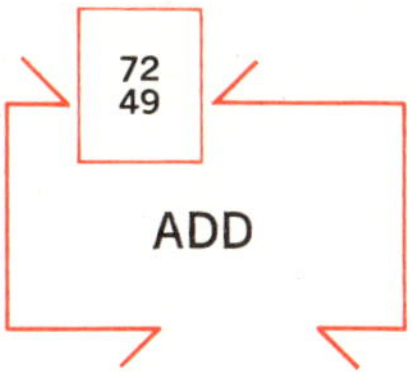

b.

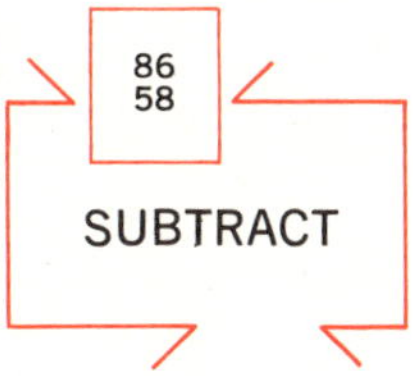

c.

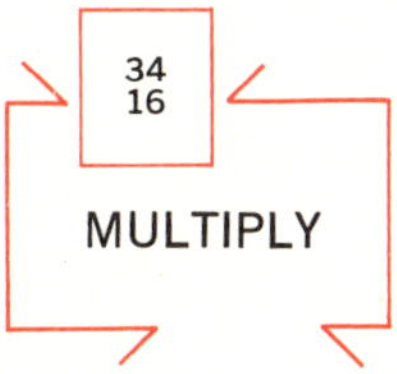

d.

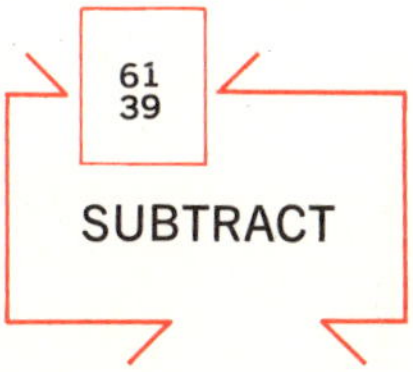

e.

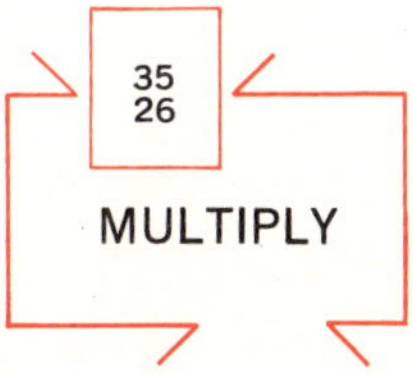

f.

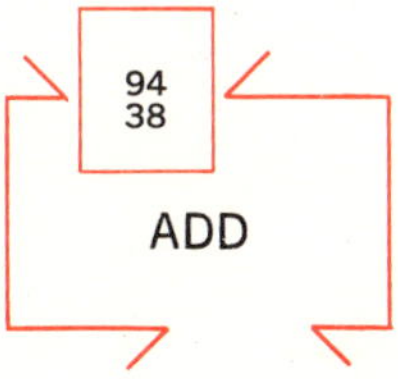

What is the output?

1.

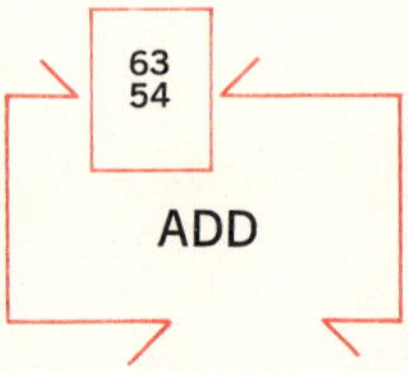

2.

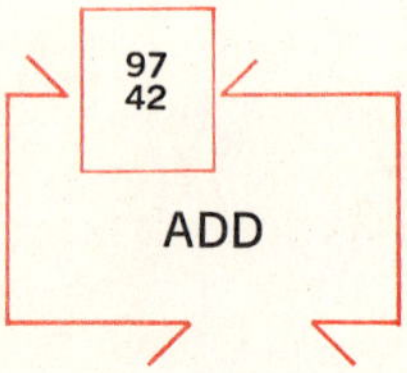

3.

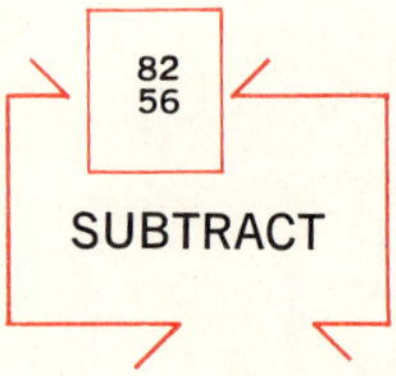

4.

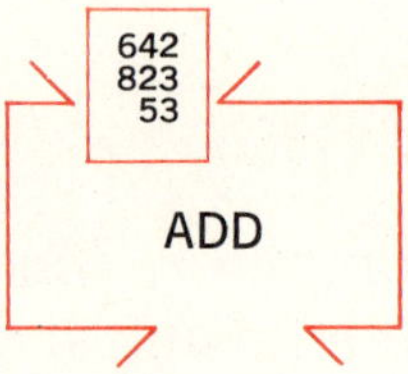

5.

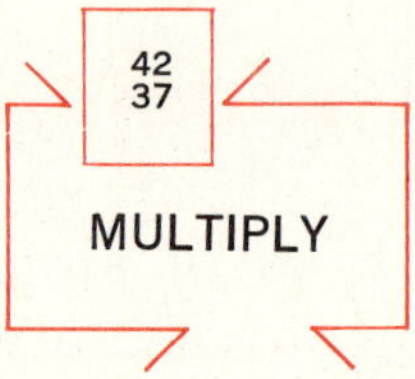

6.

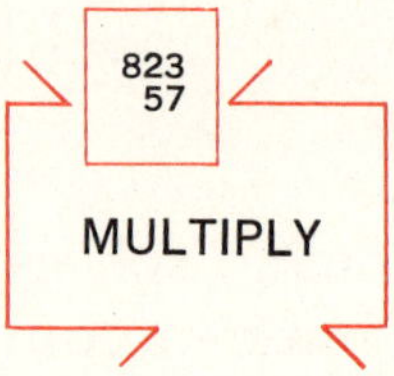

7.

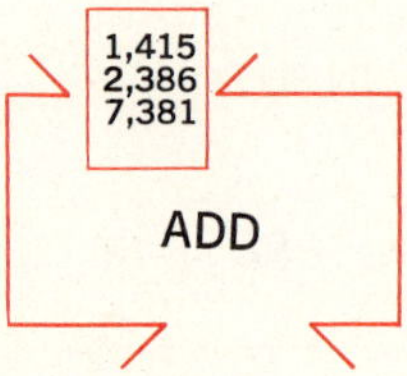

8.

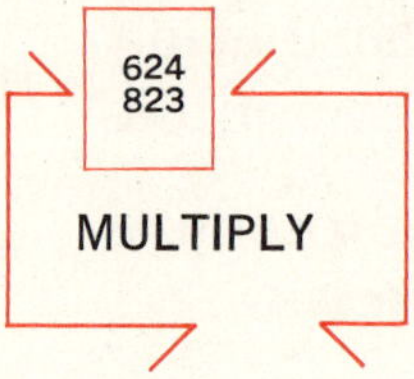

9.

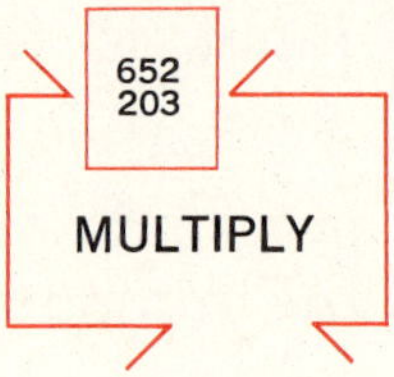

10.

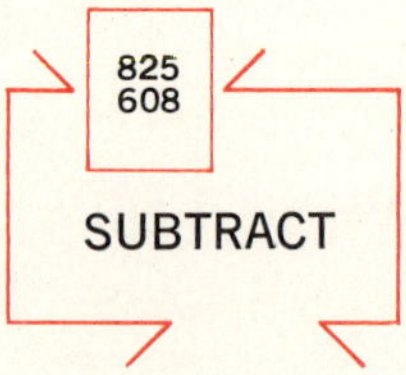

11.

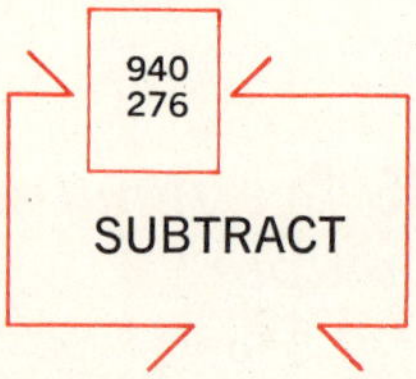

12.

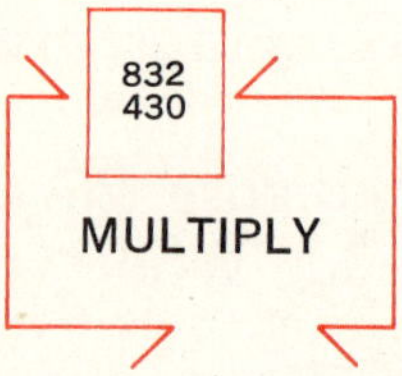

UNEXPLORED TERRITORY

Toothpick trouble! Arrange 24 toothpicks to make 9 squares. Remove 8 to leave only 2 squares.

PROBLEM SOLVING

1. Coach Rittler said that the average attendance at a basketball game was 218. What was the total attendance for the 12 games?

2. Perry Hall Junior High received a shipment of 235 books. They expect 7 more shipments of the same size.

   a. How many books did they buy in all?

   b. What was the total cost of the books?

3. Ted sold 18 boxes of Christmas candy for 75¢ a box. How much did he receive altogether?

4. At the factory, Mr. Hunt packs 50 plastic cups in each box. On an average day, he packs 125 boxes. How many cups does he handle on an average day?

5. Mr. Sanchez ordered 150,000 fliers for the national convention of math teachers. They cost 2¢ each. Find the total cost.

6. The Statue of Liberty is 305 feet tall while the Eiffel Tower in Paris is 984 feet tall. How much taller is the Eiffel Tower?

7. The attendance at Hempstead High's football games were 234, 318, 362, 291, and 283. Find the total attendance.

FLOW CHARTING

The computer operator uses a *flow chart* to outline the steps that a computer follows. This simple flow chart tells how to call a friend on the telephone.

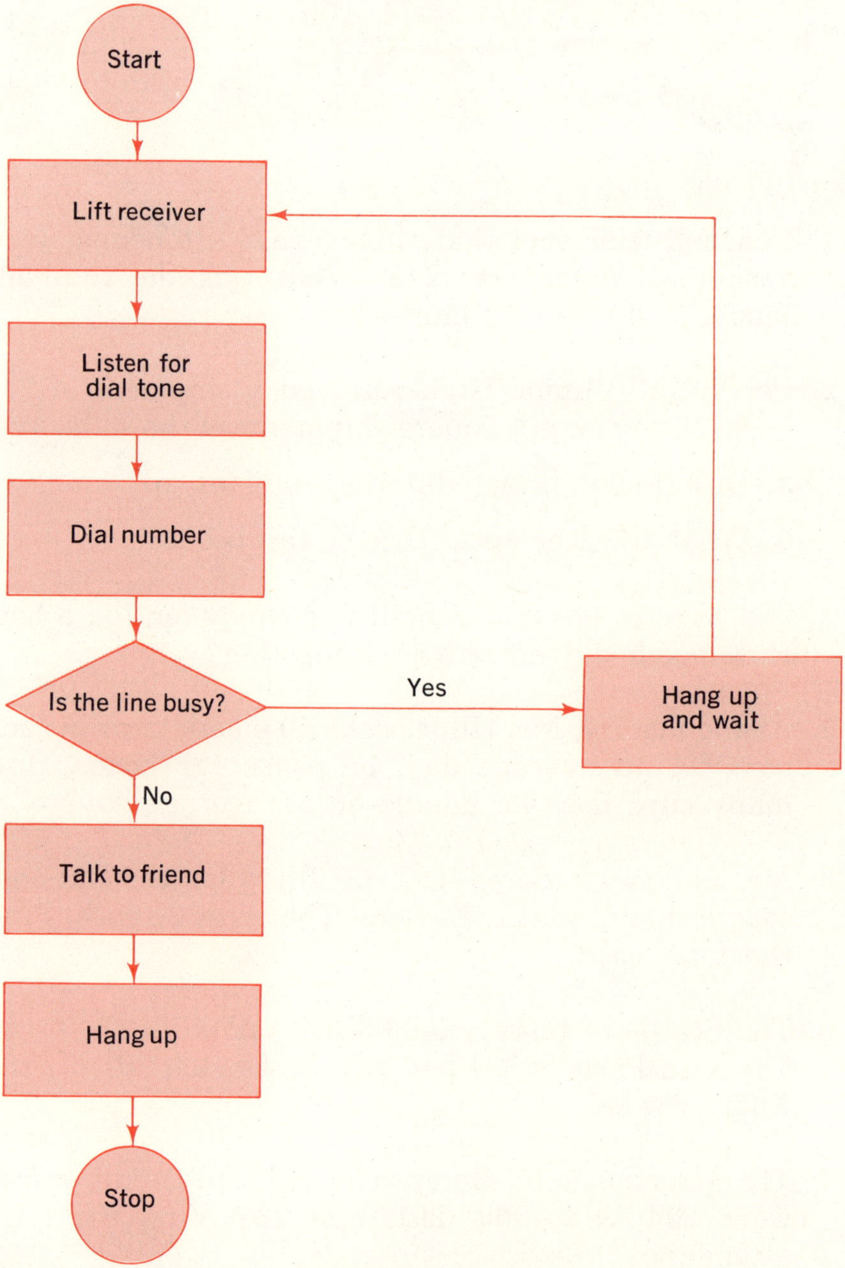

1. Instructions are given in rectangles. List the first three instructions in the flow chart on page 14.

2. Yes or no decisions are given in diamonds.

   a. What decision is shown in the chart on page 14?

   b. What happens if the answer is yes?

   c. What happens if the answer is no?

3. What is the last instruction in the chart?

4. This flow chart tells how to cross a street without a light.

   a. What is the first instruction?

   b. What decision is shown?

   c. What happens if the answer is yes?

   d. What happens if the answer is no?

   e. What is the last instruction?

5. a. List the instructions that would go in a flow chart for crossing a street with a traffic light.

   b. Which two steps would be circled?

   c. What yes or no decision should be included?

   d. Which symbol do decisions get?

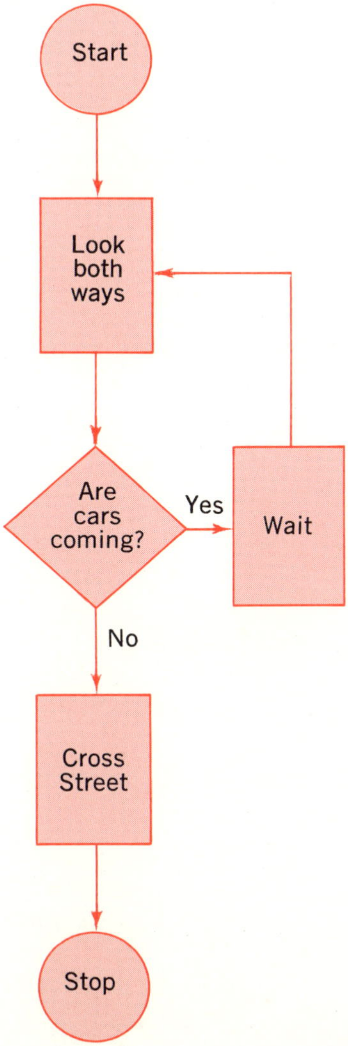

EXERCISES

1. Draw the symbol for the first and last steps in any flow chart.

2. Draw the symbol for an instruction.

3. Draw the symbol for a yes or no decision.

4. Here is a flow chart telling how to brush your teeth.

   a. What is the first instruction?

   b. What is the third instruction?

   c. What happens if your teeth are not clean?

   d. What is the last instruction?

5. a. Draw your own flow chart showing how to wash your face.

   b. What is your first instruction?

   c. What is your third instruction?

   d. What happens if your face is not clean?

   e. What is your last instruction?

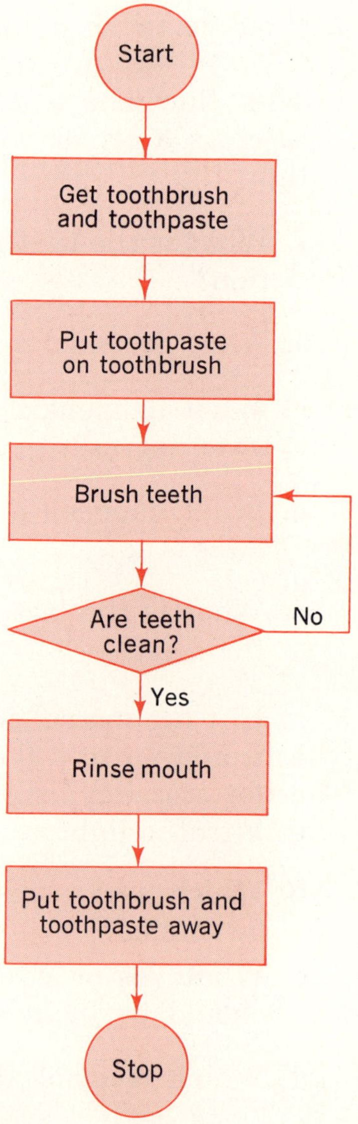

6. Here is a flow chart for finding the reciprocal of a number named by a fractional numeral.

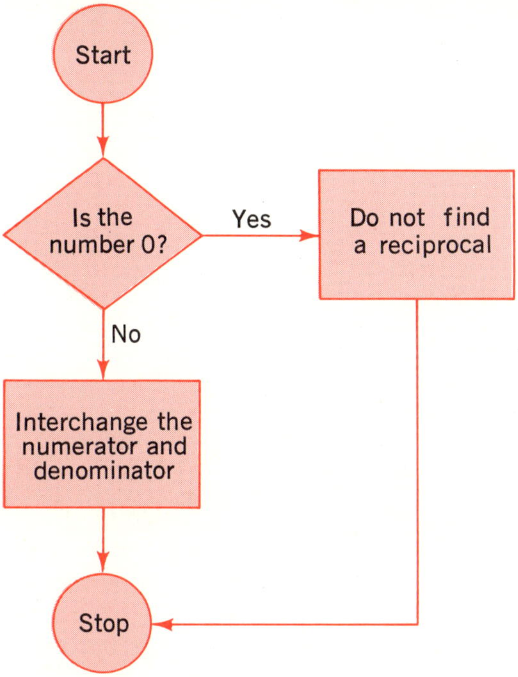

Examples

1. The reciprocal of $\frac{2}{3}$ is $\frac{3}{2}$.

2. The reciprocal of $\frac{1}{4}$ is $\frac{4}{1}$.

Use the flow chart to find the reciprocal of each.

a. $\frac{5}{6}$	**b.** $\frac{18}{9}$	**c.** $\frac{0}{3}$	**d.** $\frac{10}{35}$	**e.** $\frac{7}{5}$
f. $\frac{3}{7}$	**g.** $\frac{8}{5}$	**h.** $\frac{2}{1}$	**i.** $\frac{5}{16}$	**j.** $\frac{1}{15}$

Let's Practice

The multiples of 6 are 6, 12, 18, 24, 30, 36, 42, . . .
The multiples of 8 are 8, 16, 24, 32, 40, 48, . . .
The *Least Common Multiple* of 6 and 8 is 24.

Find the least common multiple of each pair.

1. 4, 8	**2.** 6, 9	**3.** 2, 3	**4.** 7, 11
5. 5, 13	**6.** 6, 10	**7.** 2, 8	**8.** 5, 15
9. 6, 4	**10.** 10, 15	**11.** 9, 12	**12.** 14, 21

DIVISION

Here is a flow chart showing one way to do a whole number division problem with remainders.

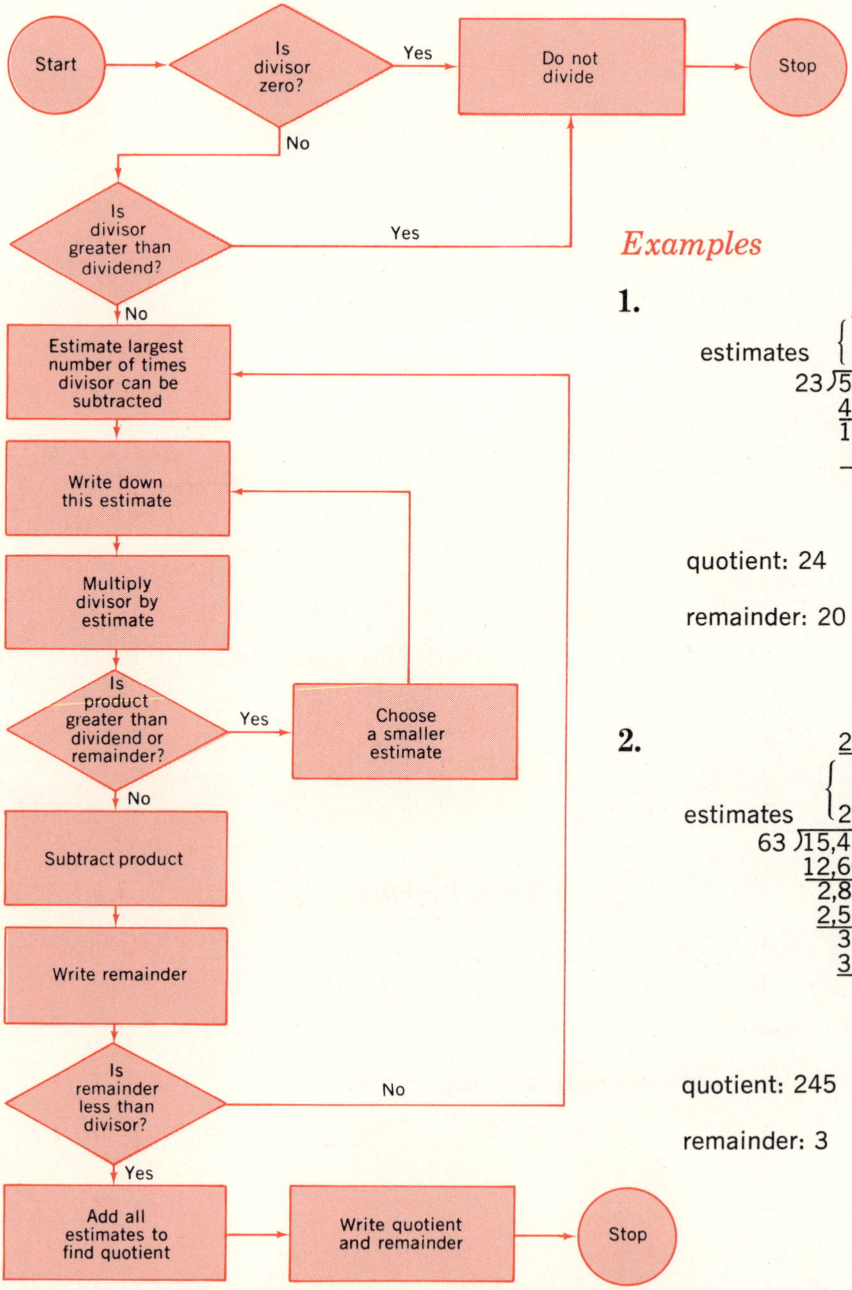

Examples

1.

```
                 24
estimates  {      4
           {     20
             23)572
                460
                112
                 92
                 20
```

quotient: 24

remainder: 20

2.

```
                 245
           {       5
           {      40
estimates  {     200
          63 )15,438
              12,600
               2,838
               2,520
                 318
                 315
                   3
```

quotient: 245

remainder: 3

1. What should you do if the divisor is zero?

2. Suppose the divisor is larger than the dividend. What happens?

3. What happens if the product of an estimate and the divisor is greater than the dividend or remainder?

4. Suppose the remainder is greater than the divisor. What is the next instruction?

5. Suppose the remainder is less than the divisor. What is the next instruction?

6. Follow the flow chart and divide.

a. $18\overline{)354}$	b. $36\overline{)495}$	c. $0\overline{)279}$
d. $72\overline{)5,942}$	e. $83\overline{)24}$	f. $997\overline{)6,582}$

EXERCISES

Pretend you are a computer. Follow the flow chart to do these divisions.

1. $9\overline{)522}$	2. $8\overline{)2,568}$	3. $43\overline{)2,276}$
4. $182\overline{)133}$	5. $54\overline{)784}$	6. $86\overline{)6,382}$
7. $24\overline{)9,045}$	8. $94\overline{)6,476}$	9. $45\overline{)56,982}$
10. $74\overline{)58,315}$	11. $83\overline{)96,108}$	12. $20\overline{)19,434}$
13. $52\overline{)32,147}$	14. $62\overline{)89,555}$	15. $44\overline{)77,777}$

Solve these problems.

16. Mrs. Weaver had 197 eggs to put into boxes which contain one dozen. How many boxes could she fill?

17. The Pert Kennel uses about 98 pounds of dog food a day. About how long will a supply of 520 pounds last?

DO YOU REMEMBER?

Find simplest fractional numerals.

1. $\frac{6}{8}$ 2. $\frac{8}{16}$ 3. $\frac{5}{20}$ 4. $\frac{4}{36}$ 5. $\frac{9}{81}$

Write mixed numerals for Exercises 6-8 and fractional numerals for Exercises 9-10.

6. $\frac{21}{16}$ 7. $\frac{15}{7}$ 8. $\frac{43}{10}$ 9. $8\frac{3}{4}$ 10. $5\frac{2}{5}$

Add and simplify.

11. $\frac{3}{8} + \frac{1}{8}$ 12. $\frac{3}{4} + \frac{1}{2}$ 13. $\frac{5}{7} + \frac{3}{4}$ 14. $7\frac{2}{3} + 2\frac{3}{5}$ 15. $3\frac{1}{9} + 7\frac{3}{8}$

16. $2\frac{1}{3} + 4\frac{3}{4} + 3\frac{5}{6}$ 17. $3\frac{1}{9} + 2\frac{2}{3} + 7\frac{4}{6}$

Subtract and simplify.

18. $\frac{7}{9} - \frac{2}{9}$ 19. $\frac{5}{6} - \frac{2}{3}$ 20. $\frac{6}{7} - \frac{2}{3}$ 21. $6\frac{7}{8} - 2\frac{1}{3}$ 22. $5\frac{1}{2} - 3\frac{5}{6}$

23. $3\frac{5}{8} - 1\frac{3}{5}$ 24. $9\frac{8}{11} - 2\frac{1}{5}$ 25. $6\frac{2}{7} - 3\frac{1}{3}$ 26. $2\frac{11}{13} - 1\frac{1}{2}$ 27. $8\frac{1}{3} - 2\frac{5}{9}$

Multiply and simplify.

28. $\frac{1}{5} \times \frac{1}{3}$ 29. $\frac{3}{4} \times \frac{5}{6}$ 30. $8 \times \frac{5}{7}$ 31. $2\frac{1}{8} \times \frac{5}{6}$

32. $3\frac{7}{8} \times 2\frac{2}{3}$ 33. $4\frac{2}{3} \times 1\frac{6}{7}$ 34. $5\frac{3}{8} \times 1\frac{5}{11}$ 35. $3\frac{9}{10} \times 6\frac{2}{3}$

Divide and simplify.

36. $\frac{1}{4} \div \frac{1}{3}$ 37. $\frac{5}{6} \div \frac{1}{4}$ 38. $\frac{9}{16} \div \frac{3}{8}$ 39. $\frac{7}{8} \div 2\frac{3}{4}$

40. $6\frac{1}{8} \div \frac{7}{16}$ 41. $7\frac{1}{4} \div 1\frac{1}{8}$ 42. $4\frac{2}{7} \div 1\frac{1}{14}$ 43. $9\frac{1}{4} \div 4\frac{1}{8}$

CHAPTER REVIEW

1. a. List the first five natural numbers.

   b. List the first five whole numbers.

2. Which of the following name whole numbers?

   a. 5 b. 0 c. $\frac{1}{2}$ d. 50

   e. 3×4 f. $5 - 8$ g. $2 \div 3$ h. $12 \div 3$

3. Name four operations on whole numbers.

4. Which operations on whole numbers always have whole number answers?

Add.

5. $25 + 42$	6. $71 + 28$	7. $63 + 18$	8. $57 + 65$
9. 324, 146, 302	10. 157, 291, 368	11. 167, 324, 489	12. 3,205, 360, 2,512
13. 1,685, 2,476, 392, 5,596	14. 9,280, 306, 409, 5,290	15. 3,699, 2,128, 3,964, 1,385	16. 18,394, 296, 9,380, 25,596

Subtract.

17. $48 - 23$	18. $92 - 31$	19. $46 - 37$	20. $304 - 127$
21. $423 - 309$	22. $230 - 102$	23. $1,346 - 986$	24. $2,581 - 1,036$
25. $3,065 - 962$	26. $5,861 - 5,843$	27. $9,606 - 2,998$	28. $38,391 - 19,396$

Multiply.

29. $\begin{array}{r} 32 \\ \times\ 25 \\ \hline \end{array}$ 30. $\begin{array}{r} 23 \\ \times\ 48 \\ \hline \end{array}$ 31. $\begin{array}{r} 16 \\ \times\ 38 \\ \hline \end{array}$ 32. $\begin{array}{r} 204 \\ \times\ 56 \\ \hline \end{array}$

33. $\begin{array}{r} 150 \\ \times\ 72 \\ \hline \end{array}$ 34. $\begin{array}{r} 1{,}322 \\ \times\ 105 \\ \hline \end{array}$ 35. $\begin{array}{r} 3{,}020 \\ \times\ 473 \\ \hline \end{array}$ 36. $\begin{array}{r} 2{,}365 \\ \times\ 250 \\ \hline \end{array}$

37. $\begin{array}{r} 5{,}695 \\ \times\ 3{,}062 \\ \hline \end{array}$ 38. $\begin{array}{r} 2{,}865 \\ \times\ 3{,}946 \\ \hline \end{array}$ 39. $\begin{array}{r} 9{,}385 \\ \times\ 2{,}006 \\ \hline \end{array}$ 40. $\begin{array}{r} 7{,}062 \\ \times\ 8{,}906 \\ \hline \end{array}$

Divide.

41. $7\overline{)350}$ 42. $8\overline{)872}$ 43. $25\overline{)375}$

44. $20\overline{)104}$ 45. $32\overline{)276}$ 46. $55\overline{)1{,}037}$

47. $82\overline{)5{,}468}$ 48. $75\overline{)3{,}247}$ 49. $55\overline{)46{,}395}$

50. $38\overline{)29{,}395}$ 51. $58\overline{)36{,}092}$ 52. $63\overline{)78{,}952}$

Solve these problems.

53. The eighth grade presented a class play. The first performance 125 people attended; the second, 175 people; and the final performance, 158 people. How many people saw the play?

54. Doug bought 5 cartons of soft drinks for his party. Three of the cartons had 6 bottles each and the others had 8 each. How many bottles did he buy?

55. The Campfire Girls are having a bake sale. They wish to wrap the cookies in packages with one dozen cookies in each. How many packages will 384 cookies fill?

Terms You Should Know

natural numbers	algorithm	output
whole numbers	input	flow chart

CHAPTER TEST

1. What whole number is not a natural number?

Which do not name whole numbers?

2. 36 − 25	3. 25 − 36	4. 0	5. 12 ÷ 5

Add.

6. 43	7. 65	8. 102	9. 4,357
54	26	347	5,879

10. 174	11. 1,695	12. 3,652	13. 19,395
395	206	2,963	2,462
268	9,007	9,596	25,386

Subtract.

14. 93	15. 86	16. 102	17. 360
41	39	57	108

18. 1,586	19. 2,695	20. 9,707	21. 59,381
859	1,769	5,968	27,296

Multiply.

22. 52	23. 216	24. 807	25. 1,347
14	37	49	208

26. 5,962	27. 3,954	28. 5,076	29. 3,506
465	2,836	3,567	4,207

Divide.

30. $8\overline{)39}$	31. $9\overline{)576}$	32. $15\overline{)102}$
33. $33\overline{)2,861}$	34. $56\overline{)39,273}$	35. $75\overline{)69,062}$

Solve this problem.

36. How many cartons will be required to package 96 tomato plants with $\frac{1}{2}$ dozen in each carton?

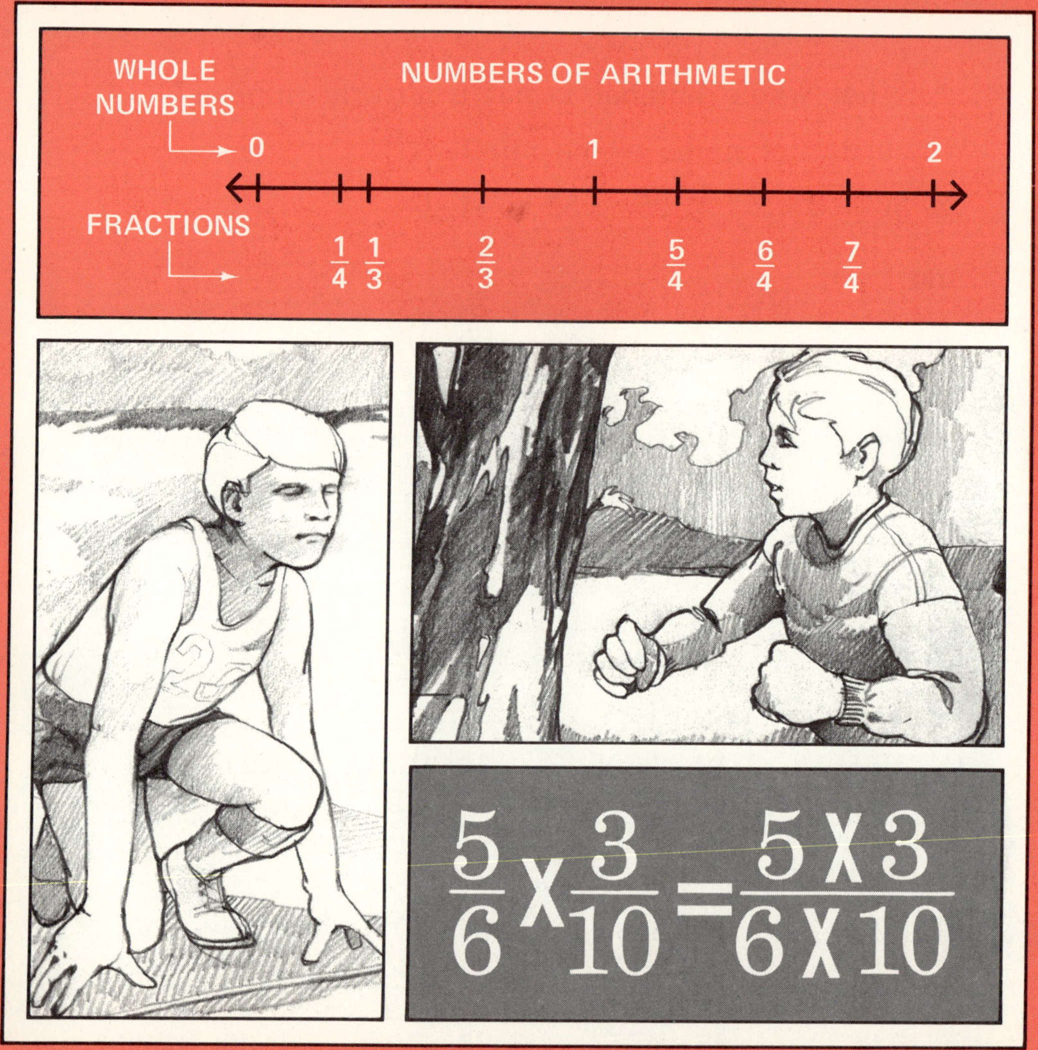

CHAPTER TWO

Fractional Numerals

NUMBERS OF ARITHMETIC

$\frac{2}{3}$, $\frac{5}{6}$, $\frac{10}{7}$, 0, 5, and 3.7 are numbers of arithmetic.

Any number that can be named by a fractional numeral with a whole number numerator and a natural number denominator is called a **number of arithmetic.**

1. Name the numerator and the denominator of $\frac{5}{6}$.

2. The number 0 can be named $\frac{0}{1}$, $\frac{0}{2}$, $\frac{0}{3}$, etc. Explain why 0 is a number of arithmetic.

3. The number 5 is a number of arithmetic since 5 can be named $\frac{5}{1}$, $\frac{10}{2}$, $\frac{15}{3}$, etc. Write three names which show that 8 is a number of arithmetic.

4. Let's find a name for 3.7 which shows that 3.7 is a number of arithmetic.

$$3.7 = 3\frac{7}{10} = \frac{37}{10}$$

Show that 6.5 is a number of arithmetic.

Both whole numbers and fractions are numbers of arithmetic.

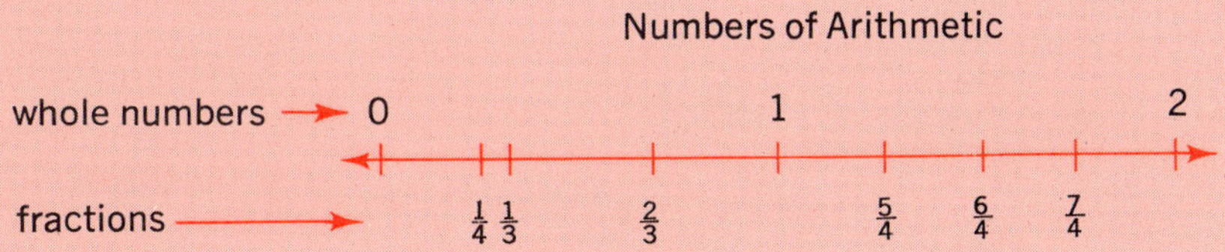

5. $\frac{2}{3}$, $\frac{4}{6}$, and $\frac{6}{9}$ are all names for two-thirds. Name the factors which are common to the numerator and the denominator of each of these names for two-thirds.

In the **simplest fractional numeral** for a number, the only common factor of the numerator and denominator is 1.

6. Here is a flow chart for finding simplest fractional numerals. Study the chart and the examples.

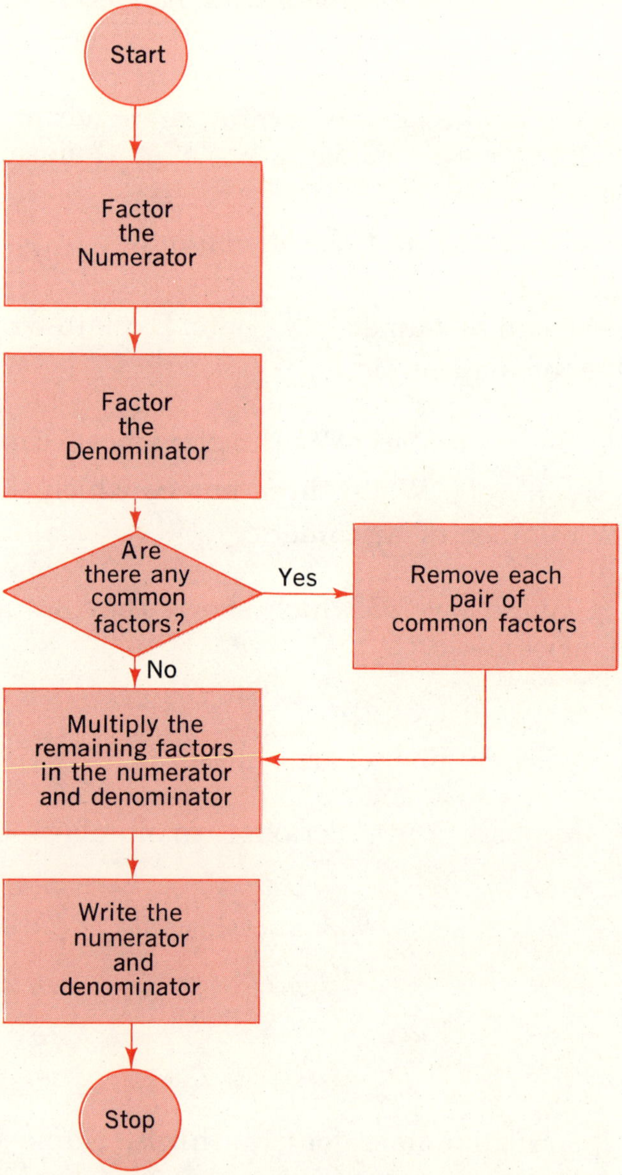

Examples

1. $\frac{5}{30}$

   $\frac{5 \cdot 1}{5 \cdot 6}$

   $\frac{1}{6}$

2. $\frac{12}{15}$

   $\frac{3 \cdot 4}{3 \cdot 5}$

   $\frac{4}{5}$

3. $\frac{18}{24}$

   $\frac{2 \cdot 3 \cdot 3}{2 \cdot 2 \cdot 2 \cdot 3}$

   $\frac{3}{4}$

7. To *simplify* a fractional numeral means to find the simplest fractional numeral. Follow the flow chart to simplify each.

a. $\frac{6}{8}$ **b.** $\frac{10}{12}$ **c.** $\frac{30}{50}$ **d.** $\frac{8}{4}$ **e.** $\frac{18}{12}$

EXERCISES

Use the flow chart to simplify each.

1. $\frac{8}{16}$	**2.** $\frac{4}{10}$	**3.** $\frac{16}{24}$	**4.** $\frac{7}{21}$	**5.** $\frac{15}{24}$
6. $\frac{6}{30}$	**7.** $\frac{18}{24}$	**8.** $\frac{42}{48}$	**9.** $\frac{25}{30}$	**10.** $\frac{24}{36}$
11. $\frac{7}{9}$	**12.** $\frac{10}{35}$	**13.** $\frac{6}{8}$	**14.** $\frac{9}{54}$	**15.** $\frac{8}{72}$
16. $\frac{15}{18}$	**17.** $\frac{4}{52}$	**18.** $\frac{16}{64}$	**19.** $\frac{5}{70}$	**20.** $\frac{3}{8}$
21. $\frac{21}{42}$	**22.** $\frac{18}{33}$	**23.** $\frac{12}{54}$	**24.** $\frac{16}{56}$	**25.** $\frac{21}{63}$
26. $\frac{25}{45}$	**27.** $\frac{13}{39}$	**28.** $\frac{17}{31}$	**29.** $\frac{38}{50}$	**30.** $\frac{9}{39}$

Let's Practice

Find the outputs.

1.

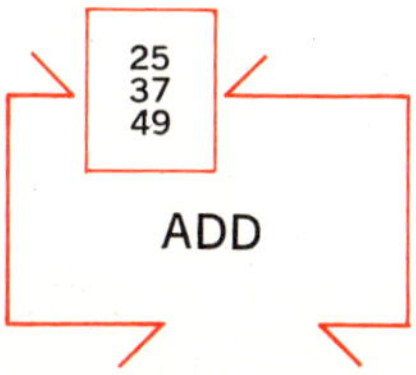

2.

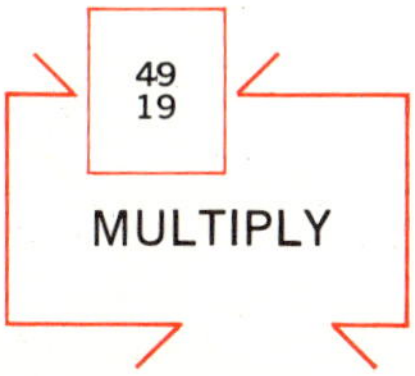

3.

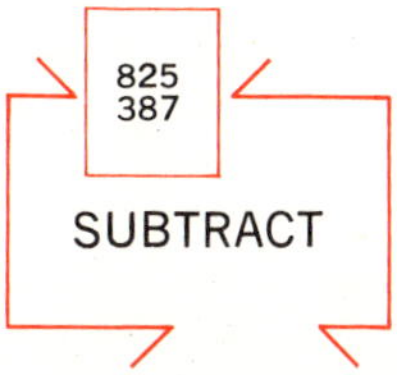

4.

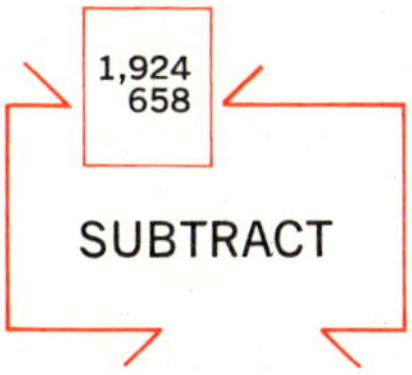

5.

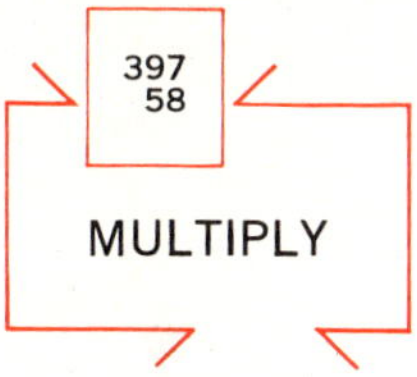

6.

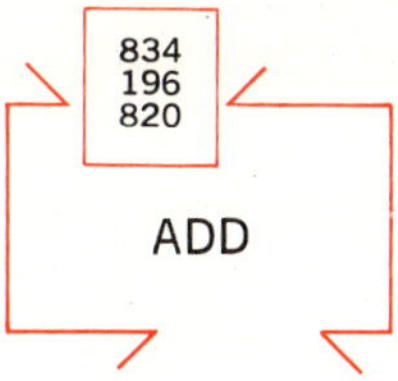

7.

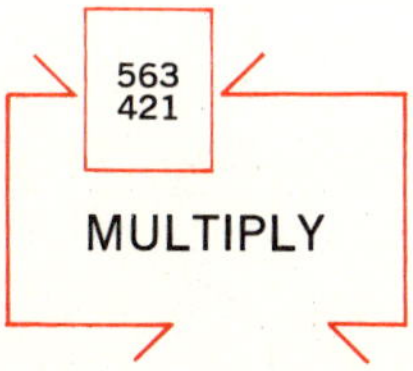

8.

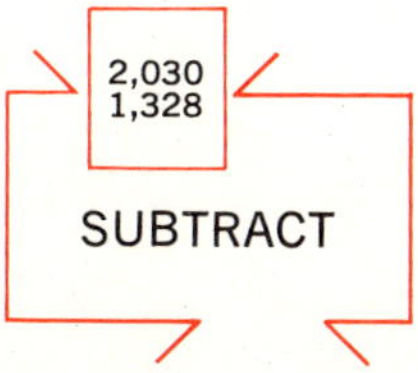

9.

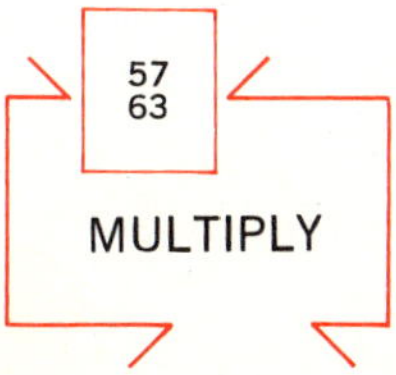

FINDING COMMON DENOMINATORS

When we add or subtract with fractional numerals, we find a common denominator.

1. Suppose we want to add $\frac{2}{3}$ and $\frac{3}{4}$. Examine these equivalent fractional numerals.

$$\frac{2}{3}, \frac{4}{6}, \frac{6}{9}, \frac{8}{12}, \frac{10}{15} \cdots$$

$$\frac{3}{4}, \frac{6}{8}, \frac{9}{12}, \frac{12}{16}, \frac{15}{20} \cdots$$

   a. Which denominator is the least common multiple of 3 and 4?

   b. Which fractional numerals would we use to add $\frac{2}{3}$ and $\frac{3}{4}$?

2. Suppose we want to subtract $\frac{1}{5}$ from $\frac{5}{6}$.

   a. What is the least common multiple of 5 and 6?

   b. What is the least common denominator for $\frac{5}{6}$ and $\frac{1}{5}$?

   c. Give the missing numerators.

$$\frac{5}{6} = \frac{\square}{30} \qquad \frac{1}{5} = \frac{\triangle}{30}$$

The least common denominator (L.C.D.) is the least common multiple of the denominators of two or more fractional numerals.

3. Find the L.C.D.

a. $\frac{1}{2}, \frac{2}{3}$	b. $\frac{1}{8}, \frac{3}{4}$	c. $\frac{1}{6}, \frac{2}{9}$
d. $\frac{1}{5}, \frac{2}{7}$	e. $\frac{3}{4}, \frac{4}{5}$	f. $\frac{3}{4}, \frac{1}{8}, \frac{5}{16}$
g. $\frac{3}{4}, \frac{1}{7}, \frac{2}{3}$	h. $\frac{2}{3}, \frac{1}{5}, \frac{5}{6}$	i. $\frac{3}{9}, \frac{2}{3}, \frac{5}{6}$
j. $\frac{1}{8}, \frac{2}{5}, \frac{9}{10}$	k. $\frac{1}{5}, \frac{3}{4}, \frac{3}{10}$	l. $\frac{3}{6}, \frac{11}{12}, \frac{5}{8}$

EXERCISES

Find the L.C.D.

1. $\frac{2}{7}, \frac{1}{3}$
2. $\frac{5}{6}, \frac{1}{8}$
3. $\frac{1}{9}, \frac{2}{5}$
4. $\frac{3}{7}, \frac{1}{12}$
5. $\frac{4}{5}, \frac{2}{6}$
6. $\frac{3}{8}, \frac{2}{3}$
7. $\frac{9}{10}, \frac{1}{5}, \frac{2}{15}$
8. $\frac{4}{9}, \frac{2}{3}, \frac{1}{4}$
9. $\frac{1}{12}, \frac{3}{16}, \frac{5}{8}$
10. $\frac{1}{8}, \frac{3}{4}, \frac{3}{16}$
11. $\frac{2}{7}, \frac{3}{9}, \frac{2}{3}$
12. $\frac{3}{4}, \frac{9}{10}, \frac{7}{15}$

Let's Practice

Find these outputs.

1. 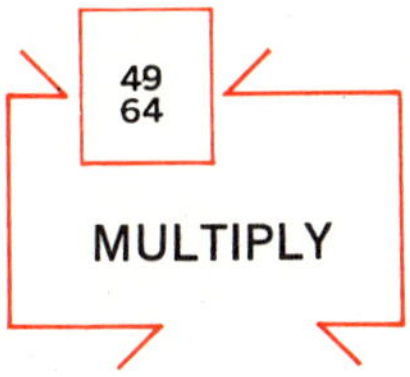

2. 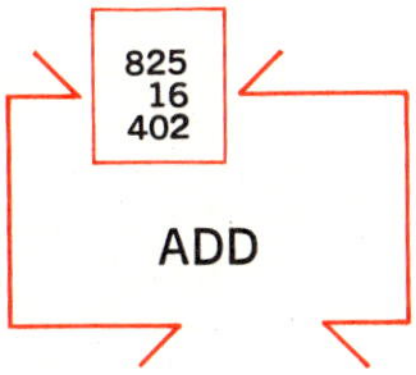

3. 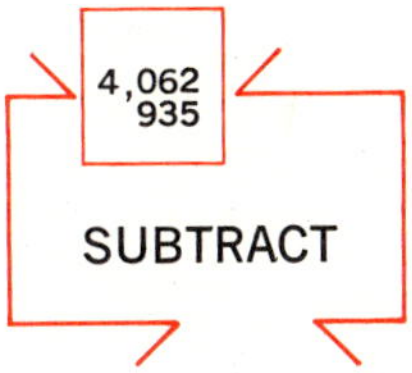

4. 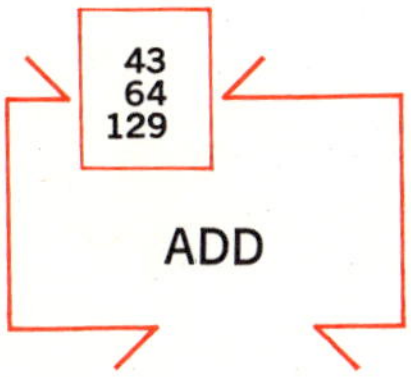

5. 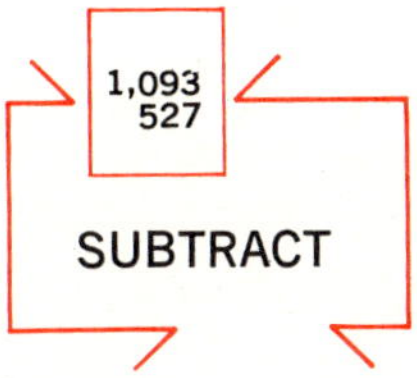

6. 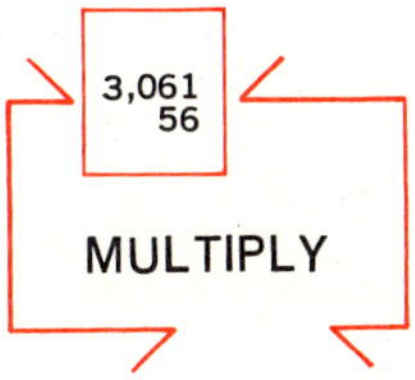

7. 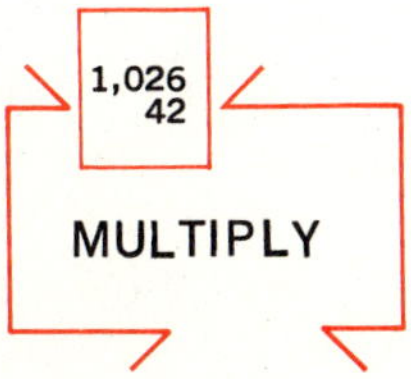

8. 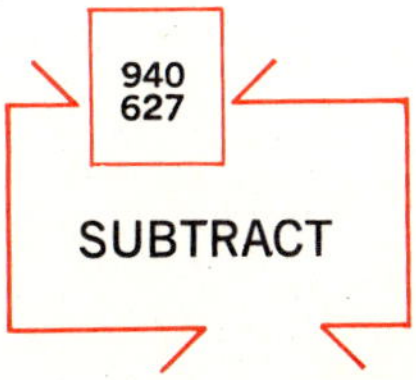

9. 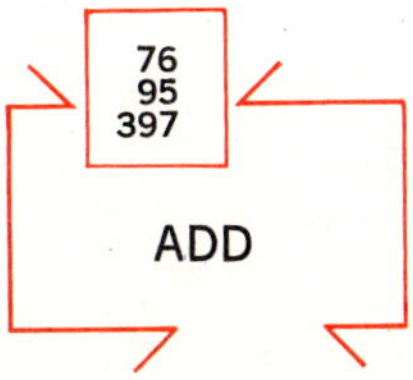

ADDITION WITH FRACTIONAL NUMERALS

Here is a flow chart for adding with fractional numerals. Study the chart and examples.

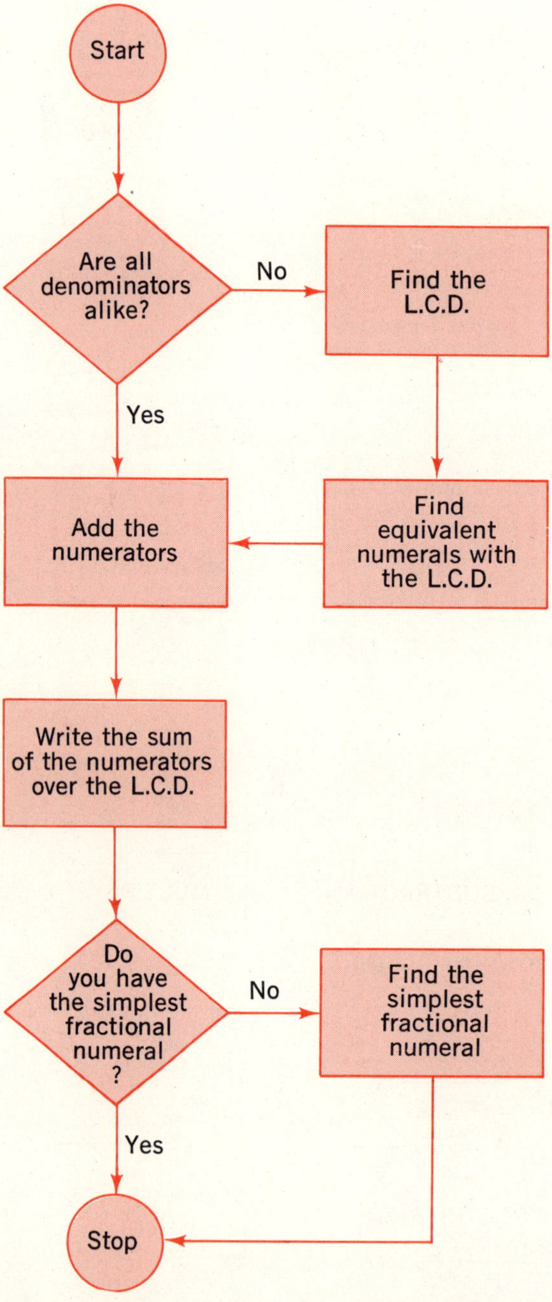

Examples

1. $\frac{1}{3} + \frac{2}{5}$

L.C.D. : 15

$\frac{1}{3} = \frac{5}{15} \qquad \frac{2}{5} = \frac{6}{15}$

$\frac{5}{15} + \frac{6}{15} = \frac{5+6}{15}$

$= \frac{11}{15}$

2. $\frac{4}{5} + \frac{7}{10}$

L.C.D. : 10

$\frac{4}{5} = \frac{8}{10}$

$\frac{8}{10} = \frac{7}{10} + \frac{8+7}{10}$

$= \frac{5 \cdot 3}{5 \cdot 2}$

$= \frac{15}{10}$

$= \frac{3}{2}$

1. Let's use the flow chart to add $\frac{1}{5}$ and $\frac{3}{4}$.

   a. What is the answer to the first yes-no decision?

   b. What is the L.C.D. for $\frac{1}{5}$ and $\frac{3}{4}$?

   c. Give the missing numerators.

   $$\frac{1}{5} = \frac{\triangle}{20} \qquad \frac{3}{4} = \frac{\square}{20}$$

   d. Complete.

   $$\frac{4}{20} + \frac{15}{20} = \frac{4 + \triangle}{20} = \frac{\square}{20}$$

   e. Do you have the simplest fractional numeral?

   f. What is $\frac{1}{5} + \frac{3}{4}$?

2. Add. Use the flow chart.

   a. $\frac{2}{3} + \frac{1}{5}$ b. $\frac{4}{7} + \frac{1}{4}$ c. $\frac{3}{8} + \frac{5}{6} + \frac{3}{4}$

EXERCISES

Use the flow chart to find these sums.

1. $\frac{1}{8} + \frac{5}{6}$
2. $\frac{2}{3} + \frac{3}{5}$
3. $\frac{2}{7} + \frac{1}{3}$
4. $\frac{1}{3} + \frac{5}{6}$
5. $\frac{4}{7} + \frac{1}{2}$
6. $\frac{8}{9} + \frac{2}{3}$
7. $\frac{9}{10} + \frac{8}{15}$
8. $\frac{1}{4} + \frac{3}{4}$
9. $\frac{5}{12} + \frac{3}{8}$
10. $\frac{4}{9} + \frac{2}{3}$
11. $\frac{1}{8}$, $\frac{1}{6}$, $\frac{3}{16}$
12. $\frac{2}{3}$, $\frac{3}{4}$, $\frac{1}{8}$
13. $\frac{5}{24}$, $\frac{1}{12}$, $\frac{5}{6}$
14. $\frac{3}{8}$, $\frac{4}{5}$, $\frac{7}{10}$
15. $\frac{8}{9}$, $\frac{3}{4}$, $\frac{13}{18}$
16. $\frac{3}{7} + \frac{4}{5} + \frac{7}{10}$
17. $\frac{5}{6} + \frac{8}{9} + \frac{2}{3}$
18. $\frac{7}{12} + \frac{8}{15} + \frac{3}{4}$

PROBLEM SOLVING

1. Carol used $\frac{2}{3}$ cup of white sugar and $\frac{1}{4}$ cup of brown sugar in baking. How much sugar did she use?

2. Each morning, David jogs $\frac{1}{2}$ mile, sprints $\frac{1}{8}$ mile, and then jogs another $\frac{1}{4}$ mile. How far does he run each day?

3. Jerry walked $\frac{3}{4}$ mile to Tim's house. Then they walked $\frac{1}{8}$ mile to the school gym. How far did Jerry walk in all?

4. Frank collected 329 baseball cards. He decided to share them evenly with his friends Don, Bob, and Fred. About how many cards will each boy have?

5. Mrs. Spinato needed $\frac{3}{4}$ cup of olive oil for salad dressing. She found that she only had $\frac{1}{3}$ cup. How much should she borrow from her neighbor?

6. Marge and Kathi packed cookies for the charity drive. In each box, they packed $\frac{3}{8}$ pound of oatmeal cookies, $\frac{1}{8}$ of a pound of sugar cookies and $\frac{1}{4}$ pound of chocolate chip cookies. How many pounds of cookies were in each box?

SUBTRACTION WITH FRACTIONAL NUMERALS

The flow chart for adding with fractional numerals can be used for subtraction. The only change is that instead of adding numerators, we subtract them.

Examples $\frac{4}{5} - \frac{1}{4} = \frac{16}{20} - \frac{5}{20} = \frac{16-5}{20} = \frac{11}{20}$

$\frac{13}{3} - \frac{5}{6} = \frac{26}{6} - \frac{5}{6} = \frac{26-5}{6} = \frac{21}{6} = \frac{7}{2}$

EXERCISES

Subtract and simplify.

1. $\frac{5}{6} - \frac{1}{6}$
2. $\frac{5}{8} - \frac{1}{4}$
3. $\frac{9}{16} - \frac{1}{3}$
4. $\frac{5}{7} - \frac{1}{2}$
5. $\frac{7}{8} - \frac{3}{4}$
6. $\frac{24}{8} - \frac{15}{8}$
7. $\frac{19}{30} - \frac{8}{15}$
8. $\frac{9}{16} - \frac{1}{4}$
9. $\frac{11}{8} - \frac{4}{32}$
10. $\frac{9}{7} - \frac{2}{4}$
11. $\frac{4}{9} - \frac{1}{5}$
12. $\frac{15}{27} - \frac{5}{9}$
13. $\frac{46}{39} - \frac{2}{3}$
14. $\frac{15}{16} - \frac{3}{4}$
15. $\frac{13}{28} - \frac{1}{7}$
16. $\frac{33}{35} - \frac{2}{7}$
17. $\frac{14}{32} - \frac{3}{16}$
18. $\frac{42}{30} - \frac{2}{3}$
19. $\frac{47}{50} - \frac{4}{5}$
20. $\frac{38}{75} - \frac{9}{25}$

Let's Practice

Copy and complete this chart.

Mixed Numeral	$2\frac{3}{5}$	$4\frac{1}{8}$	$5\frac{2}{3}$	$6\frac{1}{5}$	$9\frac{2}{7}$		
Fractional Numeral	$\frac{13}{5}$	$\frac{33}{8}$				$\frac{29}{7}$	$\frac{35}{6}$

WORKING WITH MIXED NUMERALS

1. Mr. Roy worked $2\frac{3}{4}$ hours overtime on Tuesday and $6\frac{1}{3}$ hours overtime on Saturday. Here is how he computed the total amount of overtime.

$$\begin{array}{rcl} 2\frac{3}{4} & = & 2\frac{9}{12} \\ +\ 6\frac{1}{3} & = & 6\frac{4}{12} \\ \hline & & 8\frac{13}{12} = 9\frac{1}{12} \end{array}$$

   a. Why did Mr. Roy use 12 for a denominator?

   b. Explain how Mr. Roy renamed $8\frac{13}{12}$.

2. Add. Copy and complete.

   a. $$\begin{array}{rcl} 5\frac{5}{8} & = & 5\frac{\triangle}{32} \\ +\ 2\frac{29}{32} & = & 2\frac{29}{32} \\ \hline & & 7\frac{\square}{32} = 8\frac{\triangledown}{32} \end{array}$$

   b. $$\begin{array}{r} 8\frac{9}{10} \\ +\ 4\frac{3}{5} \\ \hline \end{array}$$

   c. $$\begin{array}{r} 4\frac{1}{5} \\ +\ \frac{2}{3} \\ \hline \end{array}$$

3. Sharon lost $3\frac{5}{8}$ pounds on her diet. Diane lost $5\frac{3}{4}$ pounds. Explain how they found how much more weight Diane lost.

$$\begin{array}{rcl} 5\frac{3}{4} & = & 5\frac{6}{8} \\ -\ 3\frac{5}{8} & = & 3\frac{5}{8} \\ \hline & & 2\frac{1}{8} \end{array}$$

4. Subtract. Copy and complete.

a. $8\frac{3}{5} = 8\frac{\triangle}{15}$
$-\ 4\frac{1}{3} = 4\frac{\square}{15}$
$4\frac{\triangledown}{15}$

b. $7\frac{3}{4}$
$-\ 2\frac{1}{2}$

c. $16\frac{9}{16}$
$-\ 4\frac{3}{32}$

5. Copy and complete. Explain the steps.

$3\frac{1}{4}$
$-\ 1\frac{1}{2}$

$3\frac{1}{4} = 2 + \_\_\_\_ + \frac{1}{4}$
$= 2 + \frac{4}{4} + \_\_\_\_$
$= 2 + \frac{5}{4}$

$2 + \frac{5}{4}$
$-\ 1 + \frac{2}{4}$
$1 + \frac{3}{4} = \_\_\_\_$

6. Subtract. Copy and complete.

a. $2\frac{1}{12} = 1\frac{13}{12}$
$-\ 1\frac{3}{4} = 1\frac{\triangle}{12}$ 9
4 $\frac{\square}{12} = \frac{\triangledown}{3}$

b. $4\frac{2}{3}$
$-\ 2\frac{4}{5}$

c. $3\frac{1}{8}$
$-\ 1\frac{5}{8}$

EXERCISES

Add and simplify.

1. $8\frac{3}{4} + 5\frac{5}{6}$

2. $3\frac{1}{5} + 6\frac{3}{4}$

3. $6\frac{2}{9} + 2\frac{1}{3}$

4. $10\frac{5}{6} + 2\frac{6}{15}$

5. $7\frac{7}{9} + 3\frac{2}{5}$

6. $18\frac{10}{11} + 16\frac{2}{3}$

7. $15\frac{2}{3} + 7\frac{1}{8}$

8. $14\frac{2}{7} + 11\frac{1}{5}$

9. $17\frac{3}{4} + 12\frac{1}{6}$

10. $38\frac{5}{9} + 12\frac{3}{4}$

11. $56\frac{11}{15} + 21\frac{3}{6}$

12. $18\frac{3}{18} + 12\frac{5}{12}$

13. $88\frac{3}{11} + 12\frac{2}{3}$

14. $60\frac{1}{8} + 20\frac{1}{12}$

15. $9\frac{7}{9} + 2\frac{1}{3}$

16. $28\frac{4}{5} + 16\frac{3}{8}$

17. $4\frac{1}{2} + 3\frac{1}{4} + 5\frac{3}{8}$

18. $15\frac{1}{6} + 32\frac{2}{3} + 18\frac{3}{5}$

Subtract and simplify.

19. $16\frac{2}{3} - 5\frac{1}{3}$

20. $23\frac{5}{7} - 10\frac{1}{2}$

21. $9\frac{8}{9} - 4\frac{5}{6}$

22. $23\frac{5}{6} - 17\frac{2}{3}$

23. $9\frac{1}{3} - 4\frac{5}{6}$

24. $11\frac{1}{9} - 7\frac{3}{7}$

25. $14 - 5\frac{3}{5}$

26. $28 - 13\frac{7}{9}$

27. $17\frac{1}{3} - 16\frac{2}{5}$

28. $7\frac{3}{8} - 2\frac{2}{3}$

29. $98\frac{4}{7} - 78\frac{3}{5}$

30. $39\frac{2}{5} - 16\frac{1}{6}$

31. $78\frac{3}{4} - 16\frac{5}{6}$

32. $17\frac{8}{11} - 12\frac{1}{2}$

33. $17 - 6\frac{18}{23}$

34. $50 - 20\frac{1}{25}$

Solve these problems.

35. Mr. Irwin needed pieces of shelving $33\frac{5}{8}$ inches long and $28\frac{3}{4}$ inches long. What is the smallest piece of shelving from which these pieces can be cut?

36. Cheri fed her puppy $1\frac{2}{3}$ cups of puppy chow per day. When full grown, the dog was fed $3\frac{1}{2}$ cups per day. How much more food did the dog receive when full grown?

UNEXPLORED TERRITORY

Bull's-eye! How could you score exactly 100 using only six darts?

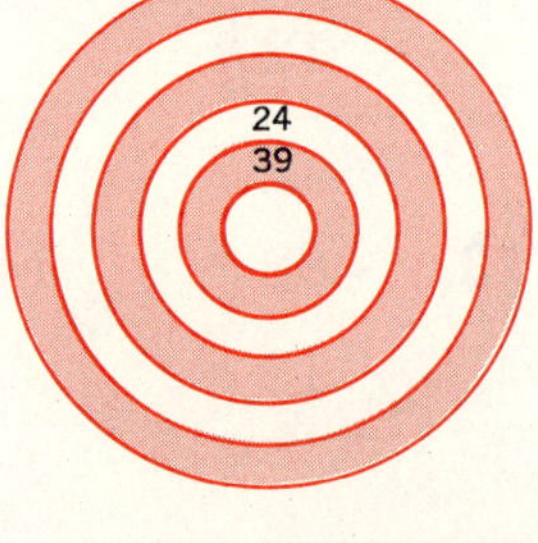

MULTIPLICATION WITH FRACTIONAL NUMERALS

This flow chart shows a way to multiply with fractional numerals.

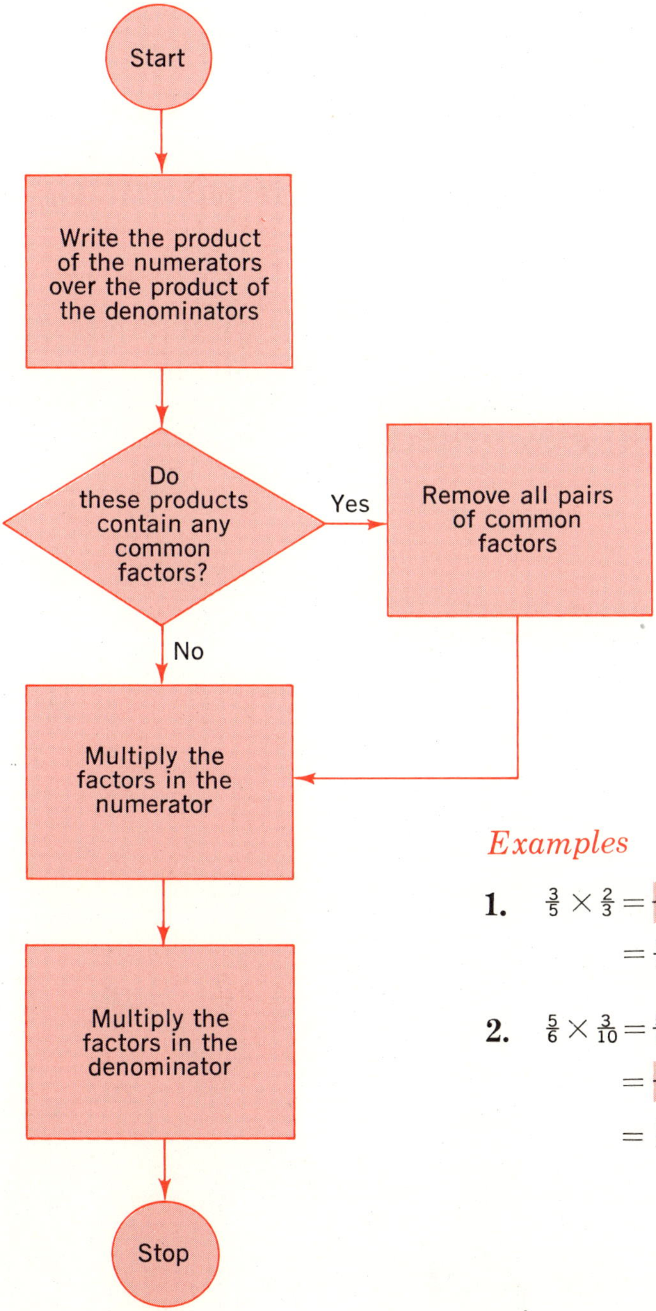

Examples

1. $\frac{3}{5} \times \frac{2}{3} = \frac{3 \times 2}{3 \times 5}$

$= \frac{2}{5}$

2. $\frac{5}{6} \times \frac{3}{10} = \frac{5 \times 3}{6 \times 10}$

$= \frac{5 \times 3 \times 1}{5 \times 3 \times 2 \times 2}$

$= \frac{1}{4}$

1. Let's use the flow chart to multiply $\frac{5}{6}$ by $\frac{2}{7}$.

   a. Write the product of the numerators.

   b. Write the product of the denominators.

   c. Copy and complete. $\frac{5 \times 2}{6 \times 7} = \frac{5 \times 2 \times 1}{\triangle \times \square \times 7}$

   d. What common factor do the numerator and the denominator contain?

   e. Write the simplest fractional numeral for the product $\frac{5}{6} \times \frac{2}{7}$.

To do a multiplication like $\frac{5}{6} \times \frac{2}{7}$, we multiply the numerators and multiply the denominators.

2. Multiply. Use the flow chart.

   a. $\frac{5}{4} \cdot \frac{3}{7}$ b. $\frac{3}{4} \times \frac{1}{9}$ c. $\frac{12}{35} \cdot \frac{5}{6}$

EXERCISES

Use the flow chart to find these products.

1. $\frac{5}{6} \times \frac{3}{5}$ 2. $\frac{8}{9} \cdot \frac{2}{5}$ 3. $\frac{10}{7} \times \frac{2}{3}$ 4. $\frac{11}{13} \cdot \frac{4}{6}$

5. $\frac{21}{5} \cdot \frac{6}{7}$ 6. $\frac{7}{8} \times \frac{5}{7}$ 7. $\frac{3}{8} \cdot \frac{5}{8}$ 8. $\frac{10}{13} \times \frac{1}{5}$

9. $\frac{5}{9} \times \frac{1}{4}$ 10. $\frac{23}{2} \cdot \frac{4}{5}$ 11. $\frac{5}{7} \cdot \frac{1}{9}$ 12. $\frac{6}{11} \cdot \frac{2}{3}$

13. $\frac{1}{9} \times \frac{9}{10}$ 14. $\frac{5}{16} \times \frac{8}{9}$ 15. $\frac{11}{15} \cdot \frac{5}{8}$ 16. $\frac{6}{7} \times \frac{9}{10}$

17. $\frac{8}{15} \cdot \frac{7}{8}$ 18. $\frac{5}{9} \cdot \frac{14}{3}$ 19. $\frac{8}{5} \times \frac{19}{16}$ 20. $\frac{14}{20} \times \frac{3}{7}$

21. $\frac{9}{13} \cdot \frac{7}{2}$ 22. $\frac{16}{5} \times \frac{5}{8}$ 23. $\frac{15}{31} \cdot \frac{2}{45}$ 24. $\frac{23}{14} \times \frac{21}{69}$

25. $\frac{6}{11} \times \frac{77}{8}$ 26. $\frac{9}{15} \cdot \frac{45}{3}$ 27. $\frac{2}{7} \cdot \frac{35}{40}$ 28. $\frac{19}{6} \times \frac{3}{20}$

Solve these problems.

25. A soup recipe requires $\frac{3}{4}$ cup of tomato juice. How much tomato juice will be needed for $\frac{2}{3}$ of a recipe?

26. The record for the 100 yard dash at Holabird Junior High was $10\frac{9}{10}$ seconds. This year, Ed Gunt set a new record of $10\frac{3}{10}$ seconds. What is the difference in the two records?

27. In planting a new lawn, Mr. Newblatt used $3\frac{1}{2}$ pounds of blue grass, $5\frac{1}{4}$ pounds of red fescue, and 2 pounds of annual rye. How many pounds of seed did he plant? What was the total cost of the seed?

28. Barbara lost $\frac{3}{4}$ pound in the first week of camp, and $\frac{7}{8}$ pound during the second week.

 a. How much more weight did she lose during the second week?

 b. How much weight did she lose in all?

Let's Practice

Every number of arithmetic except 0 has a reciprocal. Copy and complete this chart. Refer to the flow chart on page 17.

Number	$\frac{3}{4}$	0	2	$\frac{5}{7}$	$\frac{4}{5}$	$\frac{9}{10}$	8	$\frac{1}{9}$	$\frac{10}{11}$	$\frac{5}{13}$
Reciprocal	$\frac{4}{3}$	none	$\frac{1}{2}$							

DO YOU REMEMBER?

Compute.

1. $\begin{array}{r} 246 \\ 583 \\ \underline{897} \end{array}$ 2. $\begin{array}{r} 6{,}423 \\ 589 \\ \underline{8{,}346} \end{array}$ 3. $\begin{array}{r} 19{,}234 \\ 325 \\ 8{,}635 \\ \underline{21{,}658} \end{array}$ 4. $\begin{array}{r} 1{,}854 \\ 24{,}234 \\ 5{,}658 \\ \underline{19{,}826} \end{array}$

5. $\begin{array}{r} 1{,}963 \\ \underline{-\ 258} \end{array}$ 6. $\begin{array}{r} 2{,}034 \\ \underline{-\ 1{,}695} \end{array}$ 7. $\begin{array}{r} 18{,}438 \\ \underline{-\ 7{,}680} \end{array}$ 8. $\begin{array}{r} 29{,}438 \\ \underline{-\ 17{,}569} \end{array}$

9. $\begin{array}{r} 234 \\ \underline{\times\ 56} \end{array}$ 10. $\begin{array}{r} 508 \\ \underline{\times\ 219} \end{array}$ 11. $\begin{array}{r} 6{,}248 \\ \underline{\times\ 106} \end{array}$ 12. $\begin{array}{r} 9{,}569 \\ \underline{\times\ 3{,}205} \end{array}$

13. $56\overline{)827}$ 14. $29\overline{)8{,}564}$ 15. $86\overline{)9{,}720}$ 16. $234\overline{)16{,}572}$

Find simplest fractional numerals.

17. $\frac{6}{9}$ 18. $\frac{14}{56}$ 19. $\frac{18}{72}$ 20. $\frac{35}{85}$

Add and simplify.

21. $\begin{array}{r} \frac{5}{6} \\ \underline{+\ \frac{1}{5}} \end{array}$ 22. $\begin{array}{r} \frac{9}{10} \\ \underline{+\ \frac{3}{25}} \end{array}$ 23. $\begin{array}{r} \frac{8}{15} \\ \underline{+\ \frac{5}{8}} \end{array}$ 24. $\begin{array}{r} 5\frac{7}{8} \\ \underline{+\ 12\frac{9}{16}} \end{array}$

25. $\begin{array}{r} 5\frac{1}{3} \\ \underline{+\ 8\frac{2}{5}} \end{array}$ 26. $\begin{array}{r} 17\frac{3}{8} \\ \underline{+\ 6\frac{5}{6}} \end{array}$ 27. $\begin{array}{r} 12\frac{15}{16} \\ \underline{+\ 7\frac{3}{8}} \end{array}$ 28. $\begin{array}{r} 9\frac{5}{11} \\ \underline{+\ 6\frac{2}{3}} \end{array}$

Subtract and simplify.

29. $\begin{array}{r} \frac{6}{7} \\ \underline{-\ \frac{2}{3}} \end{array}$ 30. $\begin{array}{r} \frac{3}{4} \\ \underline{-\ \frac{5}{9}} \end{array}$ 31. $\begin{array}{r} 11\frac{2}{3} \\ \underline{-\ 4\frac{1}{7}} \end{array}$ 32. $\begin{array}{r} 38\frac{3}{4} \\ \underline{-\ 17\frac{4}{5}} \end{array}$

33. $\begin{array}{r} 76\frac{3}{5} \\ \underline{-\ 12\frac{2}{7}} \end{array}$ 34. $\begin{array}{r} 16\frac{4}{13} \\ \underline{-\ 14\frac{1}{2}} \end{array}$ 35. $\begin{array}{r} 28\frac{8}{9} \\ \underline{-\ 17\frac{5}{6}} \end{array}$ 36. $\begin{array}{r} 29\frac{2}{3} \\ \underline{-\ 8\frac{3}{4}} \end{array}$

DIVISION WITH FRACTIONAL NUMERALS

This flow chart shows how to divide with fractional numerals.

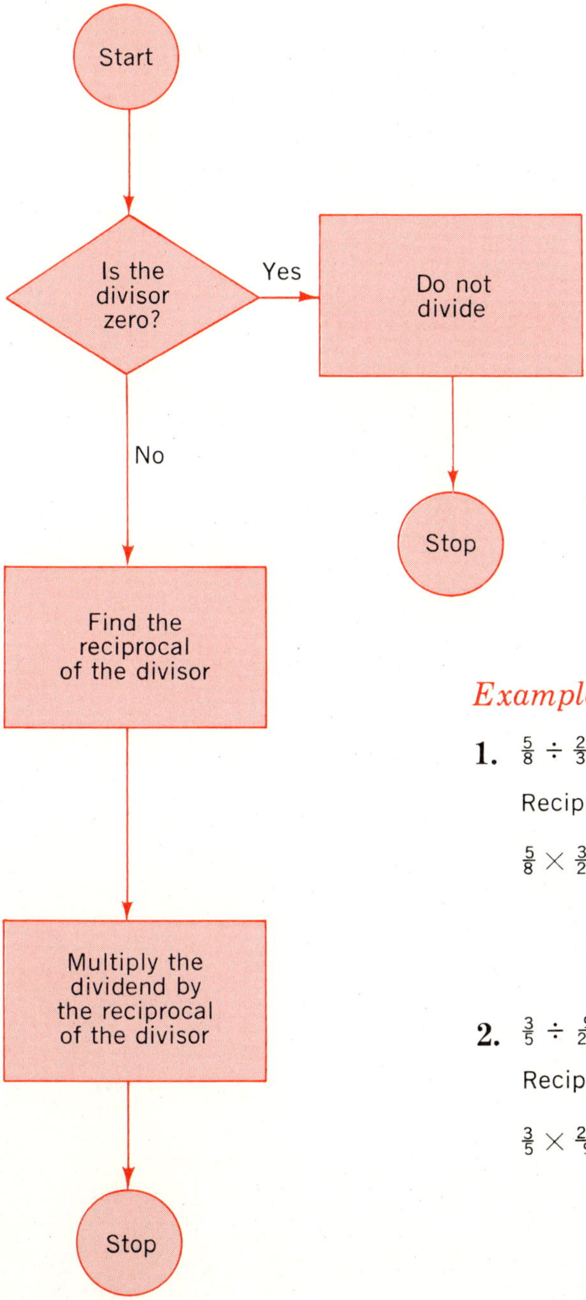

Examples

1. $\frac{5}{8} \div \frac{2}{3}$

Reciprocal of $\frac{2}{3}$ is $\frac{3}{2}$.

$\frac{5}{8} \times \frac{3}{2} = \frac{5 \times 3}{8 \times 2}$

$= \frac{15}{16}$

2. $\frac{3}{5} \div \frac{9}{20}$

Reciprocal of $\frac{9}{20}$ is $\frac{20}{9}$.

$\frac{3}{5} \times \frac{20}{9} = \frac{3 \times 20}{5 \times 9}$

$= \frac{3 \times 2 \times 2 \times 5}{3 \times 3 \times 5}$

$= \frac{4}{3}$

1. Let's use the flow chart to divide $\frac{3}{5}$ by $\frac{6}{5}$.

a. What is the answer to the first yes—no decision?

b. What is the reciprocal of $\frac{6}{5}$?

c. Copy and complete. $\frac{3}{5} \times \frac{5}{6} = \frac{3 \times 5 \times 1}{5 \times \triangle \times \square}$

d. Write simplest fractional numeral for the quotient $\frac{3}{5} \div \frac{6}{5}$.

We can find a quotient by multiplying the dividend by the reciprocal of the divisor.

2. Divide. Use the flow chart.

a. $\frac{3}{8} \div \frac{4}{5}$ b. $\frac{9}{10} \div \frac{6}{7}$ c. $\frac{4}{7} \div \frac{0}{2}$

d. $\frac{10}{11} \div \frac{5}{22}$ e. $\frac{6}{10} \div 3$ f. $15 \div \frac{5}{15}$

EXERCISES

Use the flow chart to find these quotients.

1. $\frac{3}{5} \div \frac{1}{2}$
2. $\frac{5}{7} \div \frac{3}{4}$
3. $\frac{3}{8} \div \frac{5}{7}$
4. $\frac{2}{3} \div \frac{1}{4}$
5. $\frac{7}{8} \div \frac{1}{4}$
6. $\frac{6}{5} \div \frac{1}{3}$
7. $\frac{9}{10} \div \frac{4}{5}$
8. $\frac{6}{13} \div \frac{1}{8}$
9. $8 \div \frac{1}{10}$
10. $\frac{5}{6} \div 3$
11. $\frac{8}{9} \div 5$
12. $6 \div \frac{3}{8}$
13. $\frac{5}{9} \div \frac{2}{5}$
14. $\frac{3}{10} \div \frac{7}{10}$
15. $\frac{2}{3} \div \frac{5}{6}$
16. $4 \div \frac{2}{3}$
17. $\frac{3}{8} \div 4$
18. $9 \div \frac{3}{7}$
19. $\frac{8}{9} \div \frac{7}{9}$
20. $\frac{5}{6} \div \frac{3}{5}$
21. $\frac{7}{12} \div 7$
22. $3 \div \frac{5}{8}$
23. $\frac{21}{8} \div \frac{7}{8}$
24. $\frac{2}{3} \div 10$
25. $\frac{2}{3} \div \frac{5}{8}$
26. $\frac{29}{6} \div \frac{4}{3}$
27. $\frac{5}{7} \div \frac{8}{12}$
28. $\frac{7}{10} \div \frac{3}{4}$

Solve these problems.

29. Victor had $\frac{3}{4}$ of an hour to finish a test containing 12 problems. Suppose he allows himself the same amount of time for each problem. What part of an hour can he spend on each problem?

30. Tracy was packing Christmas cookies for the church party. She put $\frac{3}{4}$ pound in each box. How many boxes can she pack completely with 8 pounds of cookies?

31. Mr. Sergi ordered $\frac{3}{4}$ of a ton of top soil. He wants to divide it evenly among his six flower beds. How much would he use for each bed?

32. Mrs. Torockio needs pieces of ruffled trim that are $\frac{3}{4}$ of a yard long. How many can she cut from 14 yards of the material.

Let's Practice

Find the measure of each angle. Use your protractor.

1.

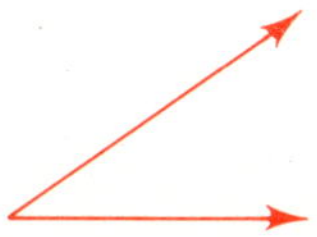

2.

3.

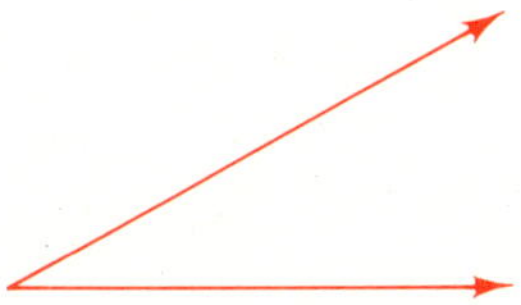

4.

5.

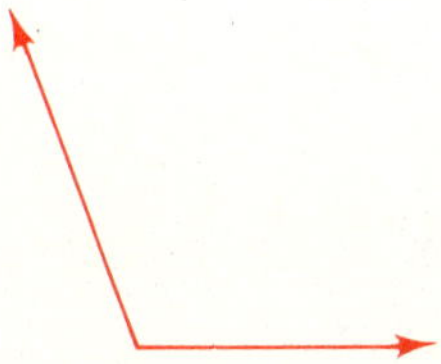

6.

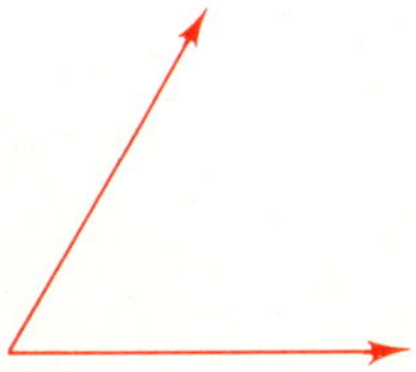

DO YOU REMEMBER?

Add and simplify.

1. $\begin{array}{r} \frac{5}{6} \\ + \frac{2}{3} \\ \hline \end{array}$
2. $\begin{array}{r} \frac{4}{5} \\ + \frac{1}{2} \\ \hline \end{array}$
3. $\begin{array}{r} \frac{6}{15} \\ + \frac{7}{15} \\ \hline \end{array}$
4. $\begin{array}{r} \frac{3}{4} \\ + \frac{1}{12} \\ \hline \end{array}$
5. $\begin{array}{r} \frac{5}{12} \\ \frac{7}{10} \\ + \frac{1}{4} \\ \hline \end{array}$
6. $\begin{array}{r} \frac{5}{16} \\ \frac{3}{5} \\ + \frac{1}{2} \\ \hline \end{array}$
7. $\begin{array}{r} 3\frac{3}{4} \\ + 5\frac{1}{4} \\ \hline \end{array}$
8. $\begin{array}{r} 24\frac{7}{8} \\ + 23\frac{3}{5} \\ \hline \end{array}$

Subtract and simplify.

9. $\begin{array}{r} \frac{7}{8} \\ - \frac{5}{8} \\ \hline \end{array}$
10. $\begin{array}{r} \frac{1}{3} \\ - \frac{1}{5} \\ \hline \end{array}$
11. $\begin{array}{r} \frac{2}{3} \\ - \frac{3}{10} \\ \hline \end{array}$
12. $\begin{array}{r} \frac{1}{3} \\ - \frac{1}{8} \\ \hline \end{array}$
13. $\begin{array}{r} 13\frac{2}{3} \\ - 5\frac{3}{5} \\ \hline \end{array}$
14. $\begin{array}{r} 16 \\ - 5\frac{3}{7} \\ \hline \end{array}$
15. $\begin{array}{r} 4\frac{1}{3} \\ - \frac{1}{5} \\ \hline \end{array}$
16. $\begin{array}{r} 5\frac{3}{4} \\ - 2\frac{1}{8} \\ \hline \end{array}$

Multiply and simplify.

17. $\frac{1}{5} \cdot \frac{3}{6}$
18. $\frac{12}{11} \cdot \frac{1}{4}$
19. $\frac{8}{3} \times \frac{3}{5}$
20. $\frac{3}{4} \cdot \frac{5}{11}$
21. $\frac{23}{7} \times \frac{5}{7}$
22. $5 \cdot \frac{6}{7}$
23. $\frac{2}{9} \times 14$
24. $\frac{1}{16} \cdot \frac{8}{9}$

Divide and simplify.

25. $\frac{2}{3} \div \frac{1}{2}$
26. $\frac{5}{6} \div \frac{3}{4}$
27. $\frac{4}{5} \div \frac{6}{9}$
28. $\frac{4}{10} \div \frac{7}{8}$
29. $5 \div \frac{6}{7}$
30. $14 \div \frac{3}{2}$
31. $\frac{5}{6} \div 4$
32. $\frac{7}{11} \div 28$

Solve this problem.

33. A post is to be $4\frac{1}{2}$ feet high with $2\frac{1}{2}$ feet beneath the ground. How long should it be?

MULTIPLICATION AND DIVISION: MIXED NUMERALS

1. When we multiply with mixed numerals, we can rename the factors.

$$
\begin{aligned}
2\tfrac{1}{4} \times 3\tfrac{2}{3} &= \frac{9}{4} \times \frac{11}{3} \\
&= \frac{9 \times 11}{4 \times 3} \\
&= \frac{3 \times 3 \times 11}{3 \times 4} \\
&= \frac{33}{4} \\
&= 8\tfrac{1}{4}
\end{aligned}
$$

   a. What fractional numeral was used for $2\frac{1}{4}$? for $3\frac{2}{3}$?

   b. Why did we rename 9 as 3×3?

   c. How was $\frac{33}{4}$ renamed?

2. Multiply and simplify.

   a. $2\frac{1}{8} \times \frac{3}{4}$ b. $\frac{7}{8} \times 2\frac{1}{5}$ c. $7 \times 2\frac{3}{4}$

   d. $7\frac{1}{2} \times 3\frac{1}{3}$ e. $6\frac{2}{3} \times 3\frac{1}{8}$ f. $\frac{5}{6} \times 3\frac{1}{6}$

3. Here is a division with mixed numerals.

$$
\begin{aligned}
2\tfrac{1}{5} \div 4\tfrac{1}{8} &= \frac{11}{5} \div \frac{33}{8} \\
&= \frac{11}{5} \times \frac{8}{33} \\
&= \frac{11 \times 8}{5 \times 33} \\
&= \frac{8 \times 11}{5 \times 3 \times 11} \\
&= \frac{8}{15}
\end{aligned}
$$

   a. What fractional numeral was used for $2\frac{1}{5}$? for $4\frac{1}{8}$?

   b. What is the reciprocal of $\frac{33}{8}$?

   c. Why did we rename 33 as 3×11?

4. Divide and simplify.

a. $3\frac{1}{4} \div 1\frac{1}{2}$ b. $5\frac{1}{6} \div 2\frac{1}{3}$ c. $1\frac{1}{8} \div 4\frac{1}{9}$

d. $12\frac{3}{5} \div 7\frac{1}{5}$ e. $16\frac{3}{4} \div 4$ f. $9\frac{1}{3} \div 6$

EXERCISES

Multiply and simplify.

1. $3\frac{1}{4} \cdot \frac{5}{8}$
2. $2\frac{1}{3} \times \frac{5}{6}$
3. $5 \times 1\frac{5}{8}$
4. $4 \cdot 2\frac{5}{6}$
5. $3\frac{1}{5} \cdot 2\frac{1}{8}$
6. $6\frac{1}{4} \times 3\frac{5}{7}$
7. $8\frac{1}{9} \times 5\frac{2}{5}$
8. $4\frac{2}{7} \cdot 3\frac{8}{9}$
9. $1\frac{5}{6} \cdot 4\frac{2}{3}$
10. $20\frac{10}{13} \times 17\frac{1}{3}$
11. $16\frac{5}{8} \cdot 9$
12. $12\frac{5}{6} \cdot 4\frac{1}{9}$

Divide and simplify.

13. $2\frac{1}{6} \div \frac{4}{5}$
14. $3\frac{5}{9} \div \frac{2}{3}$
15. $7 \div 6\frac{1}{3}$
16. $5 \div 4\frac{1}{4}$
17. $2\frac{1}{9} \div 3\frac{1}{3}$
18. $4\frac{5}{6} \div 2\frac{7}{8}$
19. $10\frac{1}{4} \div 9\frac{1}{2}$
20. $12\frac{5}{6} \div 7\frac{1}{3}$
21. $11\frac{1}{9} \div 13\frac{1}{5}$
22. $3\frac{6}{7} \div \frac{9}{14}$
23. $5\frac{2}{3} \div 6$
24. $12\frac{7}{8} \div 3\frac{1}{8}$

Solve these problems.

25. Phyllis found a recipe calling for $\frac{3}{4}$ cup of milk. She wants to make $2\frac{1}{2}$ times the recipe. How much milk should she use?

26. Mrs. Tierman selected a special tile which was $4\frac{1}{2}$ inches long. How many tiles will be needed to make a row 36 inches long?

CHAPTER REVIEW

Which is not a number of arithmetic?

1. 5 2. $\frac{2}{0}$ 3. $\frac{8}{5}$ 4. 0 5. 3.2

Find simplest fractional numerals.

6. $\frac{18}{24}$ 7. $\frac{48}{60}$ 8. $\frac{8}{11}$ 9. $\frac{30}{39}$ 10. $\frac{50}{25}$

Find the L.C.D.

11. $\frac{1}{5}, \frac{2}{3}$ 12. $\frac{9}{10}, \frac{7}{15}$ 13. $\frac{5}{12}, \frac{17}{18}$

14. $\frac{1}{4}, \frac{3}{12}, \frac{5}{24}$ 15. $\frac{9}{10}, \frac{7}{8}, \frac{5}{6}$ 16. $\frac{1}{20}, \frac{3}{4}, \frac{4}{15}$

Add and simplify.

17. $\frac{1}{6} + \frac{1}{2}$ 18. $\frac{2}{3} + \frac{1}{9}$ 19. $\frac{1}{2} + \frac{2}{5}$ 20. $\frac{5}{8} + \frac{2}{9}$

21. $\frac{2}{3}$, $\frac{4}{5}$, $\frac{1}{8}$ 22. $\frac{3}{4}$, $\frac{5}{16}$, $\frac{1}{2}$ 23. $\frac{3}{10}$, $\frac{7}{20}$, $\frac{2}{15}$ 24. $3\frac{5}{7}$, $2\frac{4}{35}$, $5\frac{3}{5}$

25. $1\frac{2}{7} + 2\frac{5}{9} + 10\frac{2}{3}$ 26. $10\frac{2}{3} + 24\frac{5}{8} + 30\frac{9}{10}$

Subtract and simplify.

27. $\frac{5}{11} - \frac{2}{11}$ 28. $\frac{9}{10} - \frac{3}{10}$ 29. $\frac{7}{10} - \frac{1}{5}$ 30. $\frac{7}{8} - \frac{5}{12}$

31. $\frac{3}{4} - \frac{2}{5}$ 32. $\frac{11}{12} - \frac{7}{10}$ 33. $\frac{24}{25} - \frac{3}{50}$ 34. $5\frac{60}{69} - 2\frac{2}{3}$

35. $7\frac{5}{8} - 2\frac{5}{12}$ 36. $12\frac{1}{3} - 5\frac{7}{9}$ 37. $28\frac{2}{7} - 15\frac{5}{9}$

Multiply and simplify.

38. $\frac{3}{4} \times \frac{1}{5}$ 39. $\frac{7}{10} \cdot \frac{3}{4}$ 40. $\frac{3}{4} \cdot \frac{2}{5}$

41. $\frac{1}{2} \times \frac{4}{5}$ 42. $\frac{3}{5} \cdot \frac{5}{6}$ 43. $\frac{9}{10} \times \frac{5}{6}$

44. $1\frac{1}{2} \cdot 1\frac{1}{3}$ 45. $6\frac{2}{3} \times 2\frac{1}{10}$ 46. $12 \cdot \frac{5}{8}$

47. $\frac{2}{3} \times 5$ 48. $2\frac{8}{9} \cdot 4\frac{1}{2}$ 49. $5\frac{5}{8} \cdot 3\frac{1}{9}$

Divide and simplify.

50. $\frac{3}{4} \div \frac{4}{5}$ 51. $\frac{1}{2} \div \frac{1}{8}$ 52. $\frac{1}{16} \div \frac{1}{2}$

53. $\frac{5}{6} \div \frac{10}{11}$ 54. $\frac{3}{10} \div \frac{9}{20}$ 55. $\frac{9}{16} \div \frac{3}{4}$

56. $1\frac{1}{2} \div 1\frac{1}{5}$ 57. $2\frac{4}{5} \div 2\frac{1}{10}$ 58. $6\frac{2}{3} \div 6$

59. $18 \div 1\frac{1}{2}$ 60. $3\frac{1}{8} \div 3\frac{3}{4}$ 61. $5\frac{5}{9} \div 4\frac{1}{6}$

Solve these problems.

62. Mrs. Smith made dance costumes that called for $4\frac{1}{4}$ yards of red ribbon, $5\frac{1}{2}$ yards of white ribbon, and $7\frac{3}{4}$ yards of blue ribbon. How many yards of ribbon did she use?

63. Sharon weighed $135\frac{1}{4}$ pounds. She went on a diet and brought her weight down to $118\frac{1}{2}$ pounds. How many pounds did she lose?

Terms You Should Know

mixed numeral	numbers of arithmetic
denominator	least common denominator
numerator	simplest fractional numeral

CHAPTER TEST

Which is not a number of arithmetic?

1. $\frac{5}{8}$ 2. $\frac{7}{3}$ 3. 4 4. $\frac{3}{0}$ 5. $\frac{0}{3}$

Find simplest fractional numerals.

6. $\frac{28}{48}$ 7. $\frac{25}{5}$ 8. $\frac{42}{28}$

Find the L.C.D.

9. $\frac{20}{21}, \frac{13}{14}$ 10. $\frac{9}{25}, \frac{2}{3}, \frac{7}{15}$

Add and simplify.

11. $\begin{array}{r} \frac{5}{8} \\ + \frac{1}{4} \\ \hline \end{array}$ 12. $\begin{array}{r} \frac{9}{10} \\ \frac{1}{3} \\ \frac{3}{5} \\ \hline \end{array}$ 13. $\begin{array}{r} 1\frac{2}{3} \\ 5\frac{5}{9} \\ 8\frac{5}{6} \\ \hline \end{array}$

Subtract and simplify.

14. $\begin{array}{r} \frac{11}{14} \\ - \frac{5}{7} \\ \hline \end{array}$ 15. $\begin{array}{r} \frac{8}{25} \\ - \frac{2}{15} \\ \hline \end{array}$ 16. $\begin{array}{r} 10\frac{2}{21} \\ - 5\frac{5}{6} \\ \hline \end{array}$

Multiply and simplify.

17. $\frac{9}{10} \times \frac{3}{5}$ 18. $\frac{25}{27} \cdot \frac{18}{35}$ 19. $1\frac{7}{8} \times 2\frac{2}{5}$

Divide and simplify.

20. $\frac{3}{8} \div \frac{7}{8}$ 21. $\frac{24}{25} \div \frac{9}{20}$ 22. $10\frac{2}{3} \div 2\frac{2}{9}$

Solve this problem.

23. Bob is packaging string beans. The beans are to be put into $2\frac{1}{2}$-pound packages. How many packages can he get from 255 pounds of beans?

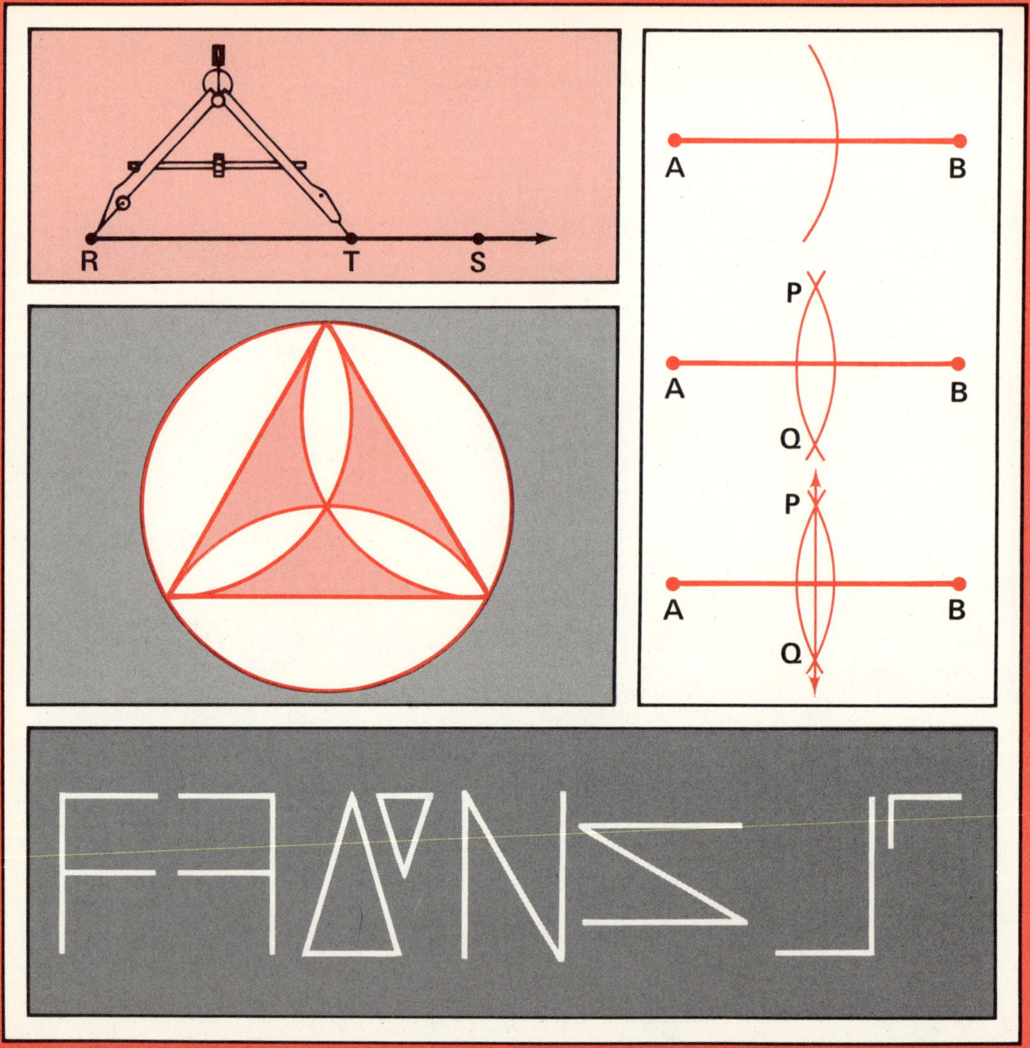

CHAPTER THREE

Congruences and Constructions

CONGRUENT FIGURES

1. *Congruent* figures look exactly alike. Which pairs of figures are congruent?

a. 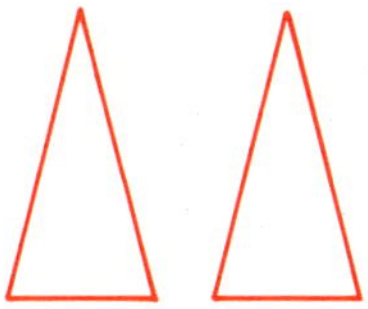
b. 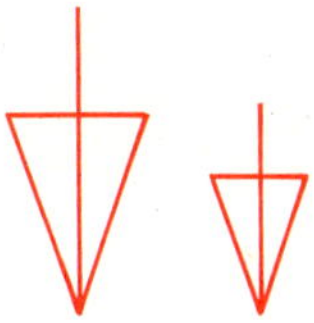
c. 

d. 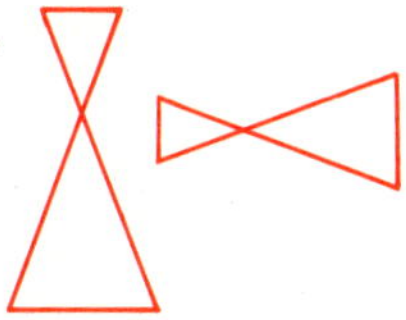
e. 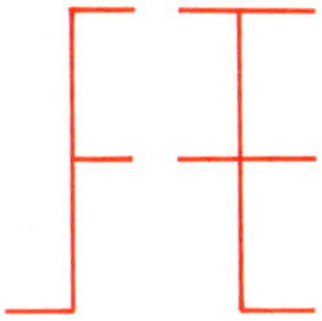
f. 

2. Congruent segments have the same length. Use your ruler. Which pairs of segments are congruent?

a. 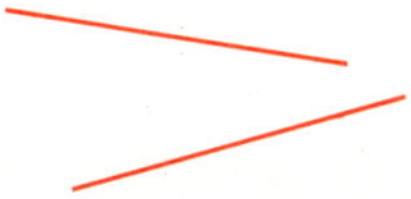
b. 
c. 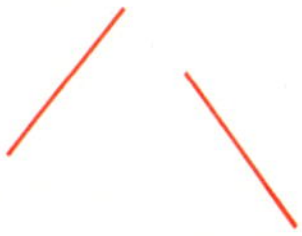

3. Congruent angles have the same measure. Which pairs of angles are congruent?

a. 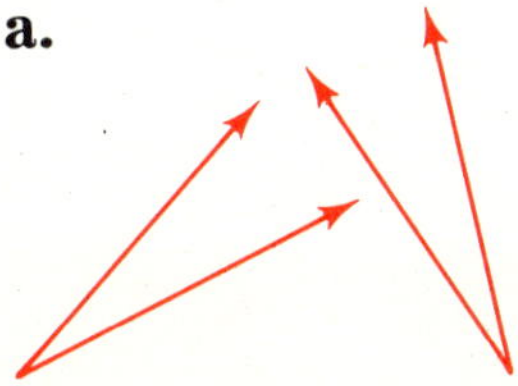
b. 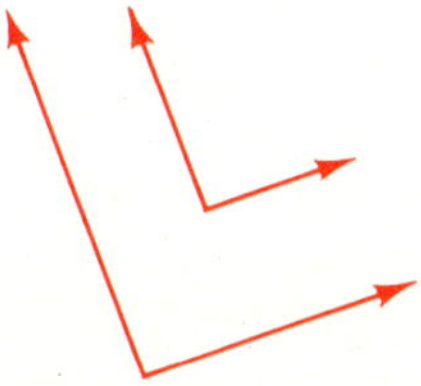
c. 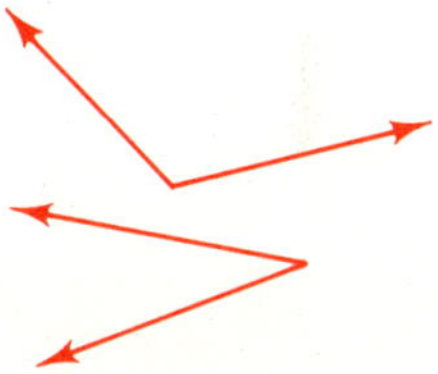

4. Congruent figures can also meet or overlap. Which pairs of figures are congruent?

a. 
b. 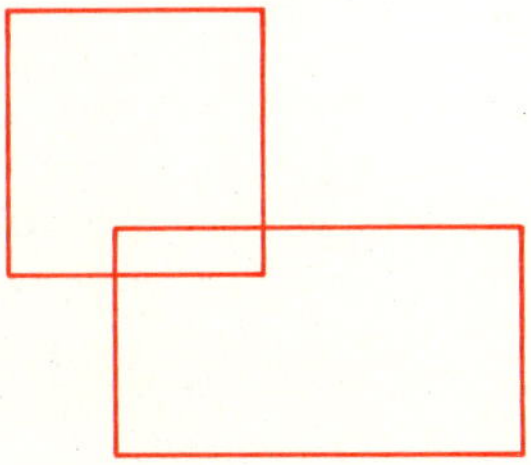
c. 

5. The symbol ≅ is read "is congruent to." We write $\triangle ABC \cong \triangle DEF$ to say "triangle *ABC* is congruent to triangle *DEF*." Read these.

a. $\triangle PQR \cong \triangle MNO$ **b.** $\overline{AB} \cong \overline{CD}$

EXERCISES

1. Which pairs of figures are congruent?

a. 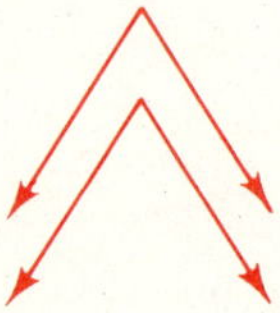

b. 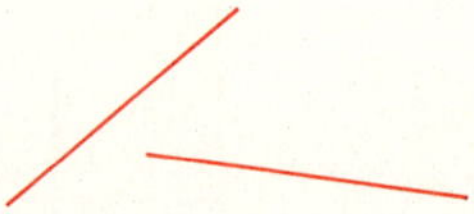

c. 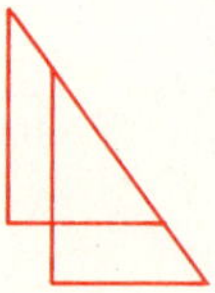

d. 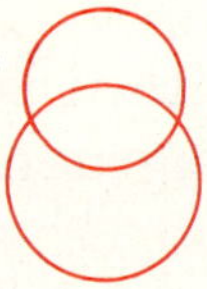

e. 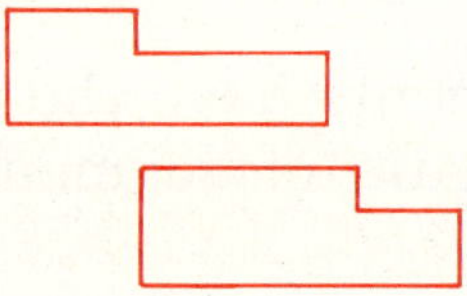

f. 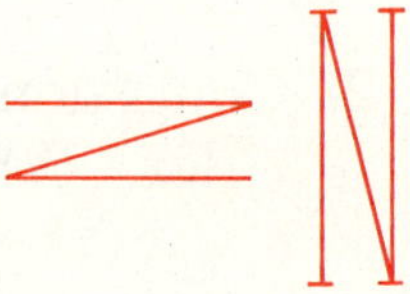

2. Draw a 2-inch line segment. Draw another segment congruent to it.

3. Draw an angle. Draw another angle congruent to it.

Let's Practice

Copy and complete this chart.

	Geometric Figure	*How Named*	*How Drawn*
1.	Segment	$\overline{AB}$	
2.	Ray		
3.			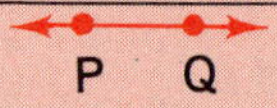
4.	Angle	$\angle ABC$ or $\angle B$	
5.		$\triangle ABC$	
6.	Plane	Plane *b*	

REFLECTIONS AND ROTATIONS

A mirror will show a reflection of you. Triangles *PQR* and *ABC* are reflections of each other. Line *l* is called a line of reflection.

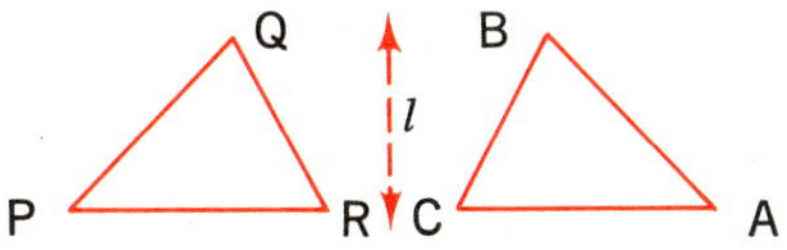

1. Copy. Draw a line of reflection for each.

a.

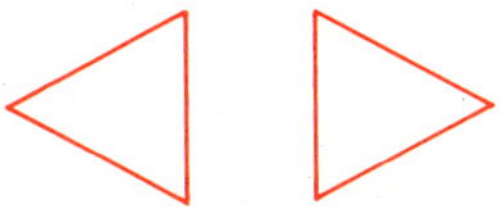

b.

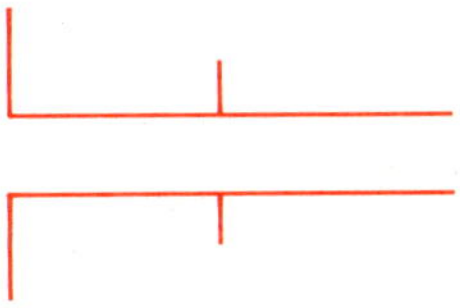

2. Copy. Draw a reflection of each.

a.

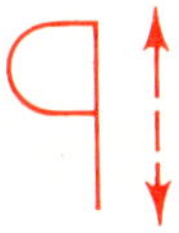

b.

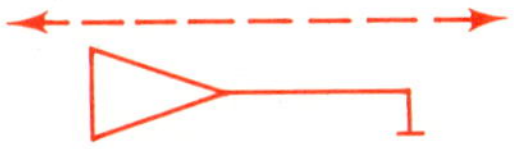

3. If figures can be made to look alike by reflection, they are congruent. $\triangle DEF$ can be reflected to look like $\triangle ABC$. $\triangle DEF \cong \triangle ABC$.

Draw a reflection of the red figure to show it is congruent to the black figure.

a.

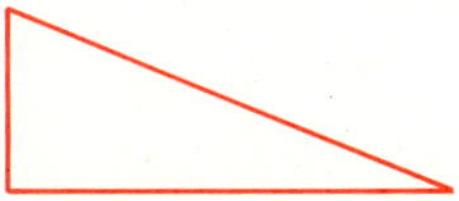

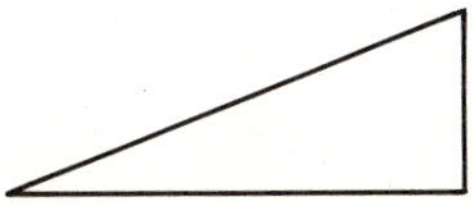

b.

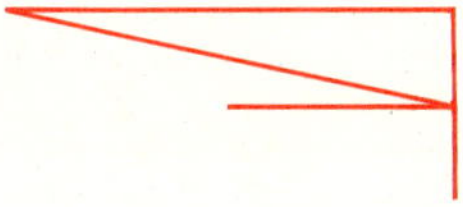

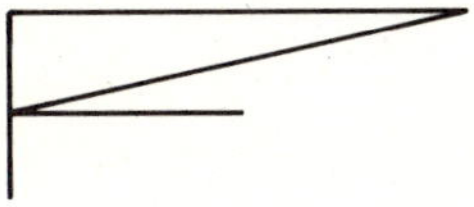

4. Which pairs of figures can be shown congruent by reflection?

a. 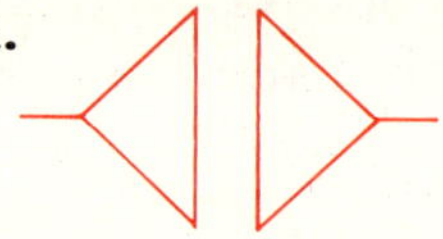 **b.** 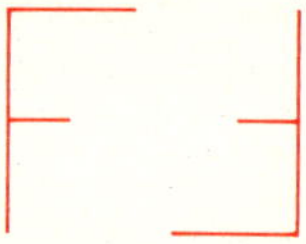 **c.** 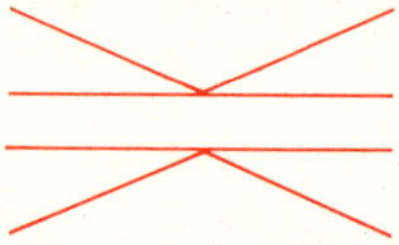

5. Sometimes we can rotate (turn) figures to make them look alike.

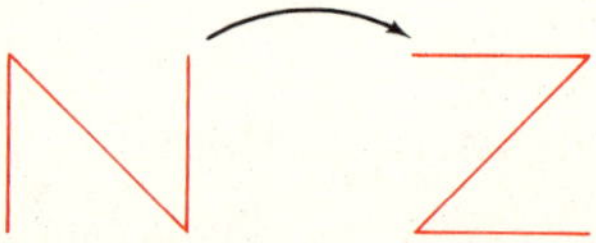

Which red figures can be rotated to look like the black figures?

a. 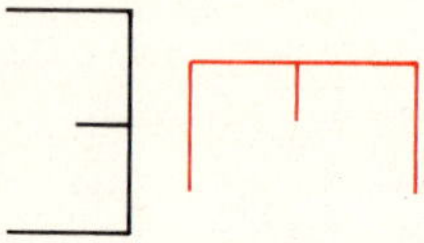 **b.** 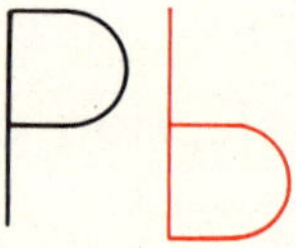 **c.** 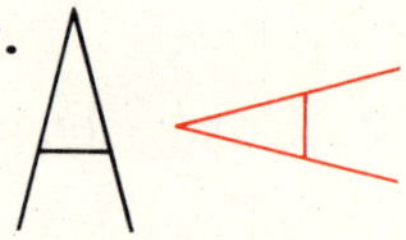

6. Which pairs of figures can be shown congruent by rotation?

a. 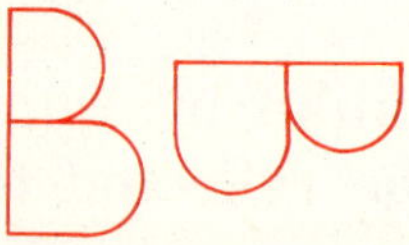 **b.** 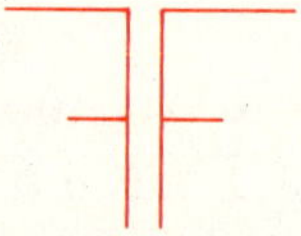 **c.** 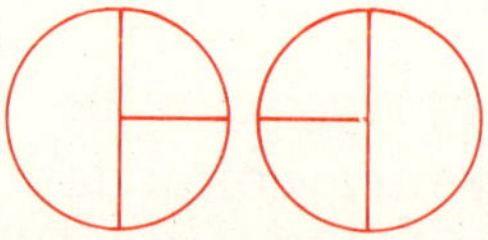 **d.** 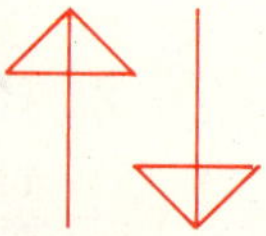

7. $\triangle ABC$ and $\triangle DEF$ can be shown congruent by *both* reflection and rotation.

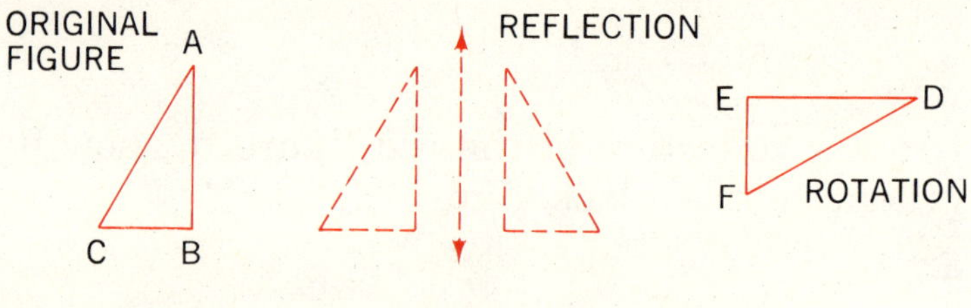

Which pairs of figures are congruent?

a. 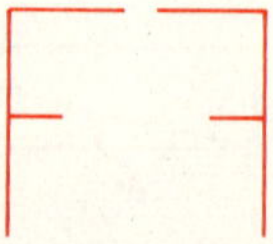 **b.**  **c.** 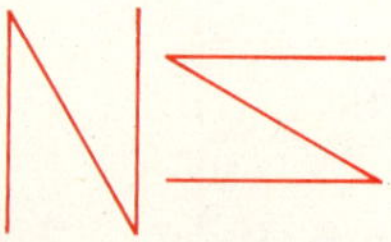 **d.** 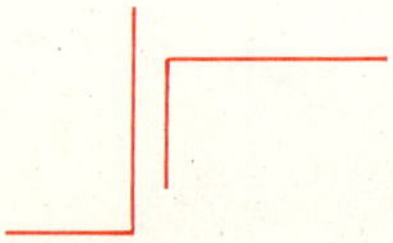

EXERCISES

Copy. Draw a reflection of each.

1. 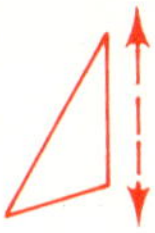

2. 

3. 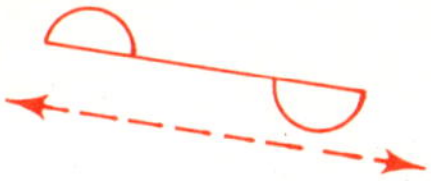

4. 

Which pairs of figures can be shown congruent by reflection?

5. 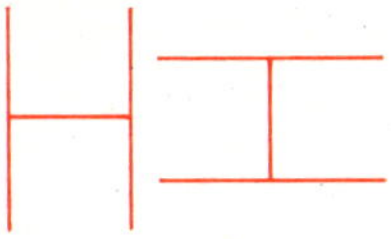

6. 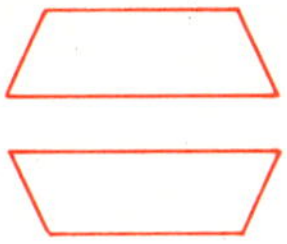

7. 

8. 

Draw figures that can be rotated to look like these.

9. 

10. 

11. 

12. 

Which pairs of figures can be shown congruent by rotation?

13. 

14. 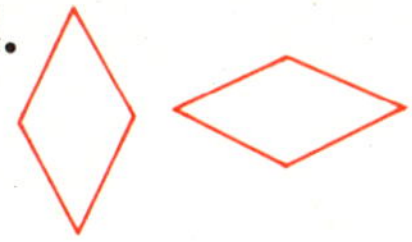

15. 

16. 

Which figures in each set are congruent to the black figure?

17. 

18. 

19. 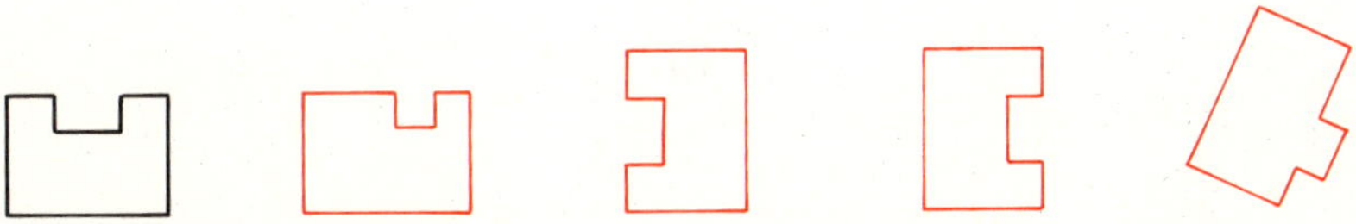

COMPARING FIGURES

In each exercise, which item is different? How is it different?

1. 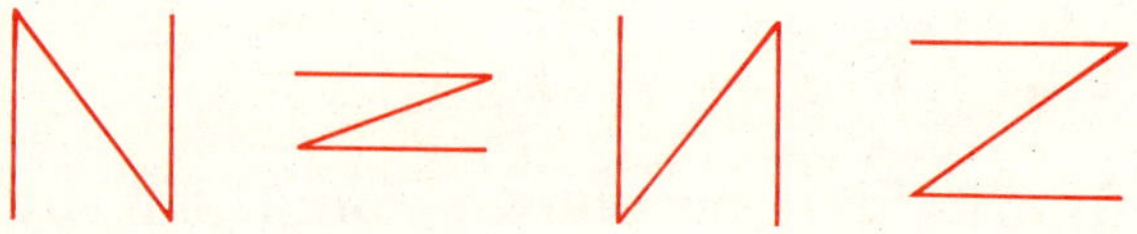

2. 

3. 

4. 

5. 

6. 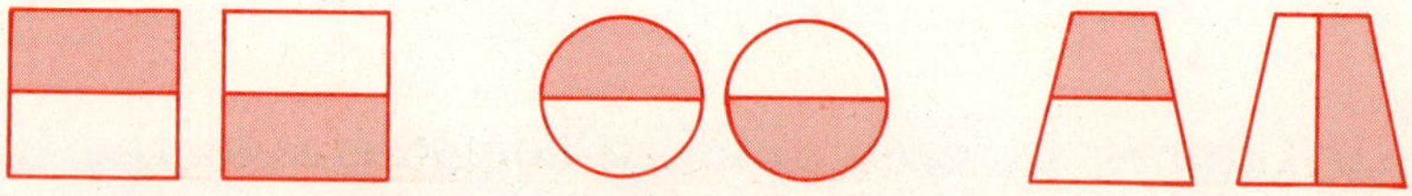

7. 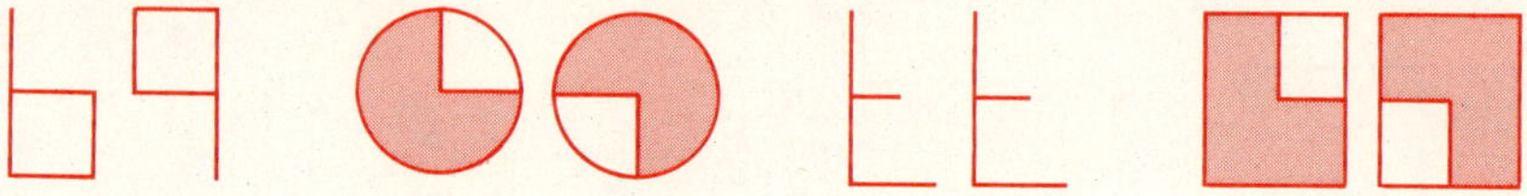

8. 

9. 

10. 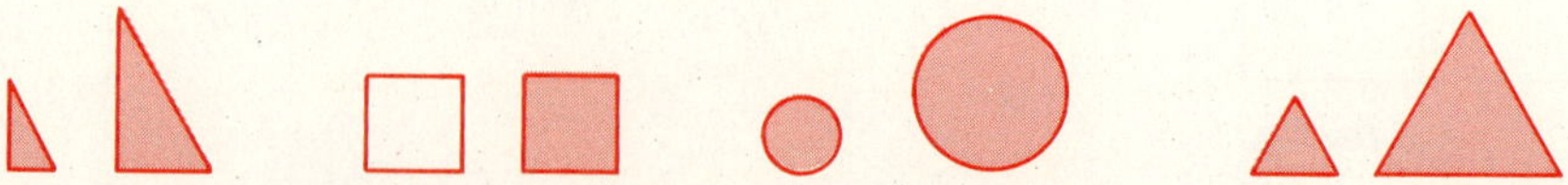

PROBLEM SOLVING

1. A wild deer loses 30% of its weight during a winter. If a deer weighs 120 pounds in the fall, what will it weigh come spring?

2. At a farm store, steel fence posts were on sale for $1.20 each. The regular price was $1.40 each. What was the per cent of discount?

3. Mr. Thrift sells gasoline for 29.4¢ per gallon. He delivered 234 gallons to the Elliotts whose tank holds 250 gallons. What was their total bill?

4. Mr. Woodward drives about 125 miles per day. About how long will it take him to drive 1,500 miles.

5. Mr. Shepard uses 10 bus tokens weekly to get to and from work. The tokens cost 30¢ each. How much does Mr. Shepard spend in 4 weeks?

6. The Goldstein family lived in an all electric home. They kept a record of monthly electricity costs. Find the average monthly cost of electricity.

January	$97.32
February	$83.15
March	$79.60
April	$52.40
May	$31.08
June	$26.42
July	$24.10
August	$20.18
September	$23.14
October	$30.18
November	$57.40
December	$81.20

COMPASS AND STRAIGHTEDGE

A *straightedge* is used to draw segments. A ruler is a straightedge with a measuring scale. A *compass* is used to mark off distances.

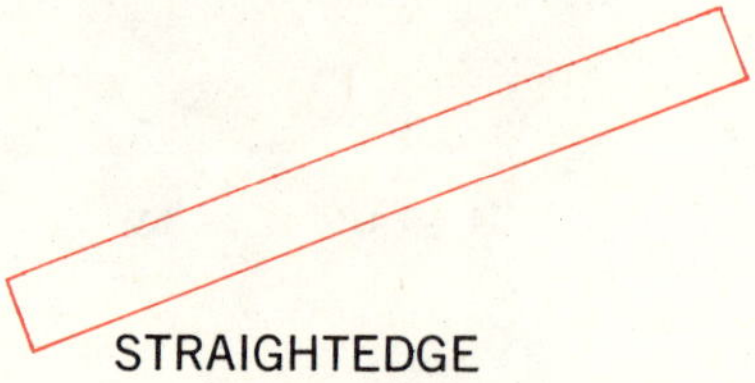

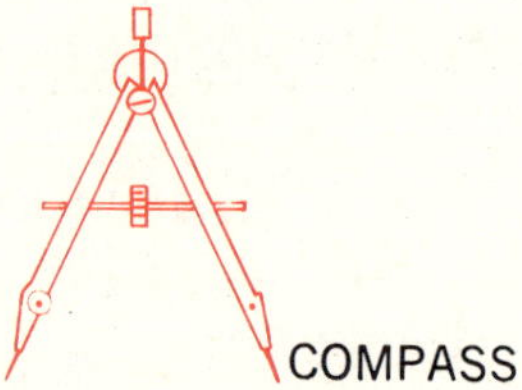

LET'S EXPLORE

1. Suppose we want to construct a segment on $\overrightarrow{RS}$ congruent to $\overline{PQ}$.

Draw a segment $\overline{PQ}$ and a ray $\overrightarrow{RS}$.

2. Place the compass so that one point is on P and the other is on Q.

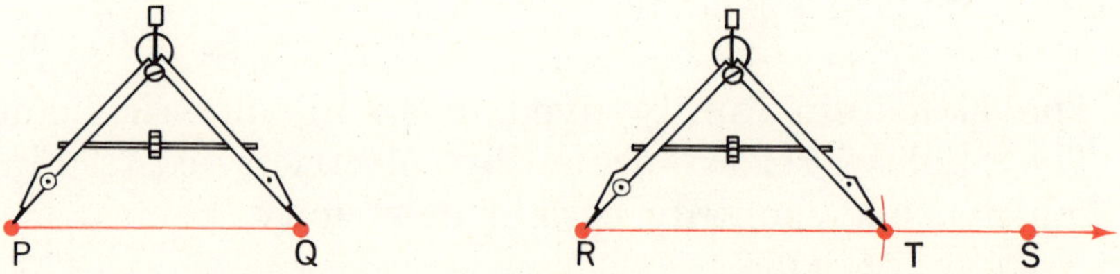

3. Pick up the compass. Do not change the distance between the points.

4. Place the metal tip on R. Then swing the compass to mark a point T on $\overrightarrow{RS}$. Then $\overline{RT} \cong \overline{PQ}$.

5. Use your compass and straightedge to construct a segment congruent to each.

a. 

b. 

In this chapter, we will make most constructions using only a compass and a straightedge.

EXERCISES

Construct a segment congruent to each.

1. 

2. 

3. 

4. 

5. 

Let's Practice

Recall that a *right* angle measures 90°. An *acute* angle measures between 0° and 90°. An *obtuse* angle measures between 90° and 180°. Tell whether each angle is right, acute, or obtuse.

1. 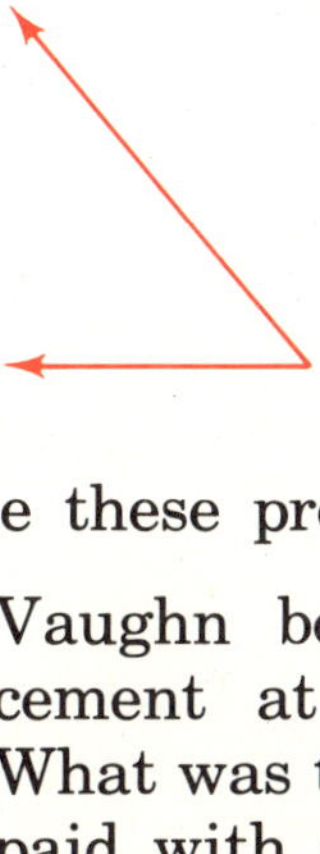

2. 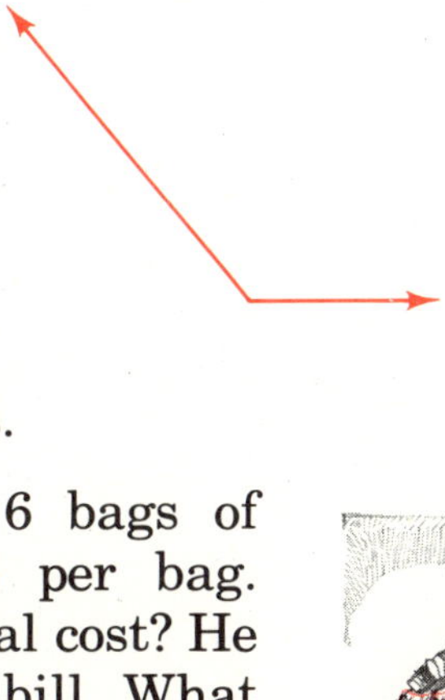

3. 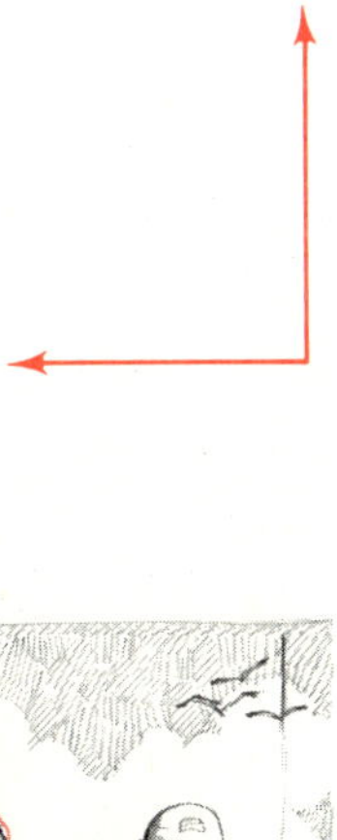

Solve these problems.

4. Vaughn bought 6 bags of cement at $2.15 per bag. What was the total cost? He paid with a $20 bill. What was his change?

5. On a fishing trip, Mark caught 12 more fish than Jeff. Jeff caught 17 fish. How many did Mark catch?

COPYING ANGLES

We can copy an angle using a compass and a straightedge.

LET'S EXPLORE

1. Draw an angle. Name it $\angle ABC$.

2. Let's construct another angle congruent to $\angle ABC$. Draw $\overrightarrow{DE}$.

3. Place the metal tip of your compass on B. Make a mark that crosses the sides of $\angle ABC$ as shown. Label the points P and Q.

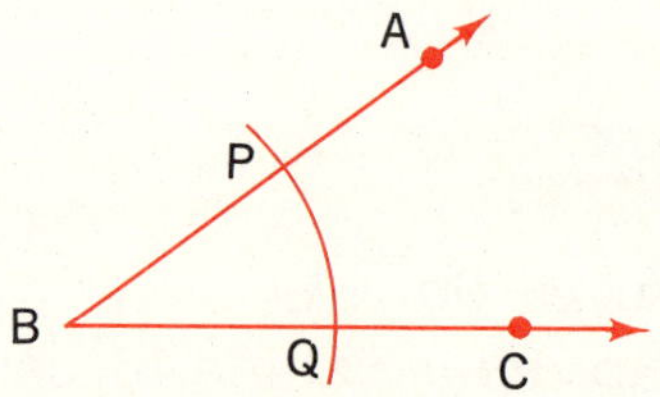

4. Do not change the compass setting. Put the metal tip on D. Make a mark that crosses $\overrightarrow{DE}$. Label the point R.

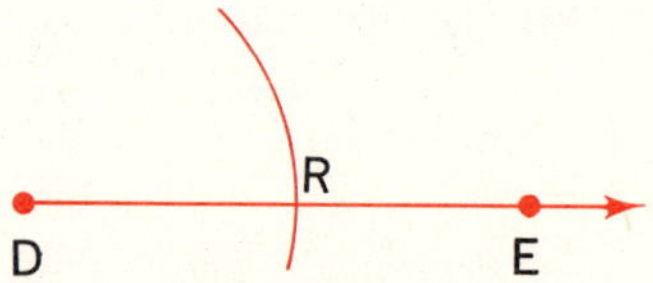

5. Place the compass points on P and Q.

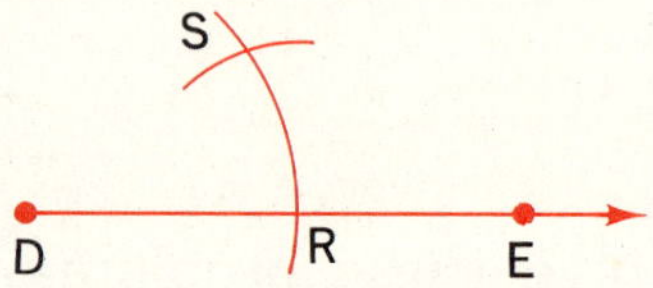

6. Do not change the compass setting. Put the metal tip on R and mark a point S.

7. Draw $\overrightarrow{DS}$. Then $\angle SDR \cong \angle ABC$.

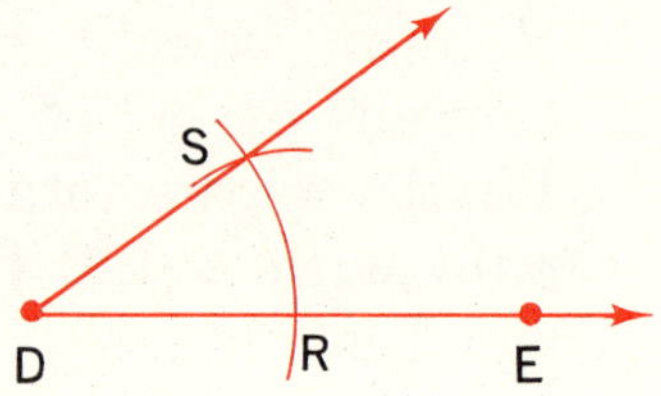

8. a. Draw three different angles.

   b. Construct an angle congruent to each of them.

9. Use your protractor to draw angles having these measures. Then construct a copy of each using your compass and straightedge.

Example 135°

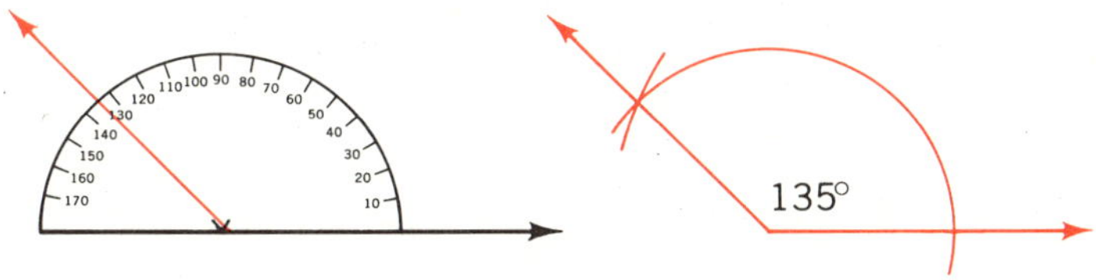

a. 30° b. 85° c. 150° d. 40° e. 170°

EXERCISES

1. Draw an acute and an obtuse angle. Construct an angle congruent to each of them.

Use a protractor to draw angles having these measures. Then construct a copy of each using your compass and straightedge.

2. 60° 3. 140° 4. 45° 5. 90° 6. 15°

7. 120° 8. 20° 9. 165° 10. 5° 11. 100°

Let's Practice

Compute and simplify.

1. $\frac{2}{7} \cdot \frac{5}{6}$ 2. $\frac{3}{5} \div \frac{2}{5}$ 3. $\frac{14}{23} \times \frac{5}{7}$ 4. $\frac{6}{7} \cdot \frac{10}{11}$

5. $\frac{3}{4} \cdot \frac{5}{6}$ 6. $\frac{2}{3} \div \frac{9}{10}$ 7. $\frac{4}{16} \div \frac{3}{8}$ 8. $\frac{8}{9} \div \frac{4}{9}$

Solve these problems.

9. Mr. Ray bought an electric organ for $1,895. The sales tax was 4%. What was the total cost?

10. Mr. Lee charges 50¢ per foot to install fences. What would he charge for installing 84 feet of fence?

BISECTING ANGLES

To *bisect* means to cut in half. Here $\angle ABC$ is bisected by $\overrightarrow{BD}$. Thus, $\angle ABD \cong \angle DBC$. Ray $\overrightarrow{BD}$ is called the *angle bisector*. We can construct an angle bisector using a compass and straightedge.

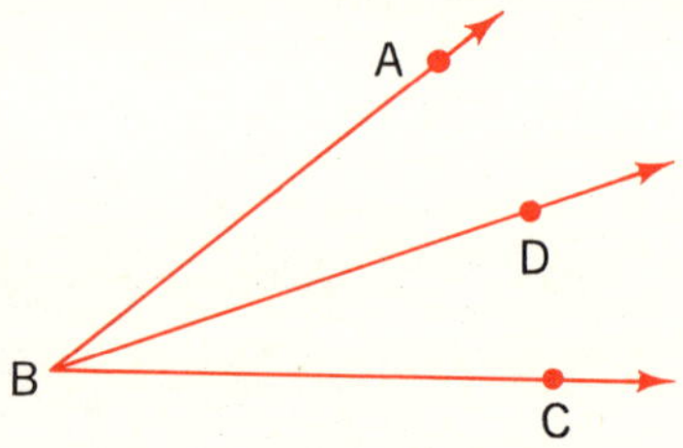

LET'S EXPLORE

1. Draw an angle. Name it $\angle ABC$.

2. Put the metal tip of the compass on B. Make marks at P and Q as shown.

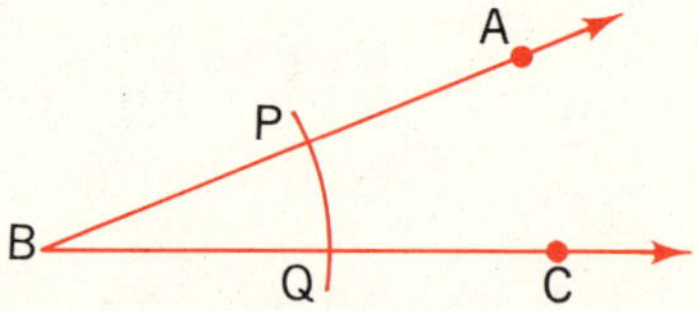

3. Put the metal tip on P. Make a mark where you think the bisector should be. Using the same opening, put the metal tip on Q and make another mark. Label the point where the marks cross D.

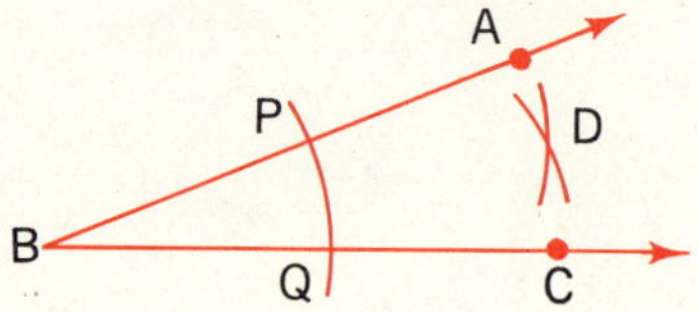

4. Draw $\overrightarrow{BD}$. $\overrightarrow{BD}$ bisects $\angle ABC$. $\angle ABD \cong \angle DBC$.

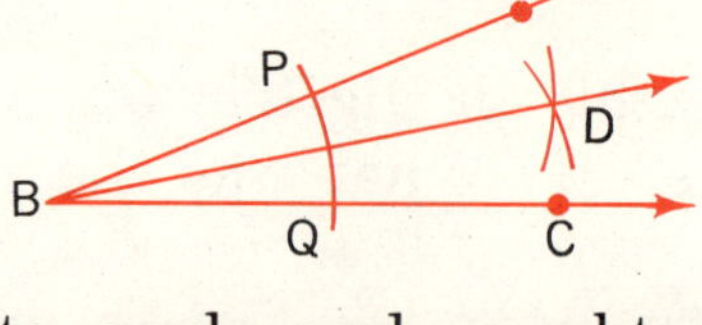

5. Draw an acute angle and an obtuse angle. Bisect each with your compass and straightedge.

6. Use your protractor to draw angles having these measures. Bisect each angle. Check your work with your protractor.

a. 50° b. 142° c. 75° d. 160° e. 80°

EXERCISES

1. Draw an acute angle, a right angle, and an obtuse angle. Bisect each with your compass and straightedge.

Use your protractor to draw angles having these measures. Bisect each angle. Check your work with your protractor.

2. 30° 3. 65° 4. 155° 5. 73° 6. 12°

7. 169° 8. 32° 9. 170° 10. 28° 11. 137°

Let's Practice

We read 24.7 as "twenty-four and seven tenths." Read these.

1. 8.3 2. 29.6 3. 54.61 4. 3.468

5. 19.82 6. .638 7. 1.5 8. 42.79

Solve these problems.

9. The length of the Saturn C-5 rocket is about 350 feet. The Saturn C-1 rocket is only $\frac{4}{7}$ as long. How long is the C-1 rocket?

10. Tickets for a play cost $3.50 for the main floor and $2.50 for the balcony. There were 1,200 main floor seats and 700 balcony seats sold. What was the total amount received for the tickets?

11. The daily attendance at Blair Junior High School is about 97%. The total enrollment is 1,263 students. About how many attend each day?

PARALLEL AND PERPENDICULAR LINES

Lines $\overleftrightarrow{AB}$ and $\overleftrightarrow{CD}$ are parallel.

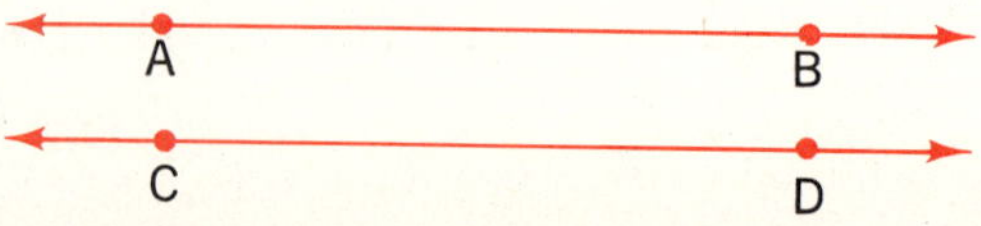

> Lines on a plane that do not meet are called **parallel** lines.

1. Draw a pair of parallel lines.

2. Segments are parallel if the lines containing them are parallel. Draw three parallel segments.

3. Which look parallel?

   a. b. c. d.

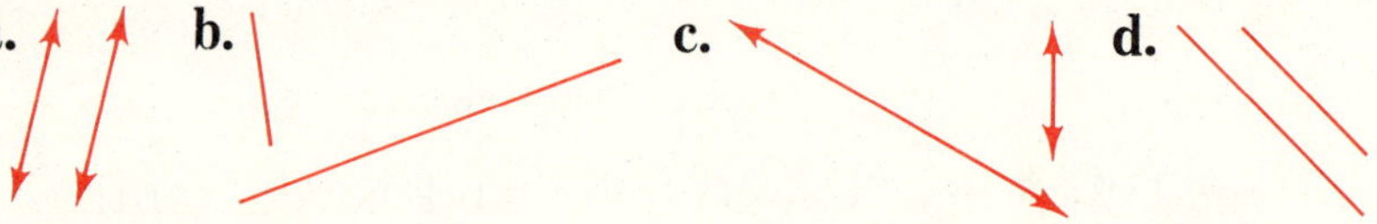

4. The symbol $\parallel$ is read "is parallel to." Read these.

   a. $\overleftrightarrow{AB} \parallel \overleftrightarrow{CD}$ b. $\overline{PQ} \parallel \overline{RS}$

> Lines or segments are **perpendicular** if they form right angles.

5. Which look perpendicular?

   a. b. c. d.

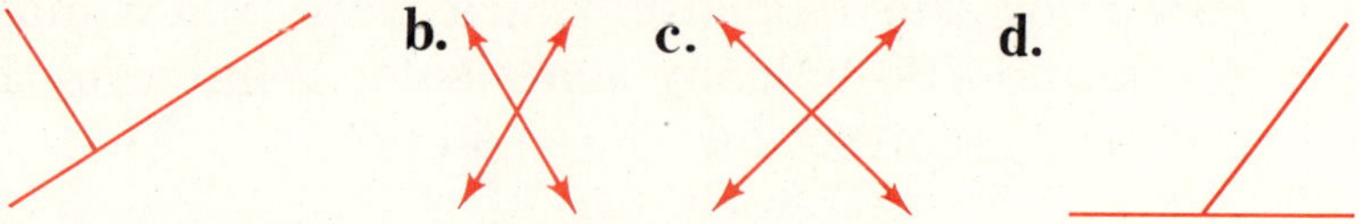

6. Use your drawing-triangle to draw some perpendicular lines and segments.

7. The symbol $\perp$ is read "is perpendicular to." Read these.

a. $\overleftrightarrow{PQ} \perp \overleftrightarrow{RS}$ b. $\overline{AB} \perp \overline{CD}$

8. Use a drawing-triangle to construct parallel lines.

a. Draw a line. Label it $\overleftrightarrow{AB}$.

b. Use your drawing-triangle to draw a line perpendicular to $\overleftrightarrow{AB}$ at B. Label it $\overleftrightarrow{BC}$.

c. Now draw a line perpendicular to $\overleftrightarrow{BC}$ at C. Label it $\overleftrightarrow{CD}$. $\overleftrightarrow{CD} \parallel \overleftrightarrow{AB}$.

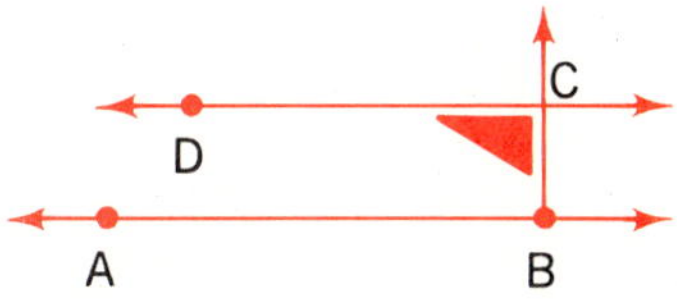

9. Draw several more pairs of parallel lines.

EXERCISES

1. Study the drawing.

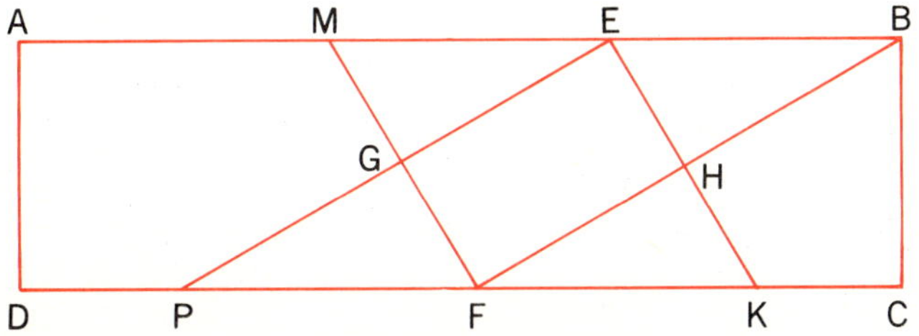

a. Which segments look parallel?

b. Which segments look perpendicular?

2. Draw some perpendicular lines. Use your drawing-triangle.

3. Draw two parallel lines. Use your drawing-triangle.

4. Read these.

a. $\overleftrightarrow{MN} \parallel \overleftrightarrow{ST}$ b. $\overline{RQ} \perp \overline{MO}$ c. $\overline{AC} \parallel \overleftrightarrow{PT}$

DO YOU REMEMBER?

Add.

1. $\begin{array}{r} 267 \\ \underline{340} \end{array}$ 2. $\begin{array}{r} 3{,}683 \\ \underline{4{,}312} \end{array}$ 3. $\begin{array}{r} 214 \\ \underline{3{,}431} \end{array}$ 4. $\begin{array}{r} 1{,}248 \\ \underline{634} \end{array}$

Subtract.

5. $\begin{array}{r} 235 \\ \underline{-\ 183} \end{array}$ 6. $\begin{array}{r} 4{,}893 \\ \underline{-\ 3{,}087} \end{array}$ 7. $\begin{array}{r} 6{,}907 \\ \underline{-\ 4{,}069} \end{array}$ 8. $\begin{array}{r} 7{,}820 \\ \underline{-\ 4{,}938} \end{array}$

Multiply.

9. $\begin{array}{r} 341 \\ \underline{\times\ 56} \end{array}$ 10. $\begin{array}{r} 803 \\ \underline{\times\ 264} \end{array}$ 11. $\begin{array}{r} 183 \\ \underline{\times\ 430} \end{array}$ 12. $\begin{array}{r} 5{,}824 \\ \underline{\times\ 1{,}635} \end{array}$

Divide.

13. $31\overline{)682}$ 14. $48\overline{)12{,}489}$ 15. $341\overline{)57{,}623}$

Compute and simplify.

16. $\frac{3}{20} + \frac{5}{18}$ 17. $\frac{2}{15} + \frac{1}{25}$ 18. $\frac{3}{5} + \frac{1}{6} + \frac{2}{3}$

19. $\frac{3}{8} + \frac{9}{11}$ 20. $\frac{7}{9} + \frac{4}{5}$ 21. $\frac{2}{7} + \frac{4}{5} + \frac{1}{10}$

22. $\frac{3}{5} - \frac{3}{20}$ 23. $\frac{7}{8} - \frac{2}{3}$ 24. $\frac{3}{4} - \frac{9}{20}$

25. $\frac{9}{11} - \frac{2}{7}$ 26. $\frac{37}{17} - \frac{1}{2}$ 27. $\frac{11}{15} - \frac{2}{9}$

28. $\frac{2}{5} \times \frac{1}{8}$ 29. $\frac{7}{10} \cdot \frac{5}{14}$ 30. $\frac{6}{7} \times \frac{2}{3} \times \frac{4}{5}$

31. $\frac{5}{11} \cdot \frac{3}{8}$ 32. $\frac{4}{13} \times \frac{3}{7}$ 33. $\frac{2}{9} \cdot \frac{6}{7} \cdot \frac{2}{3}$

34. $\frac{1}{2} \div \frac{3}{4}$ 35. $100 \div \frac{2}{3}$ 36. $\frac{3}{4} \div 6$

37. $\frac{2}{9} \div \frac{8}{12}$ 38. $500 \div \frac{5}{8}$ 39. $\frac{2}{9} \div 800$

PROBLEM SOLVING

1. Ed's math homework takes 35 minutes. English takes 40 minutes and social studies takes 45 minutes. Estimate the total time he needs for homework.

2. A professional football team's attendance for five home games was 62,584; 62,605; 58,924; 61,824; and 62,697. What was the total attendance?

3. Eddy and Frank were building models. They bought glue for 35¢ and 3 bottles of paint for 40¢ each. They paid with a $5 bill. What was their change?

4. Sal sold 9 trays of soda at the ball park. Each tray held 25 sodas.

   a. How many sodas did he sell in all?

   b. Each soda costs 30¢. How much did Sal receive for the sodas?

5. In 3 years, the conservation department planted these numbers of seedlings: 120,000; 205,000; and 168,000. How many did they plant in all?

6. The sun is about 93,000,000 miles from earth. The moon is about 240,000 miles from earth. How much farther away is the sun?

7. The Miller Trucking Company used 14 trucks to haul 225 tons of gravel. About how much gravel was hauled in each truck?

8. Cathi is planning the class field trip. There are 418 students in the class. How many 45-passenger buses will be needed?

9. There were 29,632 people at the civic center for the rock music show. The average ticket cost $3. Estimate the total amount of money received.

BISECTING SEGMENTS

In the drawing, $\overleftrightarrow{RS}$ is a *bisector* of $\overline{JL}$. We call point K the *midpoint* of $\overline{JL}$. $\overline{JK} \cong \overline{KL}$.

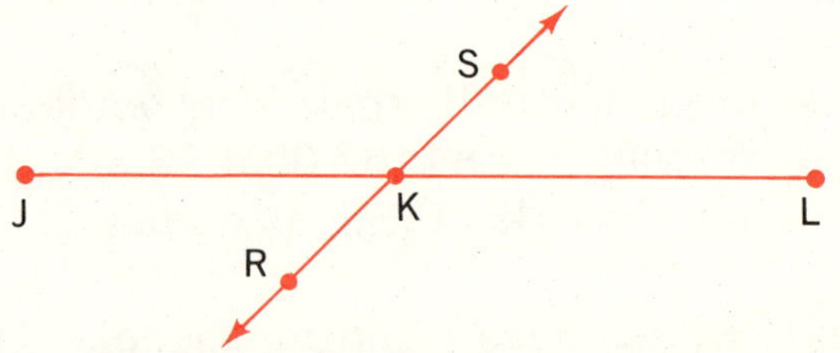

We can bisect a segment with a compass and straightedge.

LET'S EXPLORE

1. Draw a segment. Name it $\overline{AB}$.

2. Set the compass opening greater than half the length of $\overline{AB}$.

3. Put the metal tip on A. Make a mark like this.

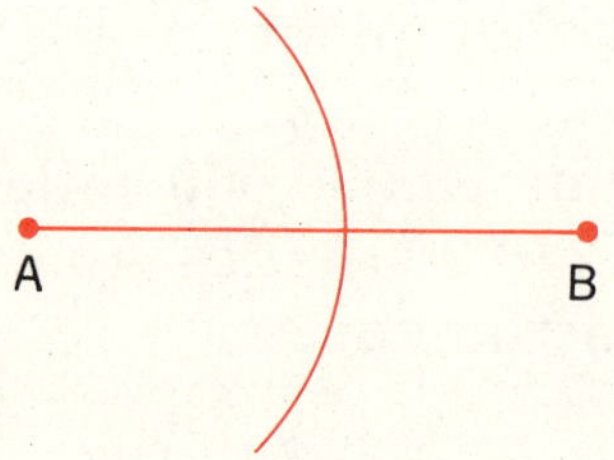

4. Keep the same compass setting. Put the metal tip on B and make another mark. Label the points where the marks cross P and Q as shown.

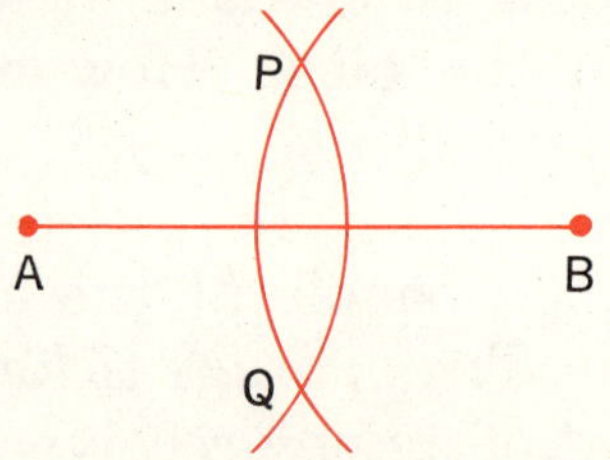

5. Draw $\overleftrightarrow{PQ}$. It will cross $\overline{AB}$ at its midpoint. $\overleftrightarrow{PQ}$ bisects $\overline{AB}$. It is also perpendicular to $\overline{AB}$.

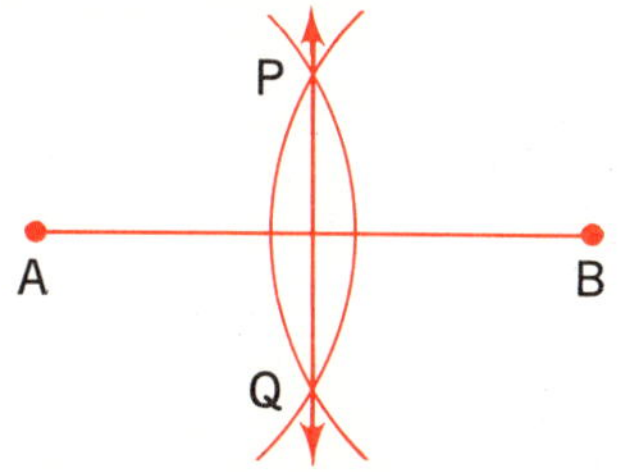

6. $\overleftrightarrow{PQ}$ is called the perpendicular bisector of $\overline{AB}$.

   a. Is there another perpendicular bisector of $\overline{AB}$?

   b. How many perpendicular bisectors does a segment have?

7. a. Draw two segments. Construct the perpendicular bisector of each.

   b. Check your work using a ruler and drawing-triangle.

EXERCISES

Copy each segment using your compass and straightedge. Then construct the perpendicular bisector of each.

1. 

2. 

3. 

4. 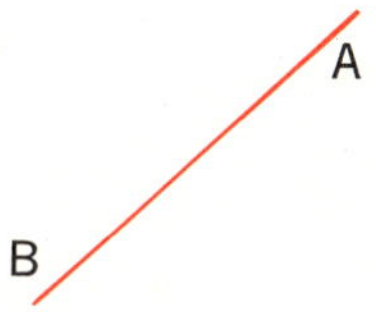

5. Draw a triangle. Construct the perpendicular bisector of each side.

Let's Practice

A decimal numeral for $12\frac{19}{100}$ is 12.19. Write decimal numerals.

1. $18\frac{33}{100}$
2. $9\frac{6}{10}$
3. $124\frac{63}{1,000}$
4. $15\frac{2}{1,000}$

CONSTRUCTING PERPENDICULARS

Now let's construct two other kinds of perpendiculars.

LET'S EXPLORE

1. Draw two points A and B. Draw $\overleftrightarrow{AB}$.

2. Let's construct a perpendicular to $\overleftrightarrow{AB}$ at point B. Put the metal tip of the compass on B. Make marks at P and Q.

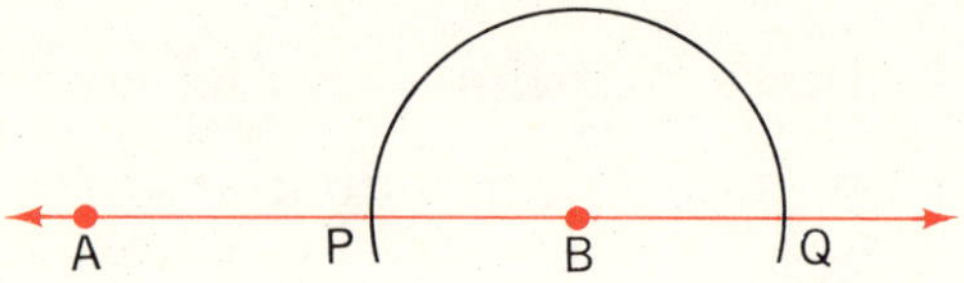

3. Set the compass opening greater than the length of $\overline{PB}$.

4. Put the metal tip on P.
Make a mark as shown.

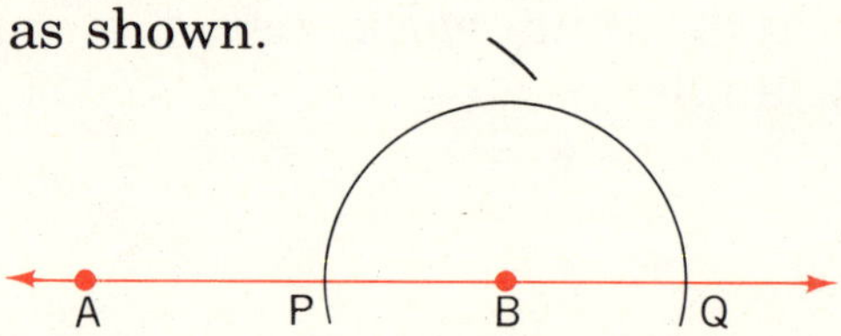

5. Keep the same compass setting. Put the metal tip on Q and make another mark. Label the point where the marks cross R as shown.

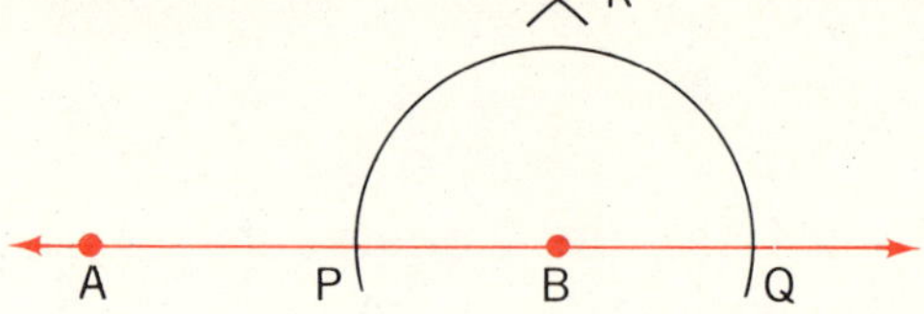

6. Draw $\overleftrightarrow{RB}$.
$\overleftrightarrow{RB} \perp \overleftrightarrow{AB}$ at B.

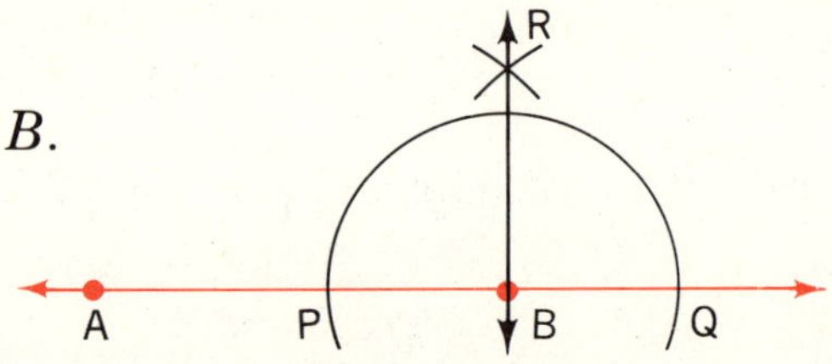

7. Now let's practice.

   a. Draw $\overleftrightarrow{RT}$.

   b. Construct a line perpendicular to $\overleftrightarrow{RT}$ at T.

8. Copy this drawing.

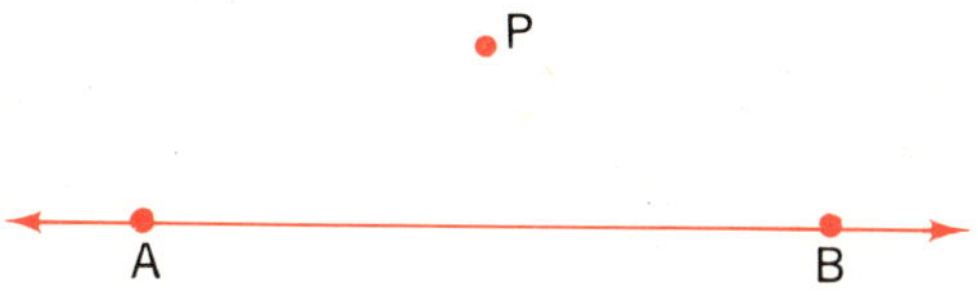

9. Let's construct a line containing P perpendicular to $\overleftrightarrow{AB}$. Put the metal tip on P. Make a mark which crosses $\overleftrightarrow{AB}$ as shown. Label the points S and T.

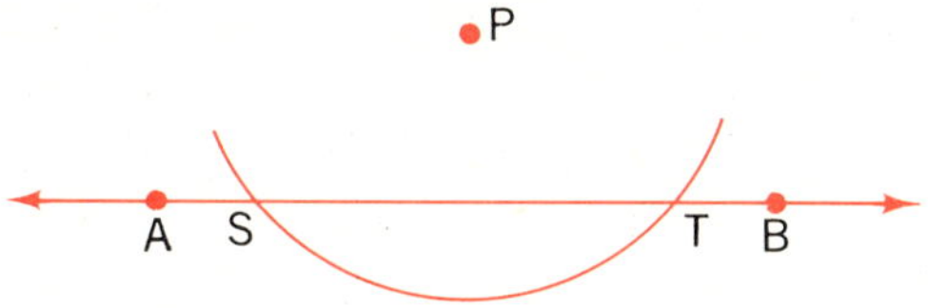

10. Put the metal tip first on S and then on T. Without changing the compass setting, make marks which cross at Q as shown.

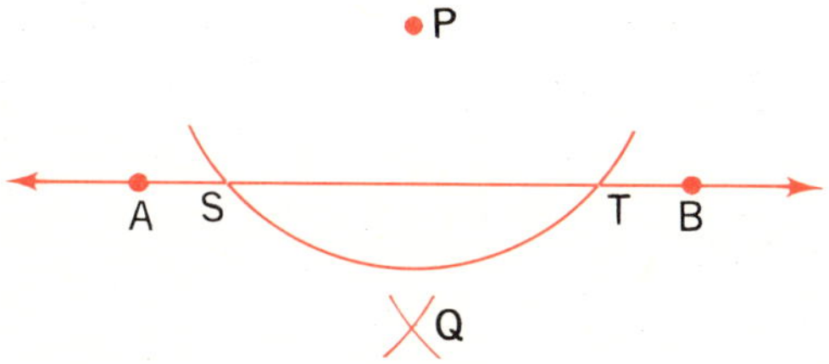

11. Draw $\overleftrightarrow{PQ}$. It will be perpendicular to $\overleftrightarrow{AB}$.

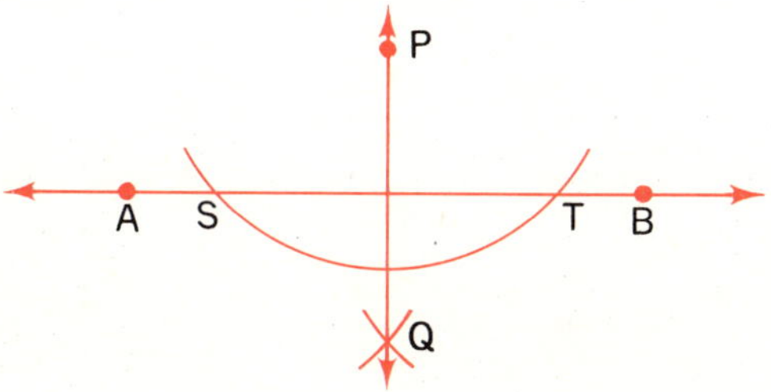

12. Now let's practice.

   a. Draw a line $\overleftrightarrow{MN}$ and a point D not on $\overleftrightarrow{MN}$.

   b. Construct a line containing D perpendicular to $\overleftrightarrow{MN}$.

EXERCISES

Copy each drawing. Construct a line containing the given point and perpendicular to the given line.

1.

2.

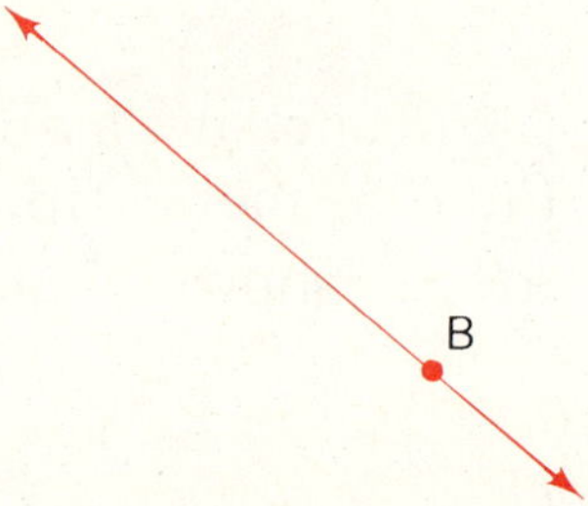

3.

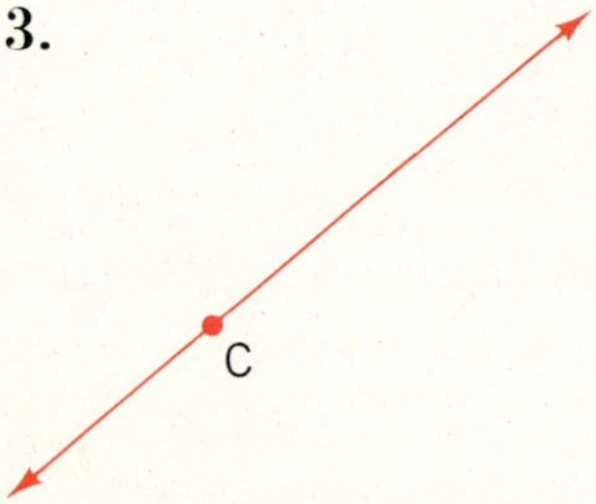

4.

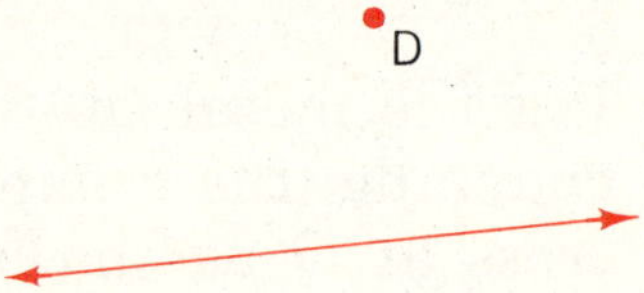

5.

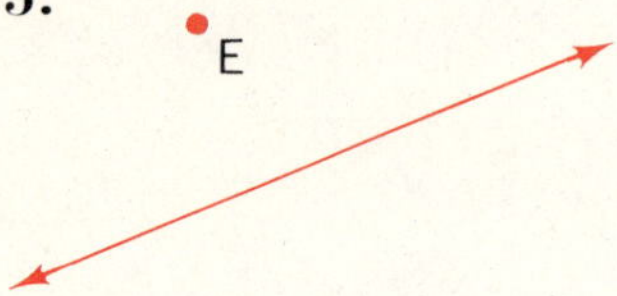

6.

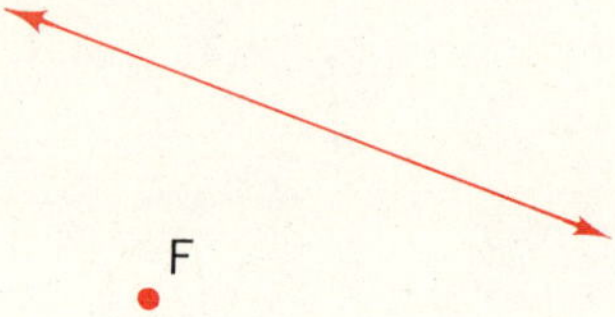

7. Draw a line and a point not on the line. Construct a perpendicular to the line from the point.

UNEXPLORED TERRITORY

Unscramble the letters to form math words.

a. rpgha

b. uernbm

c. reibcost

d. timorlgah

CONSTRUCTING PARALLEL LINES

1. **a.** Draw a pair of lines.

   b. Draw a line which crosses both of the lines. This line is called a *transversal.*

2. **a.** Draw two lines and a transversal as shown.

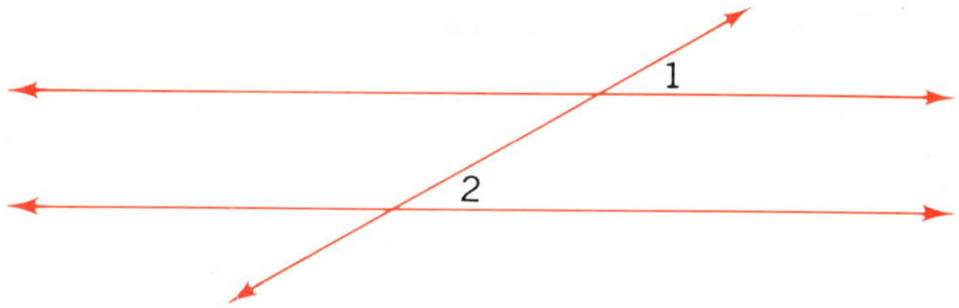

   b. Angles like ∠1 and ∠2 are called *corresponding angles.* Use your protractor. Measure ∠1 and ∠2.

3. When corresponding angles are congruent, then the lines are parallel.

   a. Use your protractor. Measure ∠3 and ∠4.

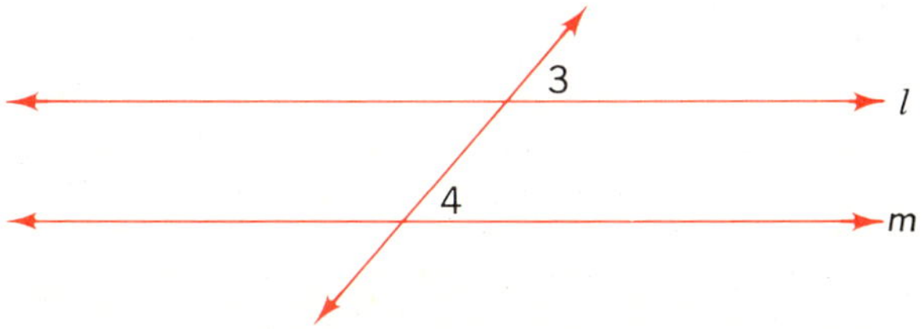

   b. Are lines l and m parallel?

LET'S EXPLORE

4. We can construct a line containing P, parallel to $\overleftrightarrow{AB}$. We will use the idea of corresponding angles. Copy this drawing.

5. Draw a line containing P which intersects $\overleftrightarrow{AB}$. Label the intersection Q.

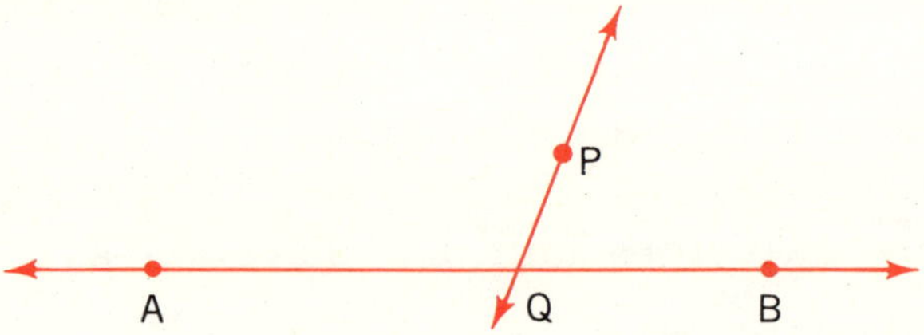

6. At P, construct an angle congruent to $\angle PQB$ as shown. Draw line m which is parallel to $\overleftrightarrow{AB}$.

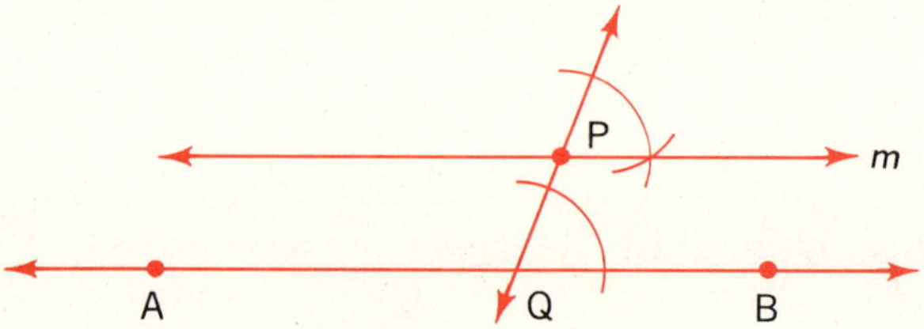

7. Copy this drawing. Construct a line containing T parallel to line l.

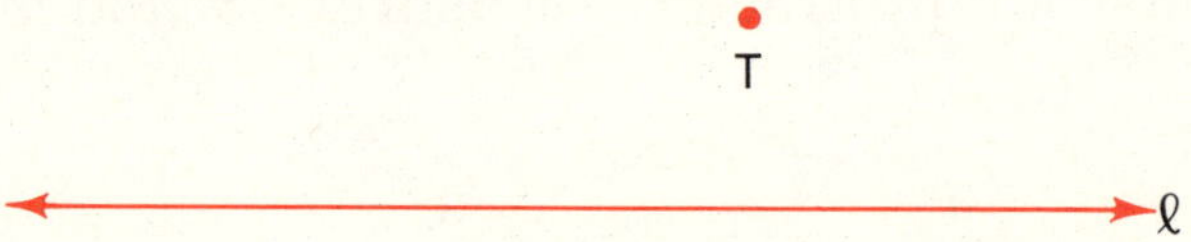

EXERCISES

1. a. Draw a pair of lines and a transversal.

   b. Label a pair of corresponding angles.

Copy each drawing. Construct a line containing the given point and parallel to the given line.

2. 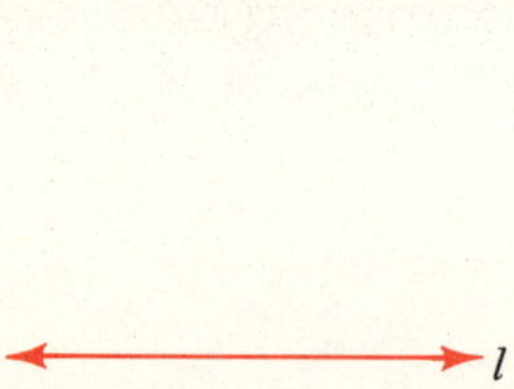

3. 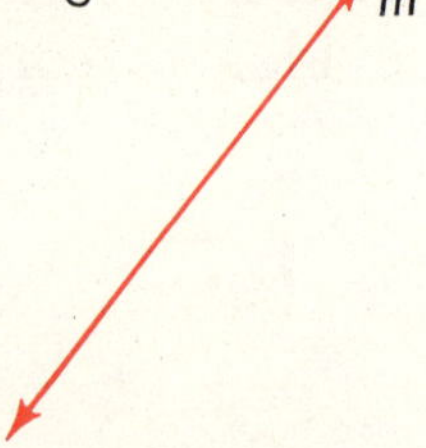

4. 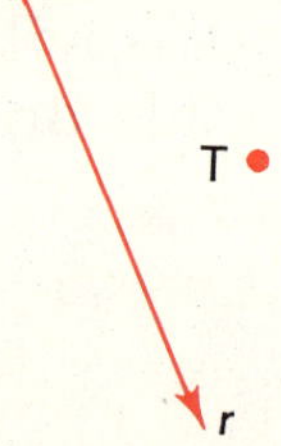

POLYGONS

A *polygon* is a closed curve made up of segments.

1. Here are several polygons.

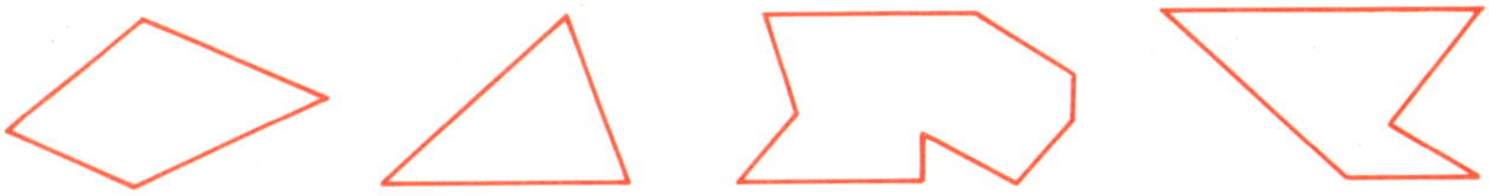

Draw two polygons.

2. Polygons are classified by the number of sides they have. Copy and complete.

Name	*Number of Sides*
Triangle	
Quadrilateral	
Pentagon	
Hexagon	
Heptagon	
Octagon	
Nonagon	
Decagon	

3. a. Draw an octagon.

b. Draw a pentagon.

UNEXPLORED TERRITORY

In the plane of the paper, how many lines can contain point P and be parallel to $\overleftrightarrow{AB}$?
How many lines can be perpendicular to $\overleftrightarrow{AB}$ at point B?

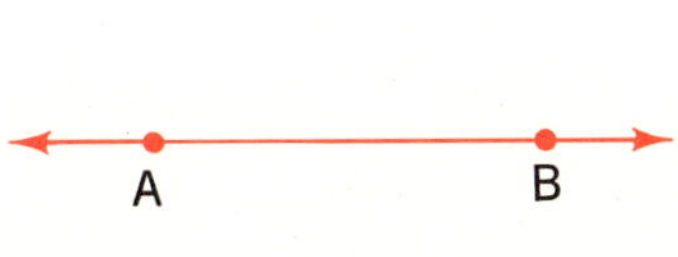

REGULAR POLYGONS AND THE CIRCLE

A *regular polygon* has all sides congruent and all angles congruent.

1. Which polygons look regular?

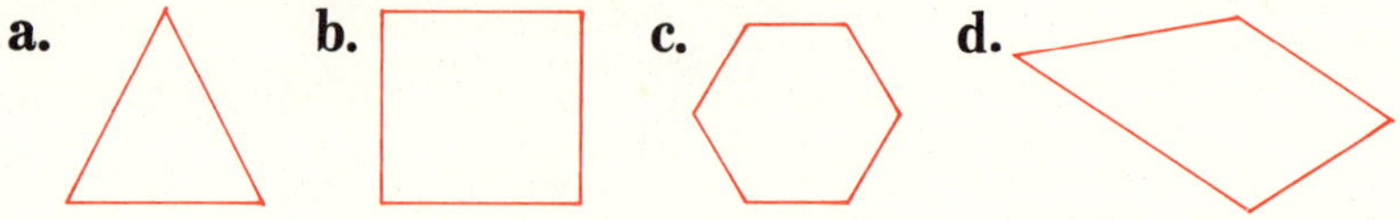

LET'S EXPLORE

2. Let's construct a regular pentagon. Draw a circle and mark its center.

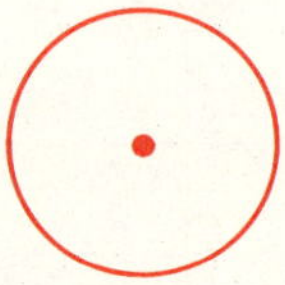

3. Remember that 360 one-degree angles can be placed around the center. Divide 360° by 5. *72°*

4. Draw 5 central angles each having a measure of 72°. (A central angle has its vertex at the center of a circle.)

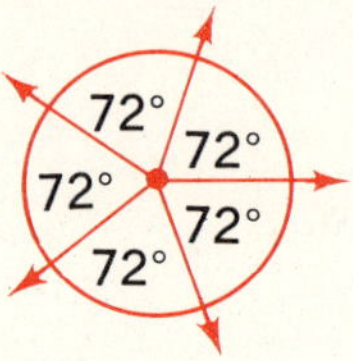

5. Connect the points where the rays cross the circle. You now have a regular pentagon.

6. You can construct any regular polygon using this procedure. Let's construct a regular hexagon.

   a. Draw a circle and mark its center.

   b. Divide 360° by 6.

   c. Draw 6 central angles each having a measure of 60°.

   d. Connect the points where the rays cross the circle. You have a regular hexagon.

EXERCISES

1. Construct a regular triangle.

2. Construct a regular quadrilateral.

3. Construct a regular octagon.

4. Construct a regular nonagon.

5. Construct a regular decagon.

Solve this problem.

6. During a price war, gasoline prices dropped 30%. The usual price was 34.9¢ per gallon. What was the new price?

UNEXPLORED TERRITORY

1. Use your compass and straightedge. Copy this design.

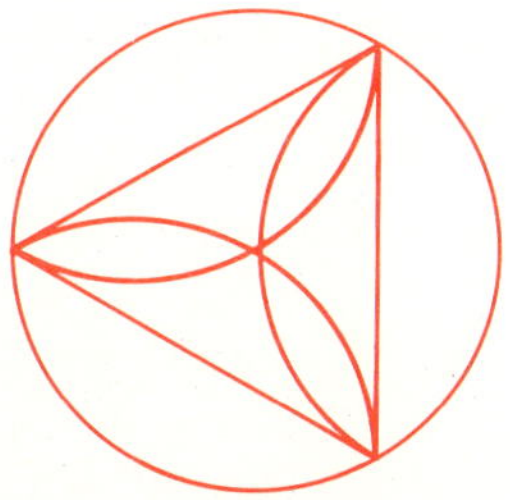

2. Make up your own design using your compass.

DO YOU REMEMBER?

Write decimal numerals.

1. $\frac{3}{10}$ 2. $\frac{53}{100}$ 3. $\frac{597}{1,000}$

4. $\frac{7}{100}$ 5. $\frac{9}{1,000}$ 6. $15\frac{927}{10,000}$

7. $18\frac{19}{10,000}$ 8. $24\frac{5,423}{100,000}$ 9. $96\frac{847}{1,000,000}$

Add.

10. $\begin{array}{r} .2 \\ .5 \\ .2 \\ \hline \end{array}$ 11. $\begin{array}{r} .4 \\ .7 \\ .8 \\ \hline \end{array}$ 12. $\begin{array}{r} 8.6 \\ 7.2 \\ .9 \\ \hline \end{array}$ 13. $\begin{array}{r} 18.7 \\ 9.23 \\ 108.359 \\ \hline \end{array}$

14. 8.3 + 19.006 + 53.42 15. .9 + 32.24 + 563.069

16. .08 + 12.3 + 8.234 17. 5.62 + 1.65 + 3.065

Subtract.

18. $\begin{array}{r} 9.7 \\ -\ 4.3 \\ \hline \end{array}$ 19. $\begin{array}{r} 8.3 \\ -\ 4.7 \\ \hline \end{array}$ 20. $\begin{array}{r} 9.06 \\ -\ 5.76 \\ \hline \end{array}$ 21. $\begin{array}{r} 8.5 \\ -\ 5.07 \\ \hline \end{array}$

22. .647 − .293 23. 4.87 − 3.98

24. 32.09 − 5.8 25. 5.639 − .806

Multiply.

26. .6 × 8 27. .23 × 9 28. .7 × .42

29. .34 × 10 30. 200 × .472 31. 9 × .006

Divide.

32. $6\overline{)12.6}$ 33. $7\overline{)35.21}$ 34. $12\overline{).0072}$

35. $5\overline{)20.5}$ 36. $.08\overline{).448}$ 37. $.002\overline{).078}$

CHAPTER REVIEW

Which figures are congruent?

1. 

2. 

3. 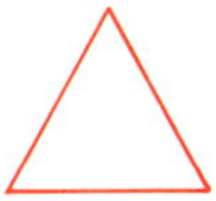

4. 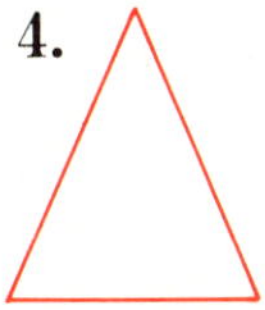

5. 

6. 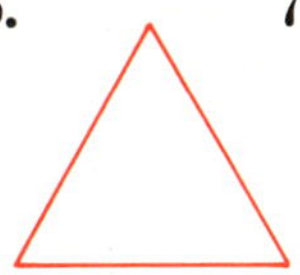

7. 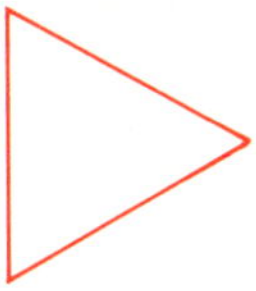

8. 

9. 

10.

Which pairs can be shown congruent by reflection?

11. 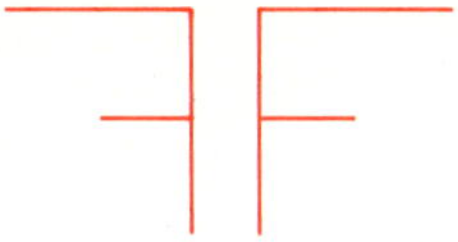

12. 

13. 

Which pairs can be shown congruent by rotation?

14. 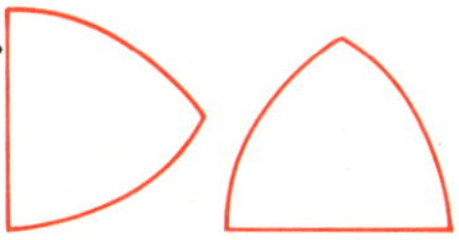

15. 

16. 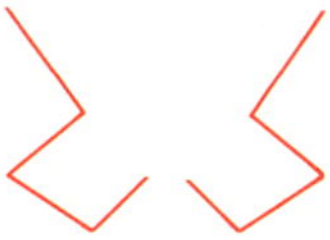

Use your straightedge and compass. Do these constructions.

17. Construct segment $\overline{CD}$ congruent to $\overline{AB}$.

18. Construct the perpendicular bisector of $\overline{CD}$.

19. Draw angles which look like these. Construct an angle congruent to each of them.

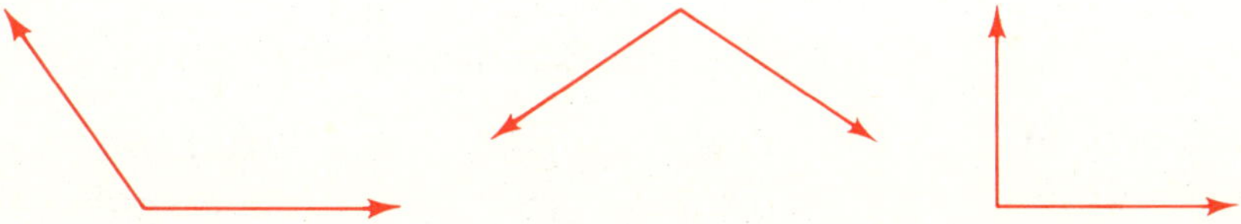

20. Draw an acute angle and an obtuse angle. Bisect the angles.

21. Draw a triangle. Construct a perpendicular from one of its vertices to the opposite side of the triangle.

22. Copy this drawing. Construct a line containing S that is perpendicular to $\overleftrightarrow{RT}$.

23. Construct a regular nonagon.

Here is a rectangle.

24. Which segments are parallel?

25. Which segments are perpendicular?

Solve these problems.

26. Mr. Rogers bought $10\frac{1}{2}$ gallons of gasoline at 39.5¢ per gallon. How much did he pay for the gasoline?

27. Robyn used $2\frac{3}{4}$ cups of flour in peanut butter cookies, $3\frac{1}{2}$ cups in chocolate chip cookies, and $2\frac{2}{3}$ cups in brownies. How much flour did she use?

28. Mrs. Weiss planted a border of pansies along her driveway. She placed the plants 4 inches apart. The driveway is 25 feet long. How many plants did she use?

Terms You Should Know

congruent	straightedge	parallel
reflection	compass	perpendicular
rotation	bisect	regular polygon

CHAPTER TEST

Tell whether each pair of figures can be shown congruent by reflection or congruent by rotation.

1. 

2. 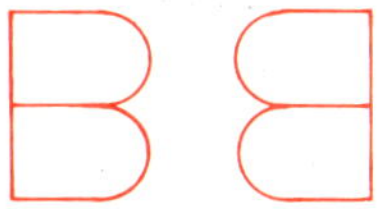

Use your straightedge and compass. Do these constructions.

3. Construct segment $\overline{IM}$ congruent to $\overline{JK}$. Bisect $\overline{IM}$.

4. Construct $\angle T$ congruent to $\angle S$. Bisect $\angle T$.

5. Copy this drawing. Construct a line containing P that is perpendicular to M.

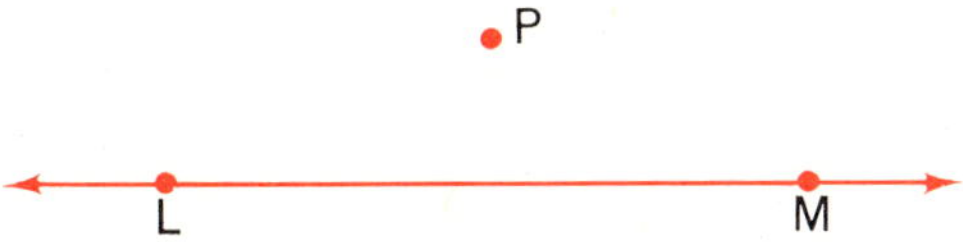

6. Construct a regular pentagon.

Study this drawing.

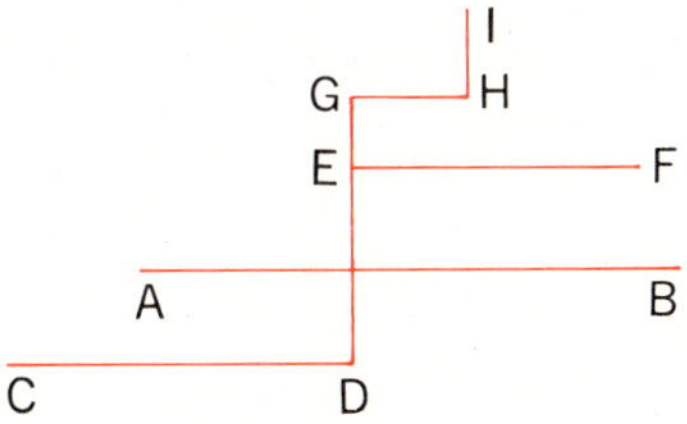

7. Which segments are parallel?

8. Which segments are perpendicular?

Solve this problem.

9. A recipe calls for $\frac{1}{2}$ teaspoon cloves, $\frac{3}{4}$ teaspoon allspice, 1 teaspoon cinnamon, and $\frac{3}{4}$ teaspoon ginger. How many teaspoons of spices are used in the recipe?

$\frac{1}{2}$	.5
$\frac{1}{4}$	.25
$\frac{3}{4}$	.75
$\frac{1}{3}$	$.3\overline{3}$

96.28341 (5 decimal places)
× 3.26 (2 decimal places)
313.8839166 (7 decimal places)

.0023 (4 decimal places)
× .4 (1 decimal places)
.00092 (5 decimal places)

CHAPTER FOUR

Decimal Numerals

DECIMAL NUMERALS

When Chet had his gas tank filled, it took $14\frac{3}{10}$, or 14.3 gallons. We call 14.3 a *decimal numeral*, or simply a *decimal.*

1. We read 14.3 as "fourteen and three tenths." Read these.

   a. 5.6 **b.** 12.7 **c.** 18.63

   d. 9.43 **e.** .581 **f.** 2.135

2. A decimal numeral for $25\frac{63}{100}$ is 25.63. Write decimal numerals.

   a. $8\frac{4}{10}$ **b.** $16\frac{25}{100}$ **c.** $19\frac{863}{1,000}$

   d. $214\frac{3}{100}$ **e.** $\frac{93}{1,000}$ **f.** $6\frac{5}{1,000}$

3. A mixed numeral for 15.253 is $15\frac{253}{1,000}$. Write mixed numerals.

   a. 13.9 **b.** 28.63 **c.** 45.573

   d. 6.07 **e.** 14.031 **f.** 62.009

4. The word name for 5.73 is "five and seventy-three hundredths." Write word names.

   a. 6.29 **b.** 10.6 **c.** 14.324

EXERCISES

Write decimal numerals.

1. $\frac{5}{10}$	2. $\frac{7}{10}$	3. $8\frac{2}{10}$	4. $9\frac{3}{10}$
5. $14\frac{24}{100}$	6. $29\frac{63}{100}$	7. $\frac{58}{100}$	8. $\frac{2}{100}$
9. $26\frac{427}{1,000}$	10. $134\frac{56}{1,000}$	11. $\frac{29}{1,000}$	12. $\frac{247}{1,000}$

Write word names.

13. .7	14. .9	15. 7.6	16. 3.8
17. .36	18. 17.26	19. 59.73	20. .06
21. 35.395	22. 137.602	23. .085	24. .356

Write simplest fractional numerals or mixed numerals.

25. .9	26. 3.7	27. 7.7	28. .09
29. .29	30. 2.59	31. 8.07	32. 27.03
33. 26.543	34. 19.063	35. 296.27	36. 83.163

Write decimals.

37. Thirty-eight and twenty-six hundredths

38. Fifty-six and nine-tenths

39. Twenty-four and two hundred thirty-six thousandths

Let's Practice

Mr. Mountcastle, a used car dealer, bought a small car for $895. He repainted it for $135 and spent $60 for an engine tune-up. Finally, he sold it for $1,290. How much profit did he make?

COMPARING NUMBERS

We can compare two numbers by writing equivalent decimals with the same number of places to the right of the decimal point.

1. Compare each pair of numbers.

Example .04, .0067

Write both with 4 places to the right of decimal point.

$.04 = .0400 \qquad .0067 = .0067$

$.0400 > .0067$

Therefore, $.04 > .0067$

a. .005, .03 b. .0032, .056 c. 4.491, 4.5

d. 5.62, 5.6188 e. 1.0, .9999 f. .16, .302

2. Tell which number is greater.

a. 2.08, 2.003 b. 21.56, 18.439 c. 38.63, 38.594

3. Which are true and which are false?

a. $.56 > .483$ b. $15.60 < 15.613$

c. $18.9004 > 18.0956$ d. $7\frac{19}{100} > 8\frac{83}{1{,}000}$

4. Write equivalent decimals with the same number of places to the right of the decimal point. Then list in order from greater to lesser.

Example .06, .483, .0056

Write each with 4 places to the right of the decimal point.

$.06 = .0600 \qquad .483 = .4830 \qquad .0056 = .0056$

$.4830 > .0600$ and $.0600 > .0056$

Therefore, $.483 > .06 > .0056$

a. .0078, .3, .296 b. .258, .031, .62

c. .64, .0064, .064 d. .5, .078, .6928

EXERCISES

Make each sentence true. Use > or <.

1. .83 ▤ .819
2. .0052 ▤ .03
3. 5.062 ▤ 5.21
4. 1.26 ▤ 1.2599
5. 2.4 ▤ 2.4378
6. 3.386 ▤ 3.4
7. .9999 ▤ 1.03
8. 18.007 ▤ 18.00659
9. .4 ▤ .403

Which are true and which are false?

10. .25 < .7
11. 2.63 > 2.599
12. 18.032 > 18.1
13. 14.0097 > 14.038
14. .814 > .08315
15. .1414 > .01414
16. 1.56998 < 1.6003
17. $\frac{45}{100}$ < .8
18. .7979 < .09797

List in order from greater to lesser.

19. 056, .5, .0572
20. .9002, .901, .9
21. .39, .3586, .0268
22. .25, .706, .0999
23. .0997, .098, .123
24. .0005, .096, .004

UNEXPLORED TERRITORY

Here is an algorithm for addition. It is called *Front-End Addition.* Try other additions using this algorithm.

	3	3	5	4
	7	3	2	9
	5	6	1	7
		9	3	4
Step 1	15			
Step 2	2	1		
Step 3		1	1	
Step 4			2	4
	17	2	3	4

DO YOU REMEMBER?

Multiply.

1. $\begin{array}{r} 80 \\ \times 8 \\ \hline \end{array}$ 2. $\begin{array}{r} 406 \\ \times 7 \\ \hline \end{array}$ 3. $\begin{array}{r} 9,008 \\ \times 6 \\ \hline \end{array}$ 4. $\begin{array}{r} 73,006 \\ \times 4 \\ \hline \end{array}$

5. $\begin{array}{r} 82 \\ \times 7 \\ \hline \end{array}$ 6. $\begin{array}{r} 2,140 \\ \times 5 \\ \hline \end{array}$ 7. $\begin{array}{r} 8,682 \\ \times 9 \\ \hline \end{array}$ 8. $\begin{array}{r} 29,702 \\ \times 8 \\ \hline \end{array}$

9. $\begin{array}{r} 56 \\ \times 29 \\ \hline \end{array}$ 10. $\begin{array}{r} 287 \\ \times 32 \\ \hline \end{array}$ 11. $\begin{array}{r} 429 \\ \times 123 \\ \hline \end{array}$ 12. $\begin{array}{r} 2,614 \\ \times 428 \\ \hline \end{array}$

13. $\begin{array}{r} 9,147 \\ \times 2,062 \\ \hline \end{array}$ 14. $\begin{array}{r} 4,720 \\ \times 3,005 \\ \hline \end{array}$ 15. $\begin{array}{r} 29,576 \\ \times 4,027 \\ \hline \end{array}$ 16. $\begin{array}{r} 47,827 \\ \times 5,624 \\ \hline \end{array}$

Divide.

17. $62\overline{)2,604}$ 18. $23\overline{)1,058}$ 19. $67\overline{)3,953}$

20. $192\overline{)14,592}$ 21. $134\overline{)6,030}$ 22. $212\overline{)8,904}$

23. $987\overline{)85,869}$ 24. $76\overline{)42,154}$ 25. $87\overline{)26,438}$

26. $326\overline{)52,829}$ 27. $431\overline{)568,294}$ 28. $258\overline{)976,434}$

29. $436\overline{)398,773}$ 30. $292\overline{)291,385}$ 31. $806\overline{)299,995}$

Solve these problems.

32. Towson High School's championship basketball team scored 2,436 points in 28 games. What was their average score per game?

33. The school board estimated that it costs $625 per child to operate the city's schools. How much does it cost for 26,829 children?

34. A recipe calls for $\frac{3}{8}$ cup olive oil, $3\frac{5}{6}$ cups tomatoes, and $2\frac{1}{3}$ cups tomato paste. How many cups are needed in all?

ADDITION

For his report on the miles to be covered in a race, Mr. Brant wrote down these mileages: 26.4 miles, 18.7 miles, and 32.61 miles. Here is the way he found the total.

$$\begin{array}{r} 26.4 \\ 18.7 \\ \underline{32.61} \\ 77.71 \end{array}$$

1. How are these additions alike?

$$\begin{array}{r} 378 \\ 459 \\ \underline{654} \\ 1{,}491 \end{array} \qquad \begin{array}{r} 3.78 \\ 4.59 \\ \underline{6.54} \\ 14.91 \end{array}$$

2. Here is the way Fran added .237 and .65.

$$\begin{array}{r} .237 \\ \underline{+\ .65} \\ .303 \end{array}$$

   a. What mistake did she make?

   b. Find the correct sum.

3. Add.

   a. $\begin{array}{r} 1.56 \\ \underline{+\ 2.298} \end{array}$ b. $\begin{array}{r} .573 \\ \underline{+\ .46} \end{array}$ c. $\begin{array}{r} .385 \\ 4.078 \\ \underline{9.64} \end{array}$ d. $\begin{array}{r} 63.54 \\ 256.4 \\ 29.825 \\ \underline{2.16} \end{array}$

When adding with decimal numerals, we add tenths to tenths, hundredths to hundredths, and so on. To help do this, we line up the decimal points.

4. Add.

   a. 26.4 + 19.56 b. 17.03 + 56.429 + .016

EXERCISES

Add.

1. .6 + .8	**2.** .9 + .4	**3.** .2 + .03	**4.** .056 + .64
5. .096 + .204	**6.** .008 + .3469	**7.** .572 + .003	**8.** .2604 + .027
9. 4.823 + 6.79	**10.** 29.008 + 2.5736	**11.** 6.428 + 2.04	**12.** 11.0004 28.465 8.93
13. 12.329 41.06 3.296	**14.** 7.004 53.2967 3.805	**15.** 87.04 156.823 24.0062	**16.** 4.3922 58.9763 825.7203
17. 828.9335 66.5 2.007	**18.** 9.72 18.426 5.3 17.0049	**19.** 56.02 9.7349 182.612 4,976.02	**20.** .30968 5.23 49.826 563.0093
21. 17.4 5.006 875.3421 16.8	**22.** 19.8304 5.62 324.98 18.0469	**23.** 97.02 5.003 124.9206 14.05	**24.** 72.564 4.2308 146.9 51.009

25. 82.005 + .16 + .0032

26. 198.04 + .5 + 28.044

27. 1.56 + .03 + 129.0356

28. 2.73 + .62 + 4.9271

29. 5 + .00721 + 47.2

30. 76.4 + 92 + 4.7258

31. 19.1 + .732 + 826.9 + 52.0032

Solve this problem.

32. In the shop, Mr. Mills cut pieces of steel rod which measured 2.5 inches, 4.75 inches, and 10.625 inches. What was the total length of the pieces he cut?

SUBTRACTION

Fred swam the 100-yard free-style race in 59.8 seconds. Dave swam it in 58.9 seconds. Here is how they found the difference in time.

$$\begin{array}{r} 59.8 \\ -\ 58.9 \\ \hline .9 \end{array}$$

1. How are these subtractions alike?

$$\begin{array}{r} 462 \\ -\ 387 \\ \hline 75 \end{array} \qquad \begin{array}{r} 4.62 \\ -\ 3.87 \\ \hline .75 \end{array}$$

2. Subtract. Check your answers by adding.

a. $\begin{array}{r} .49 \\ -\ .25 \\ \hline \end{array}$ **b.** $\begin{array}{r} 14.23 \\ -\ 11.62 \\ \hline \end{array}$ **c.** $\begin{array}{r} .824 \\ -\ .658 \\ \hline \end{array}$ **d.** $\begin{array}{r} .9034 \\ -\ .5183 \\ \hline \end{array}$

3. Barry wanted to subtract 4.597 from 12.83.

a. First he lined up the decimal points. Why?

$$\begin{array}{r} 12.83 \\ -\ 4.597 \\ \hline \end{array}$$

b. Then he placed a 0 like this. Do the subtraction.

$$\begin{array}{r} 12.830 \\ -\ 4.597 \\ \hline \end{array}$$

4. Subtract.

a. $\begin{array}{r} .4 \\ -\ .26 \\ \hline \end{array}$ **b.** $\begin{array}{r} .57 \\ -\ .386 \\ \hline \end{array}$ **c.** $\begin{array}{r} 29.04 \\ -\ 8.0354 \\ \hline \end{array}$

d. $\begin{array}{r} 46.006 \\ -\ .2835 \\ \hline \end{array}$ **e.** $\begin{array}{r} 103.29 \\ -\ .276 \\ \hline \end{array}$ **f.** $\begin{array}{r} 496.6 \\ -\ .007 \\ \hline \end{array}$

EXERCISES

Subtract. Check your answers by adding.

1. $.56 - .34$	**2.** $.2646 - .109$	**3.** $28.94 - 16.578$	**4.** $82.4 - 29.75$
5. $94.07 - 63.24$	**6.** $825.974 - 34.68$	**7.** $.423 - .0649$	**8.** $97.030 - 23.572$
9. $43.072 - 18.0631$	**10.** $8.62 - 2.003$	**11.** $943.64 - 57.82$	**12.** $108.203 - 83.5608$
13. $.86 - .0934$	**14.** $1.004 - .869$	**15.** $82.020 - 47.9$	**16.** $42.73 - 19.879$
17. $.8 - .642$	**18.** $56.008 - 2.93$	**19.** $561.8 - 2.006$	**20.** $18.004 - 9.781$

21. 42.73 − 24.856

22. 104.403 − 57.89

23. 96.034 − 18.43

24. .6749 − .048

25. 43.82 − 7.9

26. 8.3 − 5.62

27. 56.04 − 18.13

28. 104.57 − .273

29. 245.6 − 8

30. .57 − .234

31. 93.005 − 4.68

32. 732.4 − .002

Solve these problems.

33. Mr. Herman found that one route to his office was 29.4 miles long. Another route was 31.2 miles long. What was the difference in the distances?

34. Mr. Getz was making a part in the machine shop. He checked his work with a micrometer and found the size to be .268 inches. The desired size was .249 inches. How much more should he cut off?

ROUNDING NUMBERS

A number line can help us to round numbers.

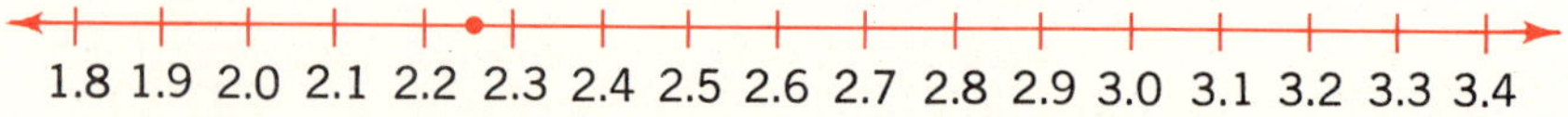

1. Locate 2.26 on the number line above.

   a. 2.26 is between what two tenths?

   b. To which tenth is it closer?

   c. What number do we get when we round 2.26 to the nearest tenth?

2. In Item 1 we rounded 2.26 to the nearest tenth.

   a. Round 2.14 to the nearest one.

   b. Round 2.14 to the nearest tenth.

3. Locate 2.15 on the number line above.

   a. 2.15 is between what two tenths?

   b. To which tenth is it closer?

4. When a number is halfway between two numbers, we often round to the greater number.

   a. Round 2.15 to the nearest tenth.

   b. Round .365 to the nearest hundredth.

5. We write $2.15 \approx 2.2$ to say "2.15 is approximately equal to 2.2." Read these.

   a. $5.64 \approx 6$ b. $129 \approx 100$ c. $4.302 \approx 4.30$

6. $8.34 \approx 8.3$ means that 8.34 has been rounded to the nearest tenth. Tell to what place each has been rounded.

   a. $5.60 \approx 6$ b. $18.29 \approx 18.3$

   c. $293.82 \approx 290$ d. $29.3235 \approx 29.324$

7. This chart shows how we can round 21.7853 in several ways. Copy and complete.

	Number	Nearest Thousandth	Nearest Hundredth	Nearest Tenth	Nearest One	Nearest Ten
	21.7853	21.785	21.79	21.8	22	20
a.	200.0251					
b.	7.8482					
c.	.92348					

EXERCISES

Tell to what place each has been rounded.

1. 6.749 ≈ 6.7
2. 100.0152 ≈ 100
3. 27.0367 ≈ 30
4. 7.2389 ≈ 7.24
5. .91346 ≈ .9
6. 18.962 ≈ 18.96

Copy and complete.

	Number	Nearest Thousandth	Nearest Hundredth	Nearest Tenth	Nearest One	Nearest Ten
	234.3708	234.371	234.37	234.4	234	230
7.	572.1347					
8.	468.41879					
9.	28.91683					
10.	30.43813					
11.	17.5983					
12.	400.2308					
13.	16.9093					
14.	134.82589					

PLACING DECIMAL POINTS BY ESTIMATING

When Peter multiplies using decimal numerals, he places the decimal point by estimating.

1. Consider 6.23×4.2.

   a. Round each factor to the nearest whole number.

   b. Multiply your answers in part a.

   c. Multiply 6.23 by 4.2. Place the decimal point so that the product is about 24.

2. Consider $.283 \times .714$.

   a. Round each factor to the nearest tenth.

   b. Multiply your answers in part a.

   c. Multiply .283 by .714. Place the decimal point so that the product is about .21.

3. Multiply. Place each decimal point by estimating.

a.	b.	c.	d.
$\begin{array}{r} 2.3 \\ \times\ 1.6 \\ \hline \end{array}$	$\begin{array}{r} 32.6 \\ \times\ 4.6 \\ \hline \end{array}$	$\begin{array}{r} .92 \\ \times\ .71 \\ \hline \end{array}$	$\begin{array}{r} 6.03 \\ \times\ .48 \\ \hline \end{array}$

EXERCISES

Round these numbers. Estimate the products.

1.	2.	3.	4.
$\begin{array}{r} 2.0734 \\ \times\ 5.9 \\ \hline \end{array}$	$\begin{array}{r} 7.869 \\ \times\ 3.2 \\ \hline \end{array}$	$\begin{array}{r} 6.2084 \\ \times\ .493 \\ \hline \end{array}$	$\begin{array}{r} .3807 \\ \times\ .5176 \\ \hline \end{array}$

5.	6.	7.	8.
$\begin{array}{r} 98.531 \\ \times\ 3.13 \\ \hline \end{array}$	$\begin{array}{r} 3.1652 \\ \times\ 4.98 \\ \hline \end{array}$	$\begin{array}{r} 56.1653 \\ \times\ .984 \\ \hline \end{array}$	$\begin{array}{r} 9.6581 \\ \times\ 764.2 \\ \hline \end{array}$

Multiply. Place each decimal point by estimating.

9.	10.	11.	12.
$\begin{array}{r} 3.72 \\ \times\ 4.61 \\ \hline \end{array}$	$\begin{array}{r} 8.93 \\ \times\ 2.34 \\ \hline \end{array}$	$\begin{array}{r} .436 \\ \times\ .83 \\ \hline \end{array}$	$\begin{array}{r} 324.8 \\ \times\ 2.6 \\ \hline \end{array}$

13.	14.	15.	16.
$\begin{array}{r} 8.462 \\ \times\ 39.63 \\ \hline \end{array}$	$\begin{array}{r} .0765 \\ \times\ 21.63 \\ \hline \end{array}$	$\begin{array}{r} .61623 \\ \times\ .92 \\ \hline \end{array}$	$\begin{array}{r} 384.5 \\ \times\ 38.6 \\ \hline \end{array}$

MULTIPLICATION

Here is a flow chart for multiplying with decimal numerals.

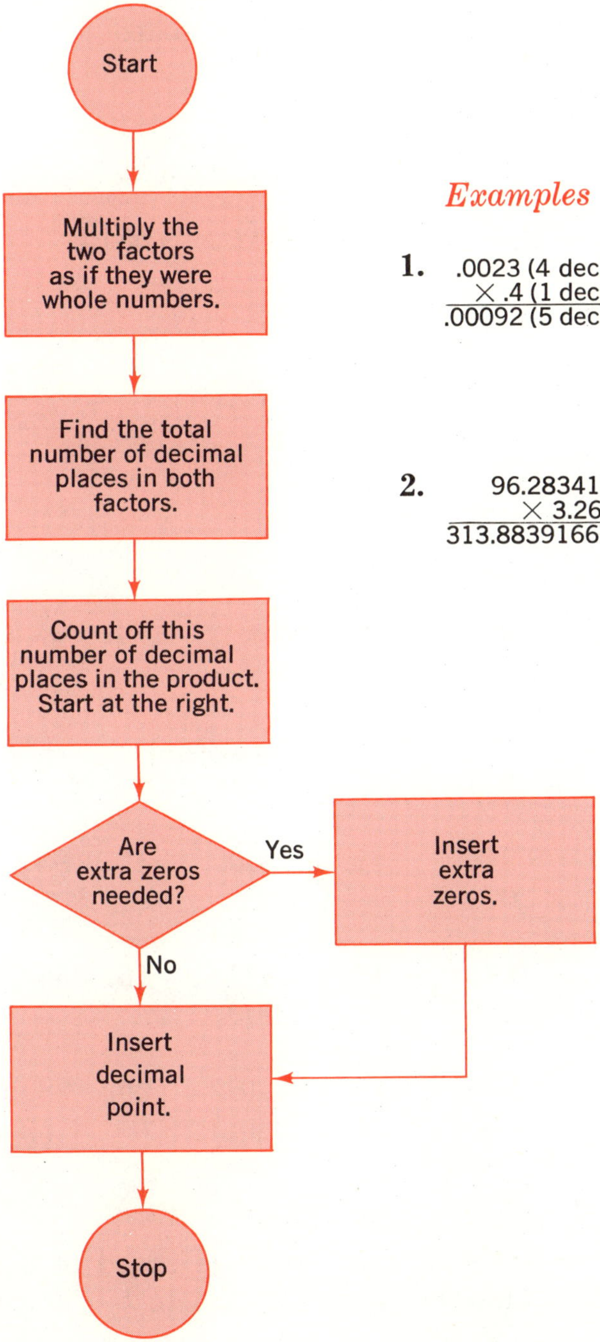

Examples

1.

```
   .0023 (4 decimal places)
   × .4  (1 decimal places)
  .00092 (5 decimal places)
```

2.

```
   96.28341 (5 decimal places)
     × 3.26 (2 decimal places)
313.8839166 (7 decimal places)
```

When multiplying with decimal numerals, the number of decimal places in the product is the total number of decimal places in the factors.

1. How many decimal places will be needed for each product?

   a. 2.43 × .56
   b. .057 × .0081
   c. 24.1 × .074
   d. 19.23 × 24.6
   e. 93.286 × .49321
   f. 6,342 × 9,581.3

2. Multiply. Use the flow chart.

   a. 8.63 × .54
   b. 19.07 × 2.43
   c. .0084 × 1.9
   d. .0642 × .073

EXERCISES

Multiply. Use the flow chart.

1. 19.43 × 2.5
2. 1.07 × .73
3. 16.54 × 29
4. 24.82 × 1.73
5. 22.75 × .043
6. 1.974 × .007
7. 634.2 × .0019
8. 489 × 2.4
9. .0034 × 17
10. 8.245 × .071
11. 10,000 × .05
12. 93.48 × 6.54
13. .0563 × 3.96
14. 9.621 × 3.854
15. 96,287 × .009
16. 386.21 × 3.56
17. 283.42 × .634
18. 124.03 × .0004
19. 83,462 × 2.443
20. 28.003 × 52.43

Solve this problem.

21. The metroliner travels at an average speed of 94.8 miles per hour. How far will it travel in 5.6 hours?

MULTIPLYING BY POWERS OF 10

1. Multiply.

a. $.2962 \times 10$	**b.** $.2962 \times 100$	**c.** $.2962 \times 1{,}000$	**d.** $.2962 \times 10{,}000$
e. 4.5983×10	**f.** 4.5983×100	**g.** $4.5983 \times 1{,}000$	**h.** $4.5983 \times 10{,}000$

> To multiply by a power of 10 such as 10, 100, or 1,000, we move the decimal point to the right. The number of places moved is the number of zeros in the standard numeral for the power of 10.

2. Multiply by moving the decimal point.

Examples

$.0589 \times 100 = 5.89$
$2.93 \times 10{,}000 = 29{,}300$
$34.2 \times 10^3 = 34{,}200$

a. $.0834 \times 100$ **b.** $.00579 \times 100{,}000$

c. $4.32 \times 1{,}000$ **d.** $.0032 \times 10{,}000$

e. 3.589×10 **f.** 12.863×10^3

EXERCISES

Multiply by moving the decimal point.

1. $.257 \times 10$
2. $.8649 \times 100$
3. $5.632 \times 10{,}000$
4. 283×10
5. $14.83 \times 1{,}000$
6. 2.938×10
7. $.0049 \times 100$
8. $.00073 \times 1{,}000$
9. 56.298×10^2
10. 1.348×10^3
11. 41.324×10^4
12. $4.386 \times 10{,}000$
13. $583.1 \times 1{,}000$
14. 561.2×10^4

DIVISION

1. Jessie said that when dividing with decimals, she estimates to place the decimal point.

$$34.2 \approx 30$$
$$30 \div 6 = 5$$

$$\begin{array}{r} 5.7 \\ 6\overline{)34.2} \\ \underline{30} \\ 4\,2 \\ \underline{4\,2} \end{array}$$

She divided as with whole numbers. The decimal point was placed to show that the quotient was about 5. Divide using Jessie's method. Check by multiplying.

a. $7\overline{)33.6}$ **b.** $8\overline{)40.48}$ **c.** $5\overline{)40.135}$

2. Tom told Jessie that in a division like $7\overline{)30.1}$, he divides as with whole numbers. Then he puts the decimal point over the decimal point in the dividend.

$$\begin{array}{r} 4\,3 \\ 7\overline{)30.1} \\ \underline{28} \\ 2\,1 \\ \underline{2\,1} \end{array} \qquad \begin{array}{r} 4.3 \\ 7\overline{)30.1} \\ \underline{28} \\ 2\,1 \\ \underline{2\,1} \end{array}$$

Use Tom's method to divide. Check by multiplying.

a. $3\overline{)17.4}$ **b.** $8\overline{).0272}$ **c.** $14\overline{)16.8}$

3. To help with division, we sometimes place zeros in the numeral for the dividend.

Example $17 \div 8$

$$\begin{array}{r} 2.125 \\ 8\overline{)17.000} \\ \underline{16} \\ 1\,0 \\ \underline{8} \\ 20 \\ \underline{16} \\ 40 \\ \underline{40} \end{array}$$

Divide. Check by multiplying.

a. $3 \div 8$ **b.** $.04 \div 8$ **c.** $7 \div 16$ **d.** $.24 \div 32$

EXERCISES

Divide. Check by multiplying.

1. $8\overline{)6.4}$	**2.** $5\overline{)4.5}$	**3.** $4\overline{)3.6}$
4. $9\overline{).819}$	**5.** $12\overline{).096}$	**6.** $13\overline{).0065}$
7. $3\overline{).0276}$	**8.** $15\overline{)2.25}$	**9.** $20\overline{)4.04}$
10. $8\overline{)28}$	**11.** $32\overline{)14.4}$	**12.** $16\overline{)1.36}$
13. $24\overline{)15}$	**14.** $28\overline{)1.82}$	**15.** $36\overline{).18}$
16. $350\overline{)6.3}$	**17.** $320\overline{).208}$	**18.** $2{,}400\overline{)3.6}$
19. $560\overline{)7.84}$	**20.** $290\overline{)5.51}$	**21.** $490\overline{)1.127}$
22. $5{,}700\overline{)1.482}$	**23.** $62\overline{)3.038}$	**24.** $74\overline{).2664}$
25. $26\overline{).2262}$	**26.** $6{,}300\overline{)51.66}$	**27.** $950\overline{)40.85}$
28. $7{,}600\overline{)20.52}$	**29.** $29\overline{)1.363}$	**30.** $82\overline{)1.558}$
31. $960\overline{)22.08}$	**32.** $8{,}300\overline{)79.68}$	**33.** $68{,}000\overline{)102}$

Solve these problems.

34. Mr. Preisinger weighed 19.2 grams of sulfur for his science class. He wished to divide this equally among 8 students. How much sulfur would each student receive?

35. In the first three games of the football season, Charlie carried the ball 44 times and gained 143 yards. What was his average gain per play?

DIVISION WITH DECIMALS

1. We can name the quotient $.4\overline{)8.32}$ as $\frac{8.32}{4}$.

   a. Multiply: $\frac{8.32}{.4} \times \frac{10}{10}$.

   b. Now the divisor is a whole number.
   Divide: $4\overline{)832.}$

2. The quotient $.36\overline{).072}$ can be named $\frac{.072}{.36}$.

   a. By what must .36 be multiplied to get 36?

   b. Multiply: $\frac{.072}{.36}$ by 1, using $\frac{100}{100}$.

   c. Divide: $36\overline{)7.2.}$

3. Divide.

Example $.027\overline{)8.64}$ $\qquad \frac{8.64}{.027} \times \frac{1{,}000}{1{,}000} = \frac{8{,}640}{27}$

$$\begin{array}{r} 320 \\ 27\overline{)8{,}640} \\ \underline{8\,1} \\ 54 \\ \underline{54} \\ 0 \\ \underline{0} \end{array}$$

a. $.09\overline{)7.2}$ b. $.007\overline{)2.17}$ c. $.25\overline{)3.75}$

d. $.009\overline{).63}$ e. $.046\overline{).001196}$ f. $.0032\overline{)1.28}$

EXERCISES

Divide.

1. $.5\overline{)16}$ 2. $.4\overline{)28}$ 3. $.3\overline{).63}$ 4. $.6\overline{).72}$

5. $.6\overline{).018}$ 6. $.7\overline{)3.5}$ 7. $7\overline{).084}$ 8. $.9\overline{).0081}$

9. $.01\overline{).046}$ 10. $.02\overline{).028}$ 11. $.02\overline{)4}$ 12. $.07\overline{).063}$

13. $.15\overline{).45}$

14. $.08\overline{)7.28}$

15. $.19\overline{)76.19}$

16. $4.1\overline{)1.558}$

17. $.056\overline{).001904}$

18. $.028\overline{)1.8004}$

19. $2.2\overline{)4.686}$

20. $.0021\overline{).00966}$

21. $.0146\overline{)3.1098}$

22. $1.3\overline{).0169}$

23. $.0159\overline{).0004134}$

24. $.0041\overline{)1.23123}$

25. $.824\overline{).0046968}$

26. $.0236\overline{).0001888}$

Solve these problems.

27. Mr. Best drove 360.8 miles and used 16.4 gallons of gasoline. How many miles per gallon did he average?

28. In his shop, Mr. Nair needed to cut pieces of aluminum stock into lengths 4.26 inches long. How many pieces can he cut from a rod 38.25 inches long?

29. Marilyn bought a packet of marigold seeds. She read on the packet that it contained 50 seeds which weighed a total of .005 ounces. About how much does a single marigold seed weigh?

UNEXPLORED TERRITORY

1. Write a three-digit numeral such that the hundreds digit minus the ones digit is greater than one.
2. Reverse the digits.
3. Subtract the two numbers.
4. Reverse the digits in the difference.
5. Add the two numbers.
6. Repeat several times. Look for a pattern.

Example

$$\begin{array}{r} 562 \\ -\ 265 \\ \hline 297 \\ +\ 792 \\ \hline 1{,}089 \end{array}$$

FRACTIONAL NUMERALS TO DECIMALS

To find a decimal numeral for $\frac{3}{8}$, we can divide the numerator by the denominator.

$$\frac{3}{8} = .375$$

$$\begin{array}{r} .375 \\ 8 \overline{)\,3.000} \\ \underline{2\;4} \\ 60 \\ \underline{56} \\ 40 \\ \underline{40} \end{array}$$

1. Find decimal numerals.

   a. $\frac{5}{8}$ b. $\frac{8}{25}$ c. $\frac{9}{40}$ d. $\frac{13}{4}$

2. Here is how Mark found a decimal numeral for $\frac{4}{9}$.

$$\begin{array}{r} .4444 \\ 9 \overline{)\,4.0000} \\ \underline{3\;6} \\ 40 \\ \underline{36} \\ 40 \\ \underline{36} \\ 40 \\ \underline{36} \\ 4 \end{array}$$

   a. Will the 4's in his quotient ever end?

   b. When could he tell that the 4's will repeat?

3. Recall that we can use a bar to show that the 4's repeat.

$$\frac{4}{9} = .4\overline{4}$$

   Find decimal numerals.

Example $\frac{5}{11}$

$$\begin{array}{r} .4545 \\ 11 \overline{)\,5.0000} \\ \underline{4\;4} \\ 60 \\ \underline{55} \\ 50 \\ \underline{44} \\ 60 \\ \underline{55} \\ 5 \end{array}$$

$$\frac{5}{11} = .45\overline{45}$$

   a. $\frac{2}{3}$ b. $\frac{9}{11}$ c. $\frac{8}{9}$ d. $\frac{5}{18}$

EXERCISES

Find decimal numerals.

1. $\frac{1}{8}$	2. $\frac{1}{12}$	3. $\frac{1}{3}$	4. $\frac{1}{5}$	5. $\frac{1}{6}$
6. $\frac{15}{4}$	7. $\frac{7}{25}$	8. $\frac{1}{7}$	9. $\frac{1}{9}$	10. $\frac{3}{5}$
11. $\frac{5}{12}$	12. $\frac{7}{9}$	13. $\frac{5}{6}$	14. $\frac{2}{9}$	15. $\frac{3}{8}$
16. $\frac{3}{4}$	17. $\frac{11}{6}$	18. $\frac{7}{11}$	19. $\frac{7}{8}$	20. $\frac{1}{11}$
21. $\frac{7}{20}$	22. $\frac{1}{15}$	23. $\frac{17}{12}$	24. $\frac{5}{16}$	25. $\frac{4}{5}$
26. $\frac{2}{5}$	27. $\frac{5}{9}$	28. $\frac{19}{6}$	29. $\frac{7}{18}$	30. $\frac{11}{6}$
31. $\frac{2}{7}$	32. $\frac{8}{11}$	33. $\frac{19}{8}$	34. $\frac{11}{20}$	35. $\frac{34}{11}$
36. $\frac{4}{15}$	37. $\frac{3}{7}$	38. $\frac{15}{16}$	39. $\frac{1}{13}$	40. $\frac{23}{9}$

UNEXPLORED TERRITORY

1. Find decimal numerals for $\frac{1}{3}$ and $\frac{2}{3}$.

2. Add.

   a. $\frac{1}{3} + \frac{2}{3}$ b. $.3\overline{3} + .6\overline{6}$

3. Find decimal numerals for $\frac{1}{6}$ and $\frac{5}{6}$.

4. Add.

   a. $\frac{1}{6} + \frac{5}{6}$ b. $.16\overline{6} + .83\overline{3}$

5. What number does $.9\overline{9}$ name?

DO YOU REMEMBER?

Add.

1. 93.4 + 186.56 + 321.824
2. 68.29 + 9.3 + 568.486
3. 5.328 + 19.64 + 365.927
4. 82.56 + 2.439 + 229.56

1.	2.	3.	4.
93.4	68.29	5.328	82.56
186.56	9.3	19.64	2.439
321.824	568.486	365.927	229.56

5. 93.42 + 65.2 + .075

6. 6.823 + .98 + 194.568

Subtract.

7.	8.	9.	10.
94.62	124.65	18.6	82.56
− 64.8	− 98.823	− 2.596	− 16.824

11. 129.62 − 84.348

12. 94.26 − 3.075

Multiply.

13.	14.	15.	16.
42.7	5.201	.841	.702
× .6	× .8	× .3	× .12

17.	18.	19.	20.
29.5	5.42	128.3	5.683
× 1.8	× .067	× 4.25	× .0064

Divide.

21. $21\overline{)88.2}$

22. $.06\overline{)4.32}$

23. $.75\overline{)337.5}$

24. $.072\overline{).36}$

25. $47.3\overline{).9933}$

26. $2.45\overline{)1.2985}$

Solve these problems.

27. Tony collected 94 soda bottles. He received a 5¢ deposit on each. How much did he receive?

28. The labor charge to remove and replace a muffler is $7.00 per hour. What would be the cost if the job took .8 hour?

PROBLEM SOLVING

1. Andy kept this record of his practice times for the 100-yard dash: 12.1 seconds, 12 seconds, 12.2 seconds, 11.8 seconds, and 11.9 seconds. Find his average time.

2. The odometer on Mr. Bodine's car read 26,803.7 before his trip and 28,246.5 after his trip. How far had he traveled?

3. Mr. Bodine kept a record of the gasoline he used on the trip.

   a. Gasoline costs 35.9 cents per gallon. Find how much he spent for gasoline.

   b. Find the average number of miles he got per gallon. (See Exercise 2.)

Day	*Gallons*
Monday	21.8
Tuesday	22.6
Wednesday	23.5
Thursday	21.9
Friday	19.5

4. Section 8B won a $50 prize for selling patron ads. They needed $2.25 per student for a field trip. There were 34 students in 8B. Do they have enough money? If not, about how much will each student have to contribute?

5. A metal washer weighs .2 ounces. The packaging department needs to package them in boxes containing 500. How much should each box weigh?

DECIMAL EQUIVALENTS

The decimal numerals for some fractions are used so often that it is important to remember them. Study this table of equivalents.

TABLE OF EQUIVALENTS

Fractional Numeral	*Decimal Numeral*
$\frac{1}{2}$	.5
$\frac{1}{4}$	.25
$\frac{3}{4}$	.75
$\frac{1}{3}$	$.3\overline{3}$
$\frac{2}{3}$	$.6\overline{6}$
$\frac{1}{5}$	.2
$\frac{2}{5}$	.4
$\frac{3}{5}$	.6
$\frac{4}{5}$	.8
$\frac{1}{8}$	.125
$\frac{3}{8}$	.375
$\frac{5}{8}$	.625
$\frac{7}{8}$	.875
$\frac{1}{6}$	$.16\overline{6}$
$\frac{5}{6}$	$.83\overline{3}$
$\frac{1}{1}$	1.00

EXERCISES

Race time! Give the decimal equivalent as fast as you can. Try not to look at the table.

1. $\frac{1}{5}$	**2.** $\frac{7}{8}$	**3.** $\frac{3}{8}$	**4.** $\frac{1}{4}$
5. $\frac{4}{5}$	**6.** $\frac{1}{8}$	**7.** $\frac{5}{6}$	**8.** $\frac{2}{3}$
9. $\frac{1}{3}$	**10.** $\frac{1}{2}$	**11.** $\frac{1}{6}$	**12.** $\frac{5}{8}$
13. $\frac{2}{5}$	**14.** $\frac{5}{6}$	**15.** $\frac{1}{1}$	**16.** $\frac{3}{4}$

Race time! Give the fractional equivalent as fast as you can.

17. $.16\overline{6}$	**18.** .625	**19.** .2	**20.** $.3\overline{3}$
21. .75	**22.** .125	**23.** .875	**24.** .8
25. $.83\overline{3}$	**26.** .25	**27.** $.6\overline{6}$	**28.** .6
29. 1.00	**30.** .375	**31.** .5	**32.** .4

Let's Practice

Solve these problems.

1. Terry ran the 100-yard dash in 11.3 seconds. Lou ran it in 10.9 seconds. How much faster was Lou?

2. The manager of Bon-Air Swim Club bought 8 cases of chlorine for $5.13 a case. What was the total cost?

3. Mr. Shelden found that his car averaged 22.8 miles per gallon at 45 miles per hour, but only 19.6 miles per gallon at 60 miles per hour. How much farther can he travel on 18.9 gallons of gasoline at 45 miles per hour than at 60 miles per hour?

DO YOU REMEMBER?

Write per cent notation.

1. .32	**2.** .57	**3.** .08	**4.** .03
5. .008	**6.** .234	**7.** 5.76	**8.** 1.862
9. $\frac{57}{100}$	**10.** $\frac{83}{100}$	**11.** $\frac{1}{4}$	**12.** $\frac{3}{8}$

Write decimal numerals.

13. 42%	**14.** 87%	**15.** 4%	**16.** 9%
17. .6%	**18.** 25.7%	**19.** .08%	**20.** 483%

Write simplest fractional numerals.

21. 87%	**22.** 7%	**23.** 40%	**24.** 75%
25. .3%	**26.** .09%	**27.** $33\frac{1}{3}$%	**28.** $6\frac{2}{3}$%

Solve these problems.

29. What is 24 per cent of 35?

30. What is 8 per cent of 56?

31. 18 is what per cent of 72?

32. 27 is what per cent of 90?

33. 13 is 5% of what number?

34. 15 is 75% of what number?

35. Mrs. Welsch got a 20% discount on a rug originally priced at $138. What did she pay for the rug?

36. Jack received $21 for selling $140 worth of soda at the ball park. What was the rate of commission?

CHAPTER REVIEW

Write decimal numerals.

1. $\frac{3}{10}$ 2. $\frac{31}{1,000}$ 3. $4\frac{9}{10}$ 4. $12\frac{13}{100}$ 5. $2\frac{513}{1,0000}$

6. Five and six tenths *5.6* 7. Two and twenty-three thousandths

Write word names.

8. .8 9. 1.2 10. .015 11. .367 12. 1.03

Write simplest fractional numerals or mixed numerals.

13. .8 14. 2.3 15. 13.3 16. 5.007 17. 81.87

Which are true and which are false?

18. $.15 > .7$ 19. $.123 < .04$ 20. $.6 < .497$

21. $.1212 > .01212$ 22. $\frac{25}{100} < .135$ 23. $1.3 > 2\frac{1}{10}$

Add.

24. $\begin{array}{r} .9 \\ +\ .8 \\ \hline \end{array}$ 25. $\begin{array}{r} .38 \\ +\ .92 \\ \hline \end{array}$ 26. $\begin{array}{r} .04 \\ +\ .153 \\ \hline \end{array}$ 27. $\begin{array}{r} .62 \\ +\ .8 \\ \hline \end{array}$

28. $\begin{array}{r} 13.476 \\ 2.58 \\ 1.023 \\ \hline \end{array}$ 29. $\begin{array}{r} 6.01 \\ 25.8 \\ 3.492 \\ \hline \end{array}$ 30. $\begin{array}{r} 1.25 \\ 15.921 \\ 8.7 \\ \hline \end{array}$ 31. $\begin{array}{r} 12.4 \\ 8.937 \\ 2.0008 \\ \hline \end{array}$

32. $1.2 + 34.1 + 5.29 + 34.083$ *74.673*

Subtract.

33. $\begin{array}{r} .53 \\ -\ .28 \\ \hline \end{array}$ *.25* 34. $\begin{array}{r} 2.563 \\ -\ 1.42 \\ \hline \end{array}$ 35. $\begin{array}{r} 15.7 \\ -\ 2.891 \\ \hline \end{array}$ 36. $\begin{array}{r} 1.342 \\ -\ .0048 \\ \hline \end{array}$

37. $\begin{array}{r} 7.3 \\ -\ 6.248 \\ \hline \end{array}$ 38. $\begin{array}{r} .0031 \\ -\ .00056 \\ \hline \end{array}$ 39. $\begin{array}{r} 25.892 \\ -\ 24.961 \\ \hline \end{array}$ 40. $\begin{array}{r} 1.2 \\ -\ .0478 \\ \hline \end{array}$

41. $82.769 - 79.68$ 42. $101.2 - 98.47$

Copy and complete.

	Number	Nearest Ten	Nearest One	Nearest Tenth	Nearest Hundredth	Nearest Thousandth
43.	132.4971					
44.	689.0394					
45.	301.0027					
46.	582.0369					
47.	100.0026					

Multiply.

48. $3.2 \times .8$ **49.** $10.02 \times .3$ **50.** 1.234×1.2 **51.** 2.891×8.7

52. 12.004×2.14 **53.** $38.53 \times .192$ **54.** $7.243 \times .00127$ **55.** $.00158 \times .0146$

56. $.382 \times 10$ **57.** $.8294 \times 10^3$

Divide.

58. $36\overline{)194.4}$ **59.** $13\overline{)10.27}$ **60.** $87\overline{)513.3}$

61. $.0064\overline{).224}$ **62.** $.0083\overline{).4897}$ **63.** $.98\overline{)0637}$

64. $1.33\overline{)319.2}$ **65.** $.255\overline{)122.4}$ **66.** $1.23\overline{)4.3542}$

Find decimal numerals.

67. $\frac{3}{4}$ **68.** $\frac{7}{8}$ **69.** $\frac{3}{5}$ **70.** $\frac{1}{2}$ **71.** $\frac{5}{8}$

72. $\frac{3}{8}$ **73.** $\frac{11}{20}$ **74.** $\frac{5}{3}$ **75.** $\frac{4}{15}$ **76.** $\frac{7}{12}$

Terms You Should Know

decimal
decimal numeral
decimal point
rounding
estimating
equivalents

CHAPTER TEST

Write decimal numerals.

1. $\frac{7}{10}$ 2. $3\frac{29}{1,000}$ 3. Twelve and five hundredths

Write word names.

4. 1.027   5. .061

Write simplest fractional or mixed numerals.

6. .085   7. 15.48

Which are true and which are false?

8. $.025 > .13$ 9. $.582 < 1.2$

Add.

10. $\begin{array}{r} .81 \\ +\ .7 \\ \hline \end{array}$ 11. $\begin{array}{r} .812 \\ +\ .16 \\ \hline \end{array}$ 12. $\begin{array}{r} 1.8 \\ .473 \\ 2.048 \\ \hline \end{array}$ 13. $\begin{array}{r} .01 \\ 2.78 \\ 24.3 \\ \hline \end{array}$

Subtract.

14. $\begin{array}{r} 1.023 \\ -\ .47 \\ \hline \end{array}$ 15. $\begin{array}{r} 1.361 \\ -\ .0472 \\ \hline \end{array}$ 16. $\begin{array}{r} 98.1342 \\ -\ 8.257 \\ \hline \end{array}$ 17. $\begin{array}{r} 125.43 \\ -\ 98.7 \\ \hline \end{array}$

18. Round 287.9264 to the nearest hundredth.

Multiply.

19. $\begin{array}{r} 34.7 \\ \times\ 25 \\ \hline \end{array}$ 20. $\begin{array}{r} 82.01 \\ \times\ 36 \\ \hline \end{array}$ 21. $\begin{array}{r} 82.37 \\ \times\ .0043 \\ \hline \end{array}$ 22. $\begin{array}{r} 1.205 \\ \times\ 1.0021 \\ \hline \end{array}$

Divide.

23. $56\overline{)2.7384}$ 24. $.052\overline{)4.5604}$ 25. $.711\overline{)396.027}$

Write decimal numerals.

26. $\frac{11}{12}$ 27. $\frac{4}{5}$

39 is 30% of what number?

$$39 = 30\% \times n$$

INTEREST = PRINCIPAL X RATE X TIME

$$I = p \times r \times t$$

CHAPTER FIVE

Working with Per Cents

PER CENTS

Bruce took a spelling quiz. He spelled 80% of the words correctly.

80% means 80 per hundred or
80 out of one hundred or

$$\frac{80}{100}.$$

1. The per cent symbol means "$\times \frac{1}{100}$."

$$n\% = n \times \frac{1}{100}$$

Write simplest fractional numerals.

Example $46\% = 46 \times \frac{1}{100}$

$$= \frac{46}{100}$$

$$= \frac{2 \times 23}{2 \times 50}$$

$$= \frac{23}{50}$$

a. 24% **b.** 6% **c.** 210% **d.** $\frac{1}{2}\%$

2. We know $\frac{1}{100} = .01$. Thus, we can also replace the per cent symbol with "$\times$ .01."

$$n\% = n \times .01$$

Write decimal numerals.

Example $82\% = 82 \times .01 = .82$

a. 93% **b.** 3% **c.** 184% **d.** 1.58%

A numeral like $n\%$ is called **per cent notation.**

$$n\% = n \times \frac{1}{100} \qquad n\% = n \times .01$$

3. Write simplest fractional numerals and decimal numerals.

a. 43% **b.** $\frac{4}{5}\%$ **c.** 1.73% **d.** 15%

EXERCISES

Write simplest fractional numerals.

1. 80%
2. 56%
3. 63%
4. 18%
5. 23%
6. 800%
7. $\frac{1}{4}$%
8. $\frac{7}{10}$%
9. 19%
10. 714%
11. $\frac{3}{4}$%
12. 362%

Write decimal numerals.

13. 43%
14. 17%
15. 64%
16. 123%
17. 2.3%
18. 17.9%
19. .03%
20. 79%
21. 525%
22. 433%
23. .056
24. .003%

Write simplest fractional numerals and decimal numerals.

25. 61%
26. 376%
27. .007%
28. .7%
29. $\frac{3}{5}$%
30. 986%
31. $\frac{2}{9}$%
32. 2%

Solve these problems.

33. Brian estimated that 20% of his time at camp was spent at the lake. What part of his time did he spend there?

34. Mrs. Stoeker said that 86% of her class had contributed to the clothing drive. Write a fractional numeral for 86%.

35. In making a dress, Mrs. Irwin estimated that she spent 15% of her time at the sewing machine. Write a fractional numeral for the part of her time spent in preparation and finishing the dress.

DO YOU REMEMBER?

Add.

1. $\begin{array}{r} 82.56 \\ 7.294 \\ \underline{563.8} \end{array}$

2. $\begin{array}{r} 293.6 \\ 3.005 \\ \underline{.7623} \end{array}$

3. $\begin{array}{r} 247.08 \\ 7.075 \\ \underline{42.78} \end{array}$

4. 84.2 + 9.63

5. 24.8 + .04 + 5.6

6. 34.82 + 2.98 + 16.004

7. .29 + 34.08 + 7.634

Subtract.

8. $\begin{array}{r} 5.007 \\ \underline{-\ 2.46} \end{array}$

9. $\begin{array}{r} 18.3 \\ \underline{-\ 11.573} \end{array}$

10. $\begin{array}{r} 83.00 \\ \underline{-\ 57.756} \end{array}$

11. .14 − .03

12. 34.6 − 8.23

13. 56 − 3.94

Multiply.

14. 4.6 × .76

15. 54.7 × 9.2

16. .056 × 8.3

17. 93.8 × 5.7

18. .073 × 95.873

19. 4.83 × 76.008

Divide.

20. $4.2\overline{)\,.0126}$

21. $.06\overline{)\,.312}$

22. $9.72\overline{)\,2.2356}$

23. $.073\overline{)\,.00584}$

24. $.56\overline{)\,.5208}$

25. $3.2\overline{)\,18.272}$

26. $16\overline{)\,1.0000}$

27. $83.2\overline{)\,.5824}$

28. $.008\overline{)\,.000012}$

Find decimal numerals.

29. $\frac{1}{8}$

30. $\frac{1}{6}$

31. $\frac{1}{4}$

32. $\frac{3}{5}$

33. $\frac{1}{12}$

34. $\frac{3}{8}$

35. $\frac{1}{9}$

36. $\frac{5}{6}$

37. $\frac{2}{9}$

38. $\frac{1}{11}$

39. $\frac{1}{7}$

40. $\frac{3}{4}$

CHANGING TO PER CENTS

We can change from a fractional numeral like $\frac{81}{100}$ to per cent notation like this.

$$\frac{81}{100} = 81 \times \frac{1}{100} = 81\%$$

We replace "$\times \frac{1}{100}$" with a % symbol.

1. Write per cent notation.

   a. $\frac{8}{100}$ b. $\frac{29}{100}$ c. $\frac{3.6}{100}$ d. $\frac{500}{100}$ e. $\frac{.26}{100}$

2. Sometimes we need to rename when the denominator is not 100.

$$\begin{aligned}\frac{4}{5} &= \frac{4}{5} \times \frac{20}{20}\\ &= \frac{80}{100}\\ &= 80 \times \frac{1}{100}\\ &= 80\%\end{aligned}$$

   Write per cent notation.

   a. $\frac{2}{5}$ b. $\frac{3}{4}$ c. $\frac{9}{20}$ d. $\frac{18}{25}$ e. $\frac{17}{10}$

3. We can change from a decimal numeral to per cent notation.

$$.54 = 54 \times .01 = 54\%$$

   We replace "× .01" with a % symbol.

   Write per cent notation.

   a. .83 b. .07 c. .042 d. 2.3 e. .006

4. We can find per cent notation for a fraction like $\frac{5}{8}$ in another way.

   a. Write a decimal for $\frac{5}{8}$.

   b. Now write per cent notation for $\frac{5}{8}$.

5. Write per cent notation.

a. $\frac{3}{8}$ b. $\frac{5}{16}$ c. $\frac{3}{4}$ d. $\frac{3}{10}$ e. $\frac{7}{20}$

6. Let's find per cent notation for $\frac{2}{3}$. First we divide to find a decimal for $\frac{2}{3}$.

$$\begin{array}{r} .66 \\ 3\overline{)2.00} \\ \underline{1\,8} \\ 20 \\ \underline{18} \\ 2 \end{array} \qquad \frac{2}{3} = .66\frac{2}{3}$$

$$\frac{2}{3} = .66\frac{2}{3} = 66\frac{2}{3} \times .01 = 66\frac{2}{3}\%$$

Write per cent notation.

a. $\frac{1}{3}$ b. $\frac{1}{6}$ c. $\frac{11}{12}$ d. $\frac{3}{7}$

EXERCISES

Write per cent notation.

Examples $\frac{3}{20} = \frac{15}{100} = 15 \times \frac{1}{100} = 15\%$

$2.5 = 250 \times .01 = 250\%$

$\frac{4}{9} = .44\frac{4}{9} = 44\frac{4}{9} \times .01 = 44\frac{4}{9}\%$

1. $\frac{6}{100}$ 2. $\frac{7}{100}$ 3. $\frac{83}{100}$ 4. $\frac{3}{10}$ 5. $\frac{57}{100}$

6. $\frac{3}{4}$ 7. $\frac{9}{10}$ 8. $\frac{3}{5}$ 9. 2.8 10. $\frac{1}{4}$

11. .28 12. $\frac{11}{25}$ 13. .96 14. .983 15. $\frac{5}{8}$

16. .03 17. .56 18. 3.4 19. .007 20. .73

21. .008 22. $\frac{7}{8}$ 23. $\frac{5}{6}$ 24. $\frac{11}{6}$ 25. $\frac{4}{25}$

MOVING THE DECIMAL POINT

Mark thought there must be a short cut for changing a decimal numeral to per cent notation. He looked at several examples.

$$.27 = 27 \times .01 = 27\%$$

$$.389 = 38.9 \times .01 = 38.9\%$$

$$2.6 = 260 \times .01 = 260\%$$

He saw that he could move the decimal point two places to the right and write the per cent symbol.

.27	.27	27%
.389	.389	38.9%
2.6	2.6	260%

1. Write per cent notation. Use Mark's method.

   a. .74 b. .59 c. .273 d. .03

   e. 6.41 f. .0723 g. 5 h. .006

2. Elaine said that by reversing Mark's method, she could change from per cent notation to a decimal numeral.

$$3\% = 3 \times .01 = .03$$

$$27.6\% = 27.6 \times .01 = .276$$

$$129\% = 129 \times .01 = 1.29$$

She moved the decimal point two places to the left and dropped the per cent symbol.

3%	3.%	.03
27.6%	27.6%	.276
129%	129.%	1.29

Write decimal numerals. Use Elaine's method.

a. 48% b. 9% c. 32% d. 46.9%

e. 243% f. 2.56% g. .8% h. 7.5%

EXERCISES

Change to percent notation by moving the decimal point.

1. .74	2. .03	3. 2.56	4. .008	5. .0023
6. .068	7. .83	8. .9	9. .7462	10. .85
11. .26	12. 42.3	13. 5.6	14. .88	15. 4.68
16. .99	17. 32.8	18. 4.396	19. .092	20. .0012
21. 1.65	22. 362.86	23. .9999	24. .51	25. .07

Change to decimal numerals by moving the decimal point.

26. 57%	27. 97%	28. 1.3%	29. 8%	30. .7%
31. .08%	32. 4.73%	33. 62%	34. 5%	35. 1.8%
36. 7.34%	37. 47%	38. .96%	39. .3%	40. .005%
41. 50.6%	42. .06%	43. .11%	44. .0067%	45. 39%
46. 8.456%	47. 99%	48. 6%	49. 2.3%	50. .0006%

Let's Practice

Solve these equations.

1. $.83 \times 40 = n$
2. $.75 \times 18 = x$
3. $.27 \times 64 = n$
4. $.4 \times 80 = y$
5. $8 \times n = 48$
6. $m \times 5 = 205$
7. $.27 \times n = 49$
8. $.83 \times y = 102$
9. $n \times .8 = 6464$
10. $.36 \times n = 1{,}440$
11. $49 \times .35 = x$
12. $100 \times .356 = y$

PER CENT OF A NUMBER

We can use decimal and fractional numerals to compute with per cents.

1. Becky got 94% on her math exam. There were 50 problems. Let's find how many she had correct.

   94% of 50 is what number?

   $94\% \times 50 = n$

   a. Write a decimal numeral for 94%.

   b. Solve this equation: $.94 \times 50 = n$.

   c. How many problems did Becky have correct?

2. In a football game, Jon threw 30 passes and completed 40% of them. Let's find how many he completed.

   40% of 30 is what number?

   $40\% \times 30 = n$

   a. Write a fractional numeral for 40%.

   b. Solve this equation: $\frac{2}{5} \times 30 = n$.

   c. How many passes did Jon complete?

3. What is 20% of 42? We write a number sentence first.

   20% of 42 is what number?

   $20\% \times 42 = n$

   We can solve in two ways.

   $.20 \times 42 = n$ $\qquad$ $\frac{1}{5} \times 42 = n$

   $8.4 = n$ $\qquad$ $8\frac{2}{5} = n$

   20% of 42 is 8.4, or $8\frac{2}{5}$.

   a. What is 25% of 80? b. What is 80% of 160?

   c. What is 60% of 8.7? d. What is 56% of 124?

EXERCISES

1. What is 15% of 80?
2. What is 24% of 16?
3. What is 1% of 62?
4. What is 9% of 100?
5. What is 12% of 900?
6. What is 83% of 1,600?
7. What is 63% of 94?
8. What is 3% of 52?
9. What is 19% of 43?
10. What is 94% of 82?
11. What is 58% of 648?
12. What is 11% of 960?
13. What is 3% of 400?
14. What is 26% of 42?
15. What is 13% of 156?
16. What is 38% of 50?

Solve these problems.

17. Mr. Weimer budgets 15% of each paycheck for savings. His check is for $164. How much does he save?

18. In the class election, Vic got about 42% of the votes. There were 268 votes. About how many votes did Vic get?

19. Mr. Levitt earned $21,000. He had to pay 32% in taxes. How much tax did he have to pay?

20. Don made 35% of his shots in a basketball game. He took 40 shots. How many shots did he make?

UNEXPLORED TERRITORY

When the day after tomorrow is yesterday, today will be as far from Saturday as today is from last Friday if today were yesterday. What day is today?

RATIO AND PER CENT

1. Eight men took 2 dogs on a hunting trip. The ratio of men to dogs is 8 to 2. We can write 8:2 or $\frac{8}{2}$ to name the ratio. What is the ratio of dogs to men?

2. Mr. Marchione, the school principal, said the ratio of teachers to students was 1:28, or $\frac{1}{28}$. What is the ratio of students to teachers?

3. Write simplest fractional numerals for these ratios.

   a. 5 to 10 **b.** $\frac{36}{12}$ **c.** 4:6 **d.** 7 to 49

4. Tom got 3 hits out of 5 times at bat. The ratio of the number of hits to times at bat can be also shown as a per cent.

$$\frac{3}{5} = \frac{3}{5} \times \frac{20}{20} = \frac{60}{100} = 60\%$$

   Write per cent notation for these ratios.

   a. 30 out of 50 **b.** 20:10 **c.** 6 to 100

   d. $\frac{60}{180}$ **e.** $\frac{6}{20}$ **f.** $\frac{16}{24}$

The ratio of some number to 100 can be shown using per cent notation. We think of 60 per 100 $\left(\frac{60}{100}\right)$ as 60%.

EXERCISES

Write simplest fractional numerals.

1. 6 to 10
2. 27:36
3. 7:35
4. 75 to 80
5. $\frac{46}{84}$
6. $\frac{24}{20}$
7. $\frac{72}{18}$
8. $\frac{18}{52}$
9. $\frac{17}{17}$
10. 80 to 120
11. 85:20
12. 23 to 41
13. 19 out of 20
14. 39 out of 50
15. .28
16. .03
17. 6 out of 25
18. 62
19. $\frac{24}{120}$
20. 90 out of 150
21. $\frac{24}{40}$
22. .08
23. $\frac{18}{75}$
24. $\frac{30}{160}$

Solve these problems.

25. During the flu epidemic, Mr. Frey said that only 21 out of his 35 students were present. What per cent of Mr. Frey's class was present?

26. At the end of the baseball season, Butch had 36 hits out of 120 times at bat. Write per cent notation for the ratio of hits to times at bat.

27. During the football season, Lou successfully kicked 24 out of 40 field goals. Write per cent notation for the ratio of successful field goals to field goals attempted.

28. Nancy got 34 problems correct out of 50 on her science test. Write per cent notation for the ratio of correct problems to total number of problems.

FINDING PER CENTS

1. Candy said that 6 out of her 24 record albums were vocals. What per cent of her albums were vocals?

   6 is what per cent of 24?

   $6 = n \% \times 24$

   a. These two sentences are related sentences.

   $$6 = n\% \times 24 \qquad n\% = 6 \div 24$$

   Divide: $24\overline{)6.00}$.

   b. Write per cent notation for .25.

   c. What per cent of the record albums were vocals?

2. Judy said that 105 out of 350 students had paid class dues. Let's find what per cent had paid their dues.

   a. Write a number sentence for "105 is what per cent of 350?"

   b. A related sentence is $n\% = 105 \div 350$.
   Divide: $350\overline{)105.00}$.

   c. Write per cent notation for .30.

   d. 105 is what per cent of 350?

3. a. Write a number sentence for "3 is what per cent of 24?"

   b. Write a related sentence.

   c. Divide: $24\overline{)3.000}$.

   d. Write per cent notation for .125.

   e. 3 is what per cent of 24?

4. 50 is what per cent of 500?

5. 6 is what per cent of 54?

EXERCISES

1. 18 is what per cent of 90?

2. 29 is what per cent of 50?

3. 93 is what per cent of 465?

4. 64 is what per cent of 16?

5. 1,328 is what per cent of 1,600?

6. 144 is what per cent of 192?

7. 16 is what per cent of 144?

8. 19 is what per cent of 57?

9. 2 is what per cent of 80?

10. 45 is what per cent of 9?

11. 81 is what per cent of 486?

12. 42 is what per cent of 6?

13. 75 is what per cent of 750?

Solve these problems.

14. A fund drive raised $4,500 for the building of the youth center. To start the center, $12,000 was needed. What per cent of the total amount needed was raised?

15. The name 24 carat describes pure gold. What per cent of gold is in a bracelet marked 18 carat?

16. The income of the city symphony orchestra was about $1,700,000. About $960,000 came from contributions. About what per cent was from contributions?

PER CENT OF WHAT NUMBER

1. There were 39 children in the village of Lineboro. This was 30% of the total population. How many people live in Lineboro?

   39 is 30% of what number?

   $39 = 30\% \times n$

   a. Write a decimal numeral for 30%.

   b. These sentences are related.

   $39 = .30 \times n \qquad n = 39 \div .30$

   Divide: $.30\overline{)39.00}$.

   c. 39 is 30% of what number?

2. a. Translate "26 is 40% of what number?" into a number sentence.

   b. Write a decimal numeral for 40%.

   c. These sentences are related sentences.

   $26 = .40 \times n \qquad n = 26 \div .40$

   Divide: $.40\overline{)26.00}$.

   d. 26 is 40% of what number?

3. a. Translate "72 is 15% of what number?" into a number sentence.

   b. Write a decimal numeral for 15% and write a related sentence.

   c. Divide: $.15\overline{)72.00}$.

   d. 72 is 15% of what number?

4. 56 is 80% of what number?

5. 23 is 23% of what number?

6. 29 is 16% of what number?

EXERCISES

1. 42 is 30% of what number?

2. 15 is 5% of what number?

3. 216 is 3% of what number?

4. 56 is 56% of what number?

5. 12.18 is 14% of what number?

6. 6.37 is 7% of what number?

7. 290 is 20% of what number?

8. 84 is 200% of what number?

9. 25 is .5% of what number?

10. 320 is 8% of what number?

11. 116 is 2% of what number?

12. 399.84 is 42% of what number?

Solve this problem.

13. Mr. Smith, a salesman, was paid 20% of his total sales. One week he received $136.00. What were his total sales?

Let's Practice

Write exponential notation.

Example $6 \cdot 6 \cdot 6 \cdot 6 = 6^4$

1. $5 \cdot 5 \cdot 5 \cdot 5 \cdot 5 \cdot 5$
2. $3 \cdot 3 \cdot 3$
3. $7 \cdot 7 \cdot 7 \cdot 7 \cdot 7 \cdot 7 \cdot 7$
4. $110 \cdot 110 \cdot 110 \cdot 110$
5. $2 \cdot 2 \cdot 2 \cdot 2 \cdot 2 \cdot 2 \cdot 2$
6. $10 \cdot 10 \cdot 10 \cdot 10 \cdot 10$

SOLVING PER CENT PROBLEMS

Let's practice solving per cent problems.

Example What is 32% of 59?

$n = 32\% \times 59$

$n = .32 \times 59$

$n = 18.88$

18.88 is 32% of 59.

$$\begin{array}{r} 59 \\ \times\ .32 \\ \hline 118 \\ 177 \\ \hline 18.88 \end{array}$$

1. What is 24% of 630?

2. What is 130% of 72?

3. What is 86% of 97?

Example 46 is what per cent of 80?

$46 = n\% \times 80$

$n\% = 46 \div 80$

$n\% = .575$

$n\% = 57.5\%$

46 is 57.5% of 80.

$$\begin{array}{r} .575 \\ 80\overline{)46.000} \\ \underline{40\ 0} \\ 6\ 00 \\ \underline{5\ 60} \\ 400 \\ \underline{400} \end{array}$$

4. 8 is what per cent of 50?

5. 18 is what per cent of 12?

6. 112 is what per cent of 160?

Example 45 is 60% of what number?

$45 = 60\% \times n$

$45 = .60 \times n$

$n = 45 \div .60$

$n = 75$

45 is 60% of 75.

$$\begin{array}{r} 75 \\ .60\overline{)45.00} \\ \underline{42\ 0} \\ 3\ 00 \\ \underline{3\ 00} \end{array}$$

7. 190 is 76% of what number?

8. 84 is 84% of what number?

9. 61.2 is 34% of what number?

EXERCISES

Write number sentences and solve.

1. 74 is what per cent of 296?
2. What is 56% of 284?
3. 32 is 40% of what number?
4. What is 15% of 8.1?
5. 56 is what per cent of 800?
6. 343 is 35% of what number?
7. 400 is 80% of what number?
8. 33.18 is what per cent of 158?
9. 2,835 is 30% of what number?
10. What is 60% of 650?
11. What is 62.5% of 72?
12. 6 is what per cent of 40?
13. What is 40% of 55?
14. 48.16 is what per cent of 56?
15. What is 15% of 2,400?
16. 15% of what number is 9?
17. 29.34 is what per cent of 2,934?
18. 2.046 is what per cent of 682?
19. 12% of what number is 858?

PER CENT OF INCREASE OR DECREASE

1. Mrs. Fitzgerald said that the price of a giant-size box of Sudsy Soap Powder increased from \$1.20 to \$1.26.

   a. How much more did the soap powder cost?

   b. We can write a number sentence to find the per cent of increase.

   Increase is what per cent of original price?

   $$6 \quad = \quad n\% \quad \times \quad \$1.20$$

   Write a related sentence for $6 = n\% \times 1.20$.

   c. Divide: $1.20\overline{)6.00}$.

   d. What is the per cent of increase in price?

2. Kenwood High School's football team won 8 games last year. This year, they won 6.

   a. How many fewer games did they win this year?

   b. Translate to a number sentence:

   Decrease is what per cent of last year's wins?

   c. Write a related sentence for $2 = n\% \times 8$.

   d. What was the per cent of decrease in games won?

3. Mr. Chenoweth, the builder, said that the price of bricks had increased from \$70 per 1,000 to \$84.70 per 1,000. Let's find the per cent of increase.

   a. How much more did the bricks cost per 1,000?

   b. Write a number sentence to find the per cent of increase.

   c. Write a related sentence.

   d. What was the per cent of increase in the price of the bricks?

To find a per cent of increase or decrease, we find the amount of increase or decrease. Then find what per cent it is of the original amount.

EXERCISES

Solve these problems.

1. Last year, Mr. Brush paid \$864 in property tax. This year he paid \$1,071.36. What was the per cent of increase?

2. Five years ago, the population of Valle was 15,200. Today's is 12,464. Find the per cent of decrease.

3. After repairing his carburetor, Mr. Mercer said his mileage increased from 18 miles per gallon to 24 miles per gallon. Find the per cent of increase.

4. When the factory signed a new contract, Mr. Spinnato worked only 35 hours per week instead of 40. What was the per cent of decrease?

5. Jack said he weighed 105 pounds last year, but now weighs $115\frac{1}{2}$ pounds. What was the per cent of increase in his weight?

6. The county government reported that its budget had increased 20.22 million dollars from last year's budget of 134.8 million dollars. Find the per cent of increase in the budget.

7. The wholesale price of a sofa was \$200. It sold for \$350 in a store. What was the per cent of increase in the sofa's cost?

8. During a semi-annual sale, a dress originally costing \$75 was priced at \$45. What was the per cent of decrease in the price?

DO YOU REMEMBER?

1. What is 24% of 57?
2. 9 is 20% of what number?
3. What is 83% of 83?
4. 7.2 is what per cent of 40?
5. 43.75 is what per cent of 70?
6. What is 3% of 143?
7. 390 is 60% of what number?
8. 2.32 is what per cent of 29?
9. 459.2 is 82% of what number?
10. What is 1.8% of 154?
11. What is 120% of 62?

Solve these problems.

12. The enrollment at Dulaney Junior High School was 1,250. The principal estimated an increase of 90 students for next year. What was the estimated per cent of increase?
13. Chuck said his father received a raise of 8%. His father's annual salary was $12,500. How much was the raise?
14. Janet was absent from school 9 days. She said that this was 5% of the year. How many days were in her school year?
15. In a free throw contest, Steve made 67 out of 75 shots. What per cent of the shots did he make?

PROBLEM SOLVING

1. Frank and Alex washed cars at a parking lot. They received $2.00 per car, but had to give the parking lot owner 75¢ for each car. How much did each boy earn when they washed 28 cars?

2. Mrs. Carr bought a 6-pound roast which cost $1.65 per pound. How much change did she receive from a $20 bill?

3. At the start of football season, Ron weighed $156\frac{3}{4}$ pounds. He lost $8\frac{1}{2}$ pounds during the season, but gained back $3\frac{1}{8}$ pounds. What is his present weight?

4. On their vacation, Don's family drove 329.7 miles the first day, 403.8 the second day, and 287.2 the third day. How far had they traveled in the three days?

5. Mr. Webb drove 832 miles for business. He receives 10¢ per mile from his employer. Mr. Webb also spent $18.35 for parking charges and toll fees. How much money will he receive for business expenses?

6. The school mile relay team ran the first quarter mile in 59.3 seconds, the second quarter mile in 60.1 seconds, the third quarter mile in 59.9 seconds, and the fourth quarter in 59.7 seconds. How long did it take the team to run the race?

INTEREST

When you borrow money from a bank, you pay *interest* for the use of the money.

1. Jackie borrowed \$40 from a bank. The *rate of interest* was 5% per year. Let's find how much interest the bank would earn in one year.

   a. What is 5% of \$40?

   b. How much interest would the bank earn?

   c. How much money would Jackie have to pay back to the bank at the end of one year?

2. The *principal* is the amount of money borrowed, loaned, or invested. Suppose you borrowed \$200 at a rate of interest of 6% per year.

   a. How much interest would you owe at the end of a year?

   b. How much money would you owe at the end of the year?

3. Sometimes money is borrowed for less than a year. Six months is the same as $\frac{1}{2}$ year. What part of a year is each of these?

   a. 4 months   **b.** 3 months

   c. 9 months   **d.** 2 months

Interest is the product of the **principal**, the **rate** of interest, and the **time** the money is used. The rate of interest is usually given per year. The time is then expressed in years.

$$\text{Interest} = \text{principal} \times \text{rate} \times \text{time}$$
$$I = p \cdot r \cdot t$$

4. Find the interest on $850 at 7% per year for 6 months.

   a. What part of a year is 6 months?

   b. Copy and complete.

$$I = p \cdot r \cdot t$$
$$I = \underline{\quad\quad} \times .07 \times \frac{1}{2}$$
$$I = \underline{\quad\quad}$$

EXERCISES

Copy and complete the chart.

	Principal	*Interest Rate Per Year*	*Time*	*Interest*
	$ 60	6%	1 year	$3.60
1.	80	7%	1 year	
2.	95	5%	1 year	
3.	800	4%	6 months	
4.	1,500	$5\frac{1}{2}$%	1 year	
5.	2,800	12%	1 year	
6.	325	8%	3 months	
7.	750	6%	4 months	
8.	1,800	4%	1 year	
9.	1,350	6%	9 months	

Solve these problems.

10. Terry borrowed $250 from a bank. The rate of interest was 4% paid twice a year. How much interest did the bank earn every six months?

11. Mr. Roth borrowed $550 from a bank for one year at 7% interest. How much will he have to pay back in all?

COMPOUND INTEREST

Interest on savings accounts is often computed in a different way.

1. Suppose you have \$200 in a savings account at 6% per year for one year.

   a. How much interest will you receive after 1 year?

   b. How much will you then have in your account?

2. a. Suppose you leave the \$212 in the account for another year. How much interest will you receive at the end of the second year?

   b. What will be the new principal?

3. Many banks do not compute interest on cents. The bank will compute your interest on \$224. How much money would you have at the end of the third year?

Interest paid on interest is called **compound interest.**

4. When interest is added every year, we say it is *compounded annually.* Find the total number of dollars in an account after 2 years if the principal is \$300, the rate is 5%, and interest is compounded annually.

5. When interest is added every half-year, we say it is *compounded semi-annually.* Suppose \$50 is deposited on July 1 in a bank which pays 4% interest compounded semi-annually.

   a. How much interest will be received at the end of six months?

   b. What will be the new principal on January 1?

   c. How much interest will be paid on the new principal after the following July 1?

   d. What will be the new principal on July 1?

6. Interest added every fourth of a year is *compounded quarterly*. Suppose $100 is deposited on October 1 in a bank which pays 5% interest compounded quarterly.

   a. How much interest will be received at the end of three months.

   b. What will be the new principal on January 1?

   c. How much interest will be paid on the new principal on April 1?

   d. What will be the new principal on April 1?

EXERCISES

Find the amount in the following accounts after the indicated time. The interest is compounded annually.

	Principal	*Rate of Interest*	*Time*
1.	$ 100	4%	1 year
2.	$1,000	5%	3 years
3.	$3,000	6%	2 years

Find the amount in the following accounts after the indicated time. The interest is compounded semi-annually.

	Principal	*Rate of Interest*	*Time*
4.	$ 300	4%	1 year
5.	$ 200	5%	$1\frac{1}{2}$ years
6.	$1,000	6%	2 years

Find the amount in the following accounts after the indicated time. The interest is compounded quarterly.

	Principal	*Rate of Interest*	*Time*
7.	$ 200	4%	1 year
8.	$1,000	8%	9 months
9.	$ 600	6%	$1\frac{1}{4}$ years

DISCOUNT SHOPPING

Jeff saw a sweater that had a 10% *rate of discount.* The *marked price* was \$16. Let's find the *sale price.*

1. We must first find the discount.

   discount = rate of discount · marked price

   $$d = r \cdot m$$

   What is the discount?

2. Now let's find the sale price.

   sale price = marked price − discount

   $$s = m - d$$

   What was the sale price of the sweater?

3. Copy and complete.

	Marked Price	*Rate of Discount*	*Discount*	*Sale Price*
	\$ 24	15%	\$3.60	\$20.40
a.	\$ 10	10%		
b.	\$ 25	20%		
c.	\$120	15%		
d.	\$ 45.60	30%		

4. Jeff's father paid \$24 for a jacket that originally sold for \$30. Let's find the rate of discount.

   a. Find the discount.

   b. We know that the formula $d = r \cdot m$ is true. Below is a related formula. Use it to find the rate of discount.

$$r = \frac{d}{m}$$

$$r = \frac{6}{30}$$

$$r = \underline{\quad\quad}\%$$

5. Copy and complete.

	Marked Price	Rate of Discount	Discount	Sale Price
a.	$52		$13	
b.	$75		$11.25	
c.	$45		$15	

EXERCISES

What will be the sale price of these items during a 20% discount sale?

1. Sweaters: $12
2. Shirts: $7.00
3. Blouses: $11
4. Ties: $3.00
5. Jackets: $35
6. Shoes: $18.00

Find the rate of discount on each of these items.

	Item	Marked Price	Sale Price
7.	Men's Suits	$96.00	$80.00
8.	Transistor radio	$15.00	$12.00
9.	Portable TV	$92.00	$69.00
10.	Bicycle	$62.00	$52.70

Solve these problems.

11. Christine found Italian sandals on sale for $14. The marked price was $20. Find the rate of discount.

12. Deena bought 3 new albums at a 25% record sale. The marked price of each album was $5.20. The sales tax was 5%. Find the total cost of the albums including tax.

COMMISSION

Ted sold soda at the ball park. He was paid a *commission* of 15% of all the money received.

$$\text{Commission} = \text{rate of commission} \cdot \text{sales}$$

$$c = r \cdot s$$

1. Suppose Ted sold $45 worth of soda.

   a. Write a decimal numeral for 15%.

$$c = r \cdot s$$
$$c = .15 \times 45$$
$$c = 6.75$$

   b. What is his commission?

2. Copy and complete.

	Sales	Rate of Commission	Commission
a.	$ 950	20%	
b.	$8,000	6%	

3. Mrs. Tinley is a real estate agent. She sold a house for $30,000 and received $900 commission. We know the formula $c = r \cdot s$ is true. Below is a related formula. Use it to find her rate of commission.

$$r = \frac{c}{s}$$
$$r = \frac{900}{30{,}000}$$
$$r = \frac{3}{100}$$
$$r = \underline{\quad}\%$$

4. Copy and complete.

	Sales	Rate of Commission	Commission
a.	$25,000		$500
b.	$ 8,500		$425

EXERCISES

Copy and complete.

	Sales	Rate of Commission	Commission
1.	$50,000		$2,000
2.	$750	15%	
3.	$85		$10.20
4.	$17,000	3%	

Solve these problems.

5. Mr. Rich sells automobiles. He receives 3% of his sales plus a $300 per month salary. How much does he receive in a month when his total sales are $25,000?

6. Fred received $235 commission when he sold $4,700 worth of merchandise. What was his rate of commission?

7. Sue sold $97 worth of magazines and received 8% commission. How much money did she receive?

8. Pete's father receives a 5% commission plus a salary. During one week his total sales were $5,500. His total pay for that week was $400.70. Find his weekly salary.

9. Mrs. Hundt works as a hair stylist. She receives $60 per week plus 15% of all her sales above $100. During one week she had sales which totaled $378.20. Find her pay for that week.

10. Mr. Matlen works in an appliance store. He works for a 12% commission. How much did he earn in a week when he sold these items?

    1 Color T.V. $529 — 4 Hair dryers $52 each
    1 Refrigerator $468 — 3 Toasters $28 each
    2 Washing Machines $340 each

PROBLEM SOLVING

1. Ron ordered the following supplies for his beauty salon, The Pampered Miss.

   3 gallons shampoo
   1 dozen combs
   1 dozen boxes hair pins
   2 dozen hair nets
   6 bottles of permanent wave lotion

Price List	
Shampoo	\$2.75 gal
Combs	\$1.25 doz
Hairpins	\$.95 box
Hair nets	\$1.25 doz
Perm. wave lotion	\$9.60 doz

   Sales tax on the purchase was 6%. What was the total purchase price?

2. Twice a year The Pampered Miss advertises permanent waves at 20% off. A permanent wave usually costs \$20. What is the savings during the sale? How much does the permanent cost during the sale?

3. Carolyn went to the Pampered Miss to have her hair washed, \$2.00; cut, \$3.00; and set, \$3.50. She tipped Ron 15% of her bill. About how much tip did he receive?

4. Mrs. Brittain, an employee of Ron's, receives a weekly salary of \$50 plus a 50% commission on all sales over \$100. During one week her daily sales slips were as follows: \$71.29, \$52.60, \$39.80, \$75.86, and \$64.34. What pay did she receive for that week?

CHAPTER REVIEW

Write simplest fractional numerals.

1. 20% **2.** 32% **3.** 29% **4.** $3\frac{1}{2}$% **5.** 124%

Write decimal numerals.

6. 52% **7.** 27% **8.** .8% **9.** 2.5% **10.** 232%

Write per cent notation.

11. $\frac{7}{100}$ **12.** $\frac{3}{25}$ **13.** $\frac{9}{40}$ **14.** $\frac{5}{16}$ **15.** $2\frac{3}{4}$

16. .04 **17.** .8 **18.** .832 **19.** 2.48 **20.** .0056

Write simplest fractional numerals for these ratios.

21. 8 to 12 **22.** 3 out of 5 **23.** $\frac{48}{50}$

24. 12 to 24 **25.** 30:6 **26.** $\frac{3}{3}$

Solve these problems.

27. What is 60% of 82?

28. 25 is what per cent of 80?

29. 18 is 50% of what number?

30. 15 is 2% of what number?

31. 45% of 200 is what number?

32. 20 is what per cent of 15?

33. What per cent of 1,000 is 130?

34. 250 is 125% of what number?

35. What is .5% of 30?

36. There are 35 students in Mrs. Kelly's mathematics class. One day 5 students were absent. What per cent of the students were present?

37. The price of tomatoes dropped from 60¢ per pound to 40¢ per pound. What was the per cent of decrease?

38. Find the interest on \$50 at 8% for 3 months.

39. Mr. Jones paid \$18 interest on \$600 for 6 months. What was the rate of interest?

40. Mr. Stevens borrowed \$1,000 at 8% interest. He repaid the loan at the end of 6 months. How much did he pay?

41. Karen bought a dress on sale at a 15% discount. The original price of the dress was \$30. Find the sale price.

42. Mrs. Kennedy bought a tablecloth marked \$5 at a sale. She paid \$3 for the tablecloth. What was the rate of discount?

43. Michael DeStefano works for a 6% commission. Find his earnings for a week in which his sales totaled \$5,000.

44. Mr. Gilpatrick works for a salary plus a commission. He earns \$100 per week plus 3% of his sales in excess of \$2,000. Find his salary for a week in which his sales totaled \$6,000.

Terms You Should Know

per cent notation	compound interest
ratio	discount
per cent of increase	rate of discount
per cent of decrease	marked price
interest	sale price
rate of interest	commission

CHAPTER TEST

Write simplest fractional numerals.

1. 60%
2. $2\frac{1}{4}\%$
3. 130%

Write decimal numerals.

4. 30%
5. .5%
6. 450%

Write per cent notation.

7. $\frac{3}{100}$
8. $2\frac{3}{50}$
9. .032

Write simplest fractional numerals for these ratios.

10. 20 to 48
11. 6:4
12. 2 out of 8

Solve these problems.

13. 24 is what per cent of 30?

14. What is 12% of 72?

15. 12 is 20% of what number?

16. 50 is what per cent of 20?

17. The price of a movie ticket was raised from $1.80 to $2.00. What was the per cent of increase?

18. Find the interest on $100 at 10% for 3 months.

19. Mr. Matthews borrowed $100 at 12%. How much will he owe after 9 months?

20. Find the sale price of a bathing suit marked $16.80 that is sold at a 25% discount.

21. Mr. Smith earns a 6% commission on all of his sales. What were his sales for a week in which his commission was $150?

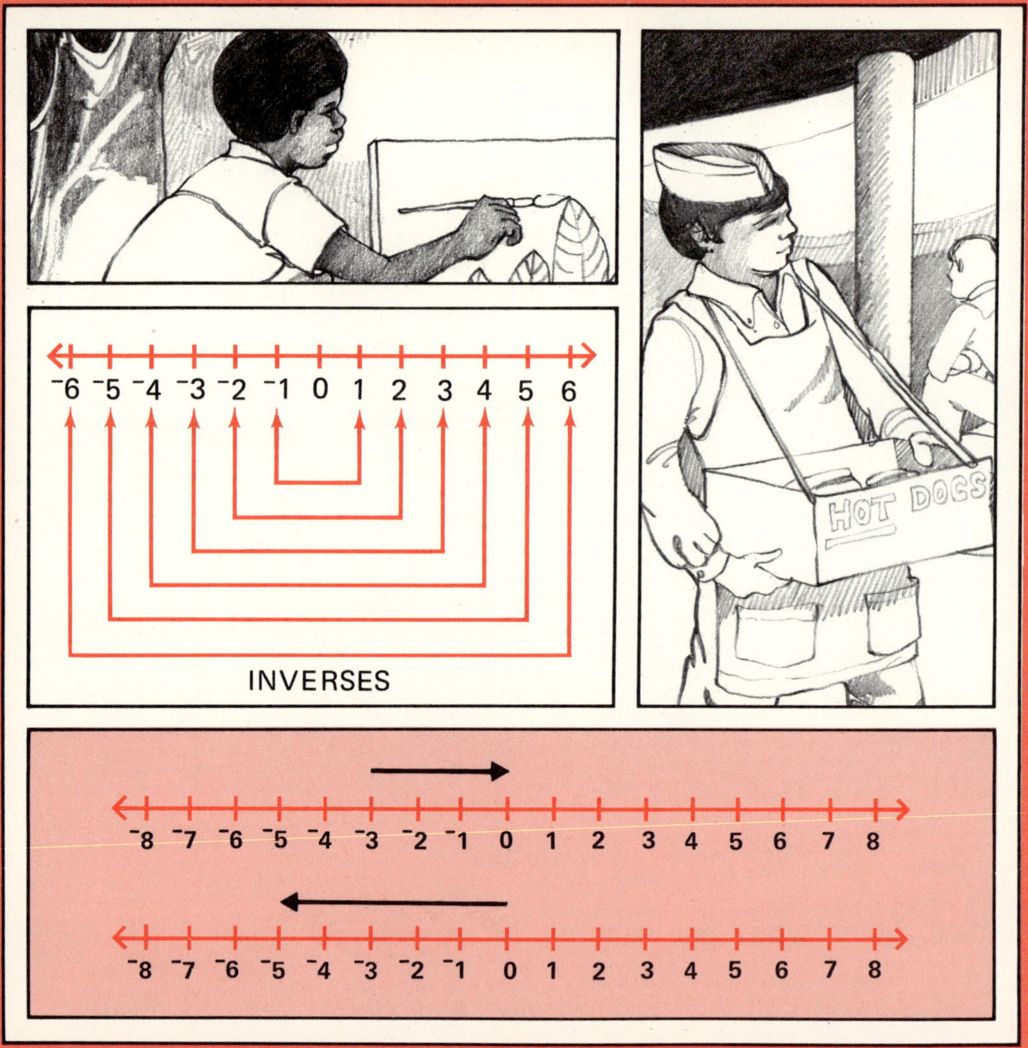

CHAPTER SIX

Integers

INTEGERS

Dick's score at the end of the first "round" of a game of shuffleboard was $^{-}18$. He was using an integer. This is the set of integers.

$$\{\ldots, {}^{-}4, {}^{-}3, {}^{-}2, {}^{-}1, 0, 1, 2, 3, 4, \ldots\}$$

1. We read $^{-}18$ as "negative eighteen." Read these.

a. $^{-}4$ **b.** $^{-}10$ **c.** $^{-}22$ **d.** $^{-}135$

2. Copy and complete this number line.

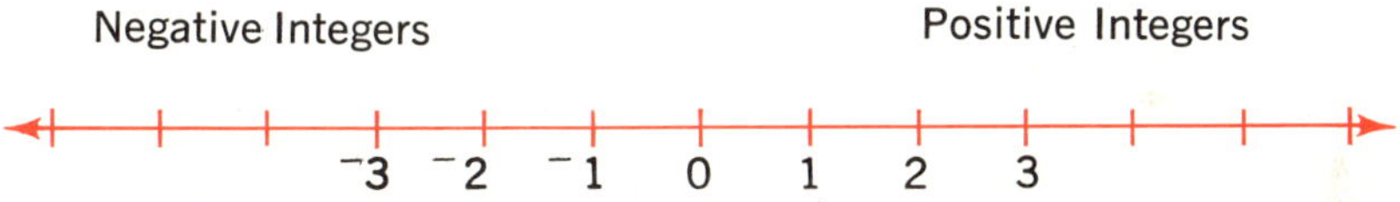

3. The greater of two integers is named farther to the right.

a. Which is greater, $^{-}7$ or $^{-}10$? Why?

b. Which is greater, $^{-}5$ or 0? Why?

c. Which is greater, 24 or $^{-}6$? Why?

4. Which are true and which are false?

a. $^{-}7 < 6$ **b.** $10 > {}^{-}2$ **c.** $0 > 14$

d. $^{-}11 > {}^{-}14$ **e.** $2 > {}^{-}7$ **f.** $16 < {}^{-}24$

5. All positive integers are greater than zero. What can we say about all negative integers?

> Positive integers are greater than zero. Negative integers are less than zero. Zero is neither positive nor negative.

6. Integers can help describe practical situations. Describe each item below with an integer.

a. 230 feet above sea level

b. 216° below zero

c. a bank withdrawal of $320

d. 12 seconds before lift off

EXERCISES

Which are true and which are false?

1. $^{-}12 < 8$	**2.** $4 > {}^{-}7$	**3.** $0 > {}^{-}73$
4. $18 < {}^{-}5$	**5.** $64 < {}^{-}124$	**6.** $26 > {}^{-}16$

Make these sentences true. Use $<$ or $>$.

7. 22 ▤ 14	**8.** $^{-}16$ ▤ 0	**9.** $^{-}43$ ▤ $^{-}18$
10. $^{-}34$ ▤ $^{-}28$	**11.** 26 ▤ $^{-}6$	**12.** $^{-}11$ ▤ 14

13. Is zero a positive integer? a negative integer?

Describe each with an integer.

14. a loss of 11 yards

15. 6 seconds before lift off

16. 194 feet below sea level

17. a bank deposit of $342

18. 27° below zero

PROBLEM SOLVING

1. Mr. Orsino said that he estimated that about 40% of the contents of his automobile cooling system was antifreeze. The system holds 12 quarts. About how much antifreeze does the system contain?

2. The eighth grade at Arbutus Middle School contains 312 students. About 18% made the honor roll. How many students made the honor roll?

3. At the middle of the Pony League season, Butch led the league in batting. He had 20 hits out of 38 times at bat. What was his batting percentage?

4. Kathy's mother gave her $12 to take her two younger sisters to the circus. Kathy's ticket cost $3.75 while her sisters' tickets cost $2.25 each. They spent $1.80 for hot dogs and popcorn. How much change did Kathy bring her mother?

5. The Ebersole family owns a house valued at $35,800. The county tax department assessed the house at 65% of its value. Find the assessed value of the house.

6. Sam and Ben were building scenery for the dramatics club play. They had to cut 12 pieces of framing wood $4\frac{1}{2}$ feet long. The wood comes in pieces 10 feet long. How many pieces should they order?

ADDING INTEGERS USING THE NUMBER LINE

Frank said he used the number line to add integers. Here is how he added $^{-}4$ and $^{-}3$.

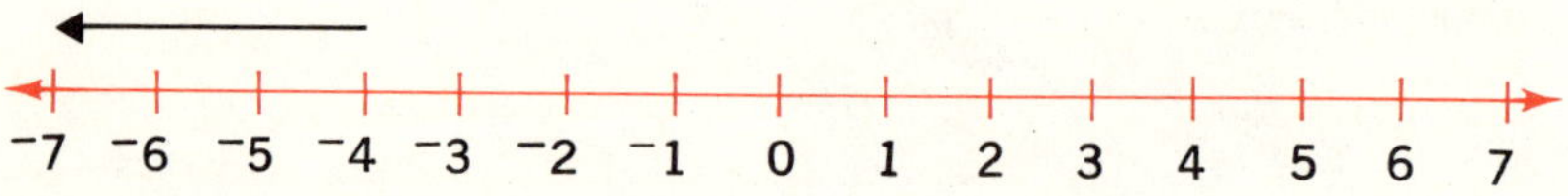

1. Frank started at the first addend, $^{-}4$.
   a. How long is the arrow for the second addend, $^{-}3$?
   b. What is the direction of the arrow?
   c. Where does the arrow stop?
   d. What is $^{-}4 + {}^{-}3$?

2. Draw a number line to show each sum.
   a. $^{-}2 + {}^{-}4$ b. $^{-}3 + {}^{-}2$ c. $^{-}2 + {}^{-}1$ d. $^{-}4 + {}^{-}5$

3. In playing a game with her friends, Lori had a score of $^{-}12$. On her next turn, she lost 6 points.
   a. What was her new score?
   b. What is $^{-}12 + {}^{-}6$?

The sum of two negative integers is a negative integer.

4. Here is how Frank added $^{-}3$ and 4 on the number line.

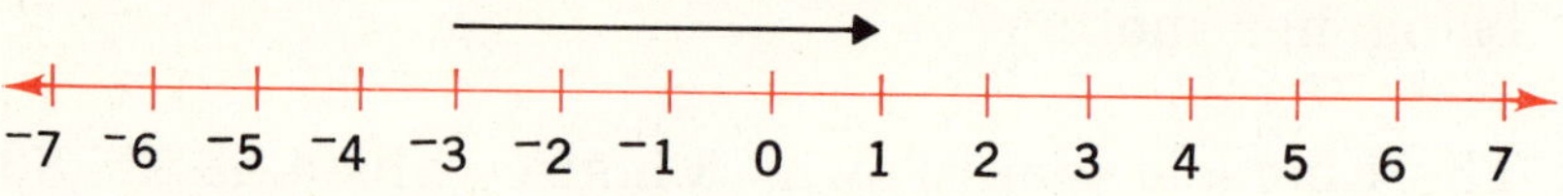

   a. Where does the arrow start?
   b. What is the direction of the arrow?
   c. How long is the arrow?
   d. Where does the arrow stop?
   e. What is $^{-}3 + 4$?

5. Draw a number line to show each sum.

a. $3 + {}^{-}6$ **b.** ${}^{-}2 + 5$ **c.** ${}^{-}3 + 3$ **d.** $0 + {}^{-}4$

6. This number line shows the addition sentence $3 + {}^{-}5 = {}^{-}2$.

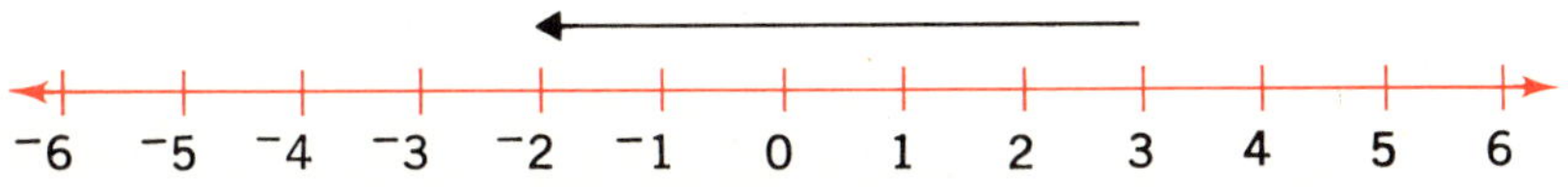

Write an addition sentence for each.

a.

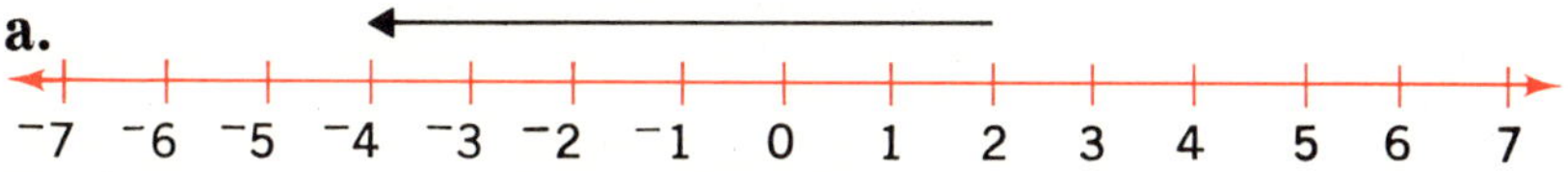

b.

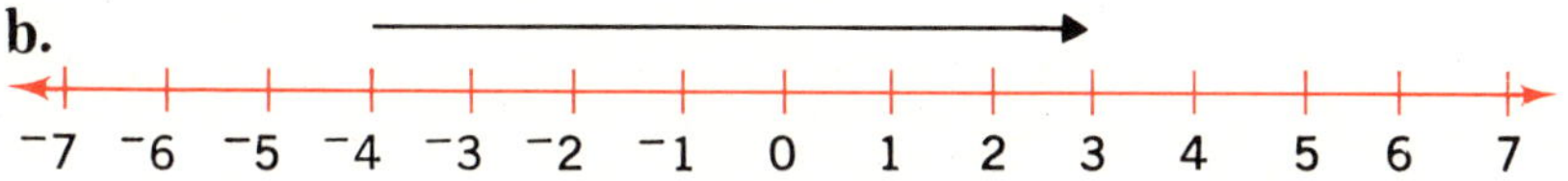

EXERCISES

Draw a number line to show each sum.

1. $8 + {}^{-}3$ 2. ${}^{-}6 + 8$ 3. ${}^{-}5 + {}^{-}4$ 4. $0 + {}^{-}4$

5. $10 + {}^{-}11$ 6. $4 + {}^{-}5$ 7. $12 + {}^{-}7$ 8. ${}^{-}3 + 11$

Write an addition sentence for each.

9.

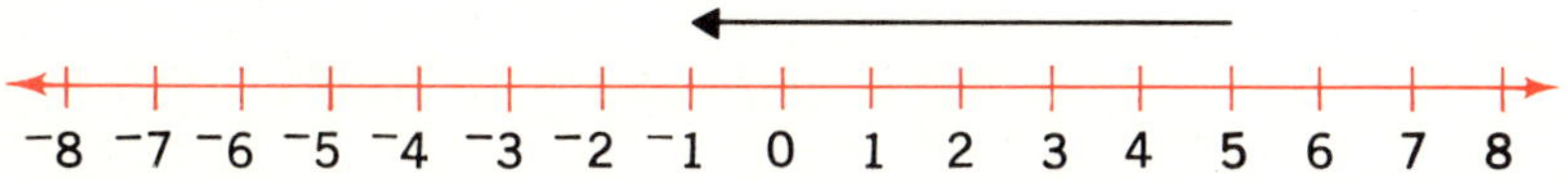

10.

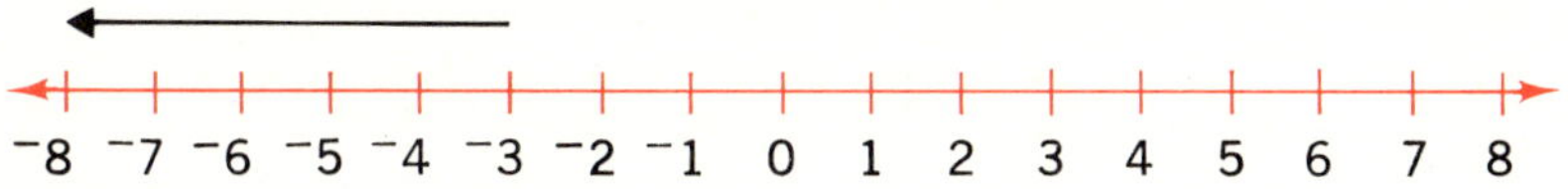

11.

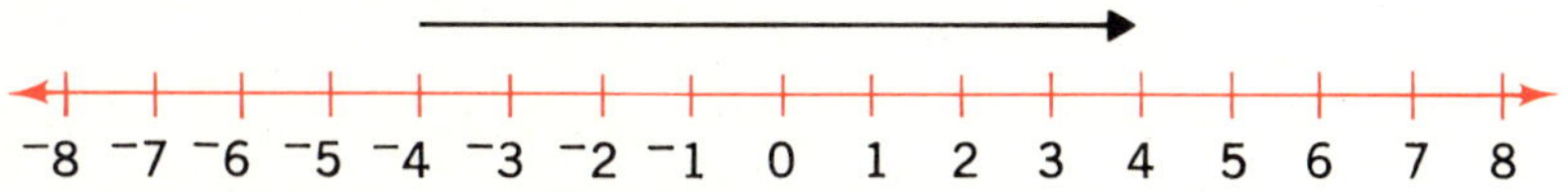

12.

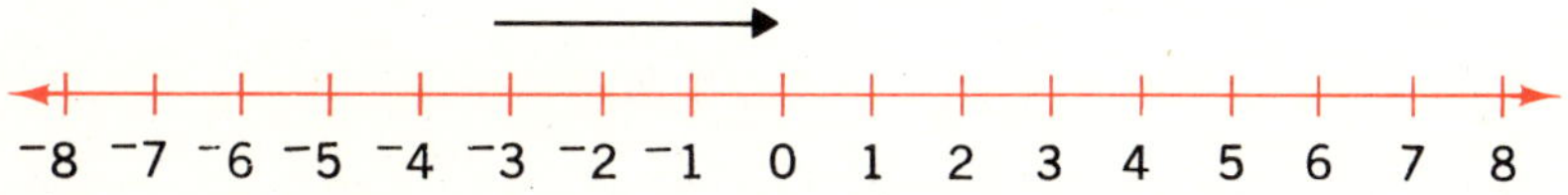

13.

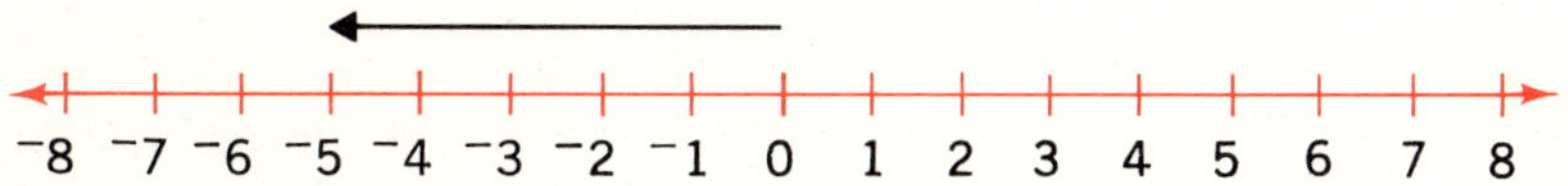

ADDITION OF INTEGERS

1. Every integer has an *inverse*. The inverse of 6 is $^{-}6$. The inverse of $^{-}3$ is 3. Zero is its own inverse.

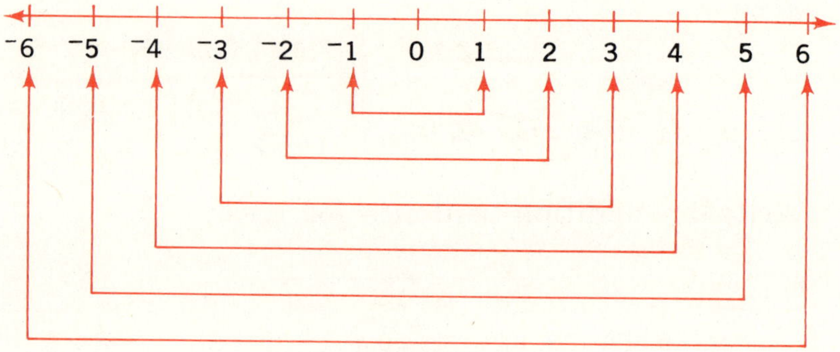

INVERSES

Name the inverse of each.

a. $^{-}4$ **b.** $^{-}14$ **c.** 0 **d.** 5

e. 98 **f.** $^{-}39$ **g.** $^{-}61$ **h.** 432

2. Add using the number line. Look for a pattern.

a. $4 + {}^{-}4$ **b.** $10 + {}^{-}10$ **c.** $^{-}6 + 6$

3. What is the sum of an integer and its inverse?

The sum of an integer and its inverse is zero.

4. Recall that addition of integers is commutative. Use number lines to show that these are true.

a. $^{-}4 + {}^{-}2 = {}^{-}2 + {}^{-}4$ **b.** $^{-}3 + 5 = 5 + {}^{-}3$

5. Addition of integers is also associative. Use number lines to show that these are true.

a. $^{-}2 + ({}^{-}5 + {}^{-}1) = ({}^{-}2 + {}^{-}5) + {}^{-}1$

b. $3 + ({}^{-}2 + {}^{-}4) = (3 + {}^{-}2) + {}^{-}4$

Addition of integers is commutative and associative.

6. We can use inverses and the associative property to help us add integers. Study each step in this addition.

$$
\begin{aligned}
5 + {}^{-}7 &= 5 + ({}^{-}5 + {}^{-}2) \\
&= (5 + {}^{-}5) + {}^{-}2 \\
&= 0 + {}^{-}2 \\
&= {}^{-}2
\end{aligned}
$$

a. How did we rename $^{-}7$ in the first step?

b. What is the sum $5 + {}^{-}7$?

7. Add. Use the method in Item 6.

a. $3 + {}^{-}8$ b. $^{-}6 + 9$ c. $^{-}12 + 5$

d. $4 + {}^{-}10$ e. $^{-}7 + 3$ f. $10 + {}^{-}2$

EXERCISES

Add.

1. $^{-}2 + {}^{-}6$
2. $0 + 4$
3. $^{-}3 + 8$
4. $9 + {}^{-}7$
5. $8 + {}^{-}3$
6. $^{-}3 + {}^{-}9$
7. $^{-}6 + {}^{-}2$
8. $4 + {}^{-}7$
9. $3 + {}^{-}7$
10. $^{-}2 + {}^{-}1$
11. $^{-}4 + 1$
12. $^{-}10 + 9$
13. $0 + {}^{-}4$
14. $^{-}5 + 5$
15. $^{-}7 + 1$
16. $11 + 16$
17. $^{-}6 + 14$
18. $^{-}15 + {}^{-}6$
19. $11 + {}^{-}9$
20. $^{-}14 + {}^{-}19$
21. $^{-}20 + 6$
22. $17 + {}^{-}17$
23. $^{-}15 + {}^{-}6$
24. $23 + {}^{-}5$
25. $40 + {}^{-}8$
26. $^{-}22 + {}^{-}9$
27. $^{-}16 + 25$
28. $^{-}20 + 6$
29. $62 + {}^{-}18$
30. $84 + {}^{-}62$
31. $^{-}156 + {}^{-}53$
32. $73 + {}^{-}14$
33. $56 + 82$

SUBTRACTION OF INTEGERS

The integer that makes the sentence $8 - 3 = m$ true also makes the related sentence $3 + m = 8$ true.

1. A related sentence for $7 - 4 = x$ is $4 + x = 7$. The dashed arrow shows the number x on the number line.

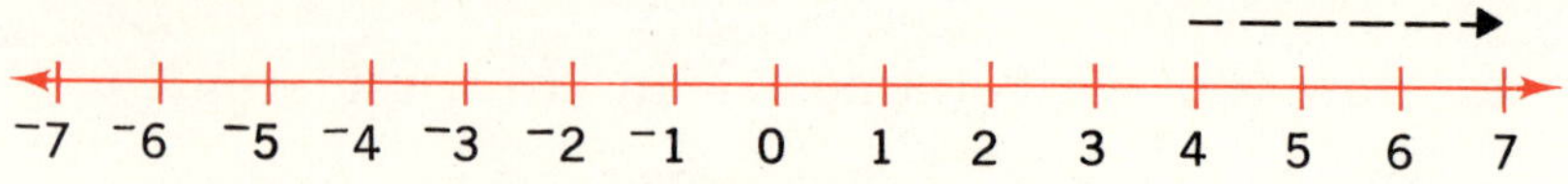

   a. How long is the dashed arrow?

   b. In what direction is the dashed arrow pointed?

   c. Solve: $7 - 4 = x$.

2. **a.** Write a related sentence for $3 - 5 = m$.

   b. The dashed arrow shows the number m. Solve: $3 - 5 = m$.

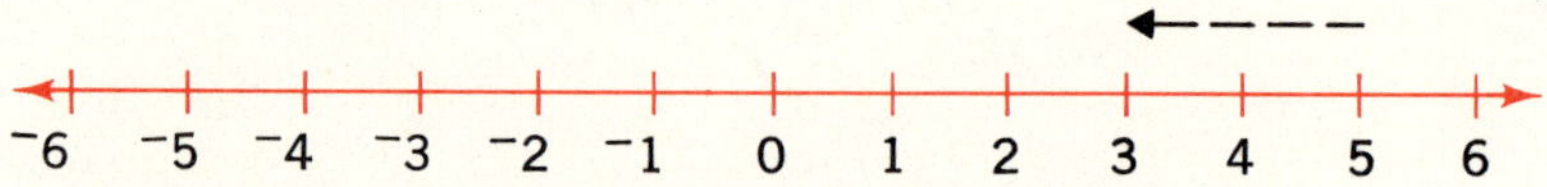

3. **a.** Suppose we want to find this difference: $^-3 - 4$. Write a related sentence for $^-3 - 4 = x$.

   b. The dashed arrow shows the number x. Solve: $^-3 - 4 = x$.

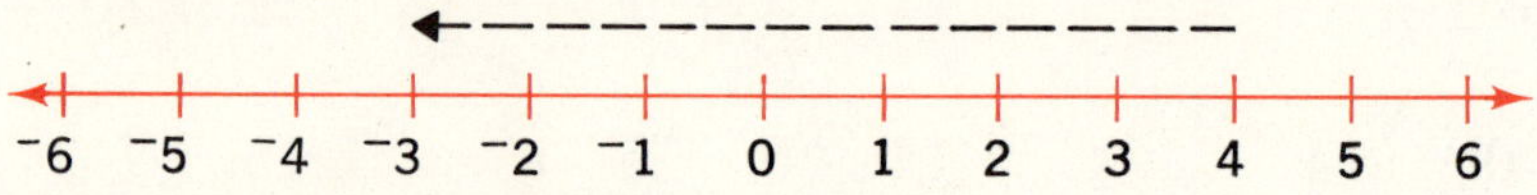

4. Write a related sentence. Then draw a number line to solve each sentence.

   a. $2 - 5 = x$ **b.** $^-3 - 6 = y$ **c.** $^-4 - 6 = n$

5. **a.** Write a related sentence for $^-6 - {}^-4 = m$.

   b. What number is shown by the dashed arrow?

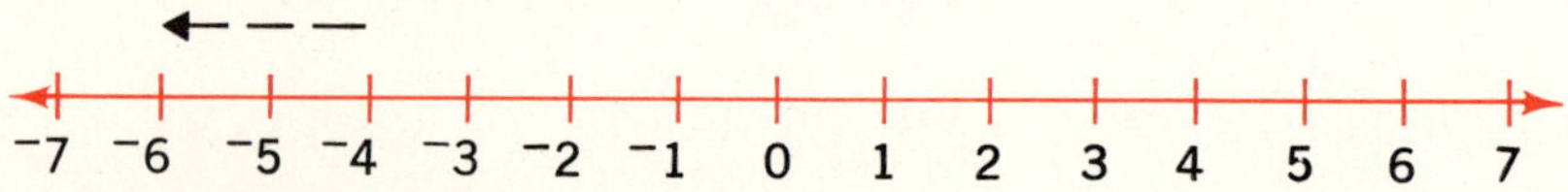

   c. What is $^-6 - {}^-4$?

To subtract integers, we can write a related addition sentence. Then we solve the addition sentence.

6. Subtract. Use number lines.

a. $5 - 7$ **b.** $2 - 6$ **c.** $^-4 - 7$

d. $^-3 - {}^-8$ **e.** $6 - {}^-3$ **f.** $^-6 - {}^-8$

EXERCISES

Subtract. Use number lines.

1. $3 - 6$ **2.** $4 - 8$ **3.** $0 - 18$

4. $0 - 9$ **5.** $^-2 - 3$ **6.** $^-6 - 9$

7. $^-10 - 4$ **8.** $^-8 - {}^-2$ **9.** $^-11 - {}^-4$

10. $0 - {}^-9$ **11.** $^-6 - {}^-10$ **12.** $^-5 - {}^-7$

13. $^-4 - {}^-7$ **14.** $^-9 - {}^-9$ **15.** $0 - {}^-5$

Solve these problems.

16. In an experiment, the temperature of a solution changed from $^-4°$ Celsius to 13° Celsius. What was the difference in temperature?

17. In a test run, a submarine dived to $^-185$ feet. Then it rose to $^-110$ feet. What was the difference in depth?

18. Mr. Uzel received a check for \$595 and a bill for \$638. How much ahead or behind is he?

19. On successive hands of a game, Pat scored 18, $^-21$, 25, $^-13$, and 29. What was her final score?

INVERSES IN SUBTRACTION

We can use inverses to make subtraction of integers easier.

1. Name the inverse of each.

   a. $^{-}10$ b. 14 c. 0 d. $^{-}6$

   e. $^{-}123$ f. 42 g. $^{-}84$ h. 5

2. Add or subtract. Look for a pattern.

   a. $6 - 3$ b. $3 - 7$ c. $^{-}4 - 6$
   $6 + {}^{-}3$ $3 + {}^{-}7$ $^{-}4 + {}^{-}6$

   d. $^{-}5 - {}^{-}3$ e. $0 - 8$ f. $^{-}9 - 0$
   $^{-}5 + 3$ $0 + {}^{-}8$ $^{-}9 + 0$

3. a. Is $4 - 3 = 4 + {}^{-}3$ a true sentence?

   b. How are 3 and $^{-}3$ related?

4. a. Is $^{-}6 - {}^{-}4 = {}^{-}6 + 4$ a true sentence?

   b. How are $^{-}4$ and 4 related?

5. How can we use inverses to subtract?

> We can subtract an integer by adding its inverse.

6. Subtract by adding the inverse.

Example $2 - 6 = 2 + {}^{-}6$
$= {}^{-}4$

   a. $3 - 8$ b. $^{-}6 - 9$ c. $^{-}8 - {}^{-}11$

EXERCISES

Name the inverse of each.

1. $^{-}19$ 2. 34 3. $^{-}28$ 4. 0

5. $^{-}296$ 6. 500 7. $^{-}60$ 8. $^{-}1{,}285$

Subtract by adding the inverse.

9. ${}^{-}6 - {}^{-}5$ 10. $8 - {}^{-}10$ 11. $5 - {}^{-}6$

12. $0 - 4$ 13. ${}^{-}5 - {}^{-}2$ 14. $3 - {}^{-}1$

15. ${}^{-}7 - 14$ 16. $9 - {}^{-}6$ 17. ${}^{-}12 - 2$

18. $0 - {}^{-}5$ 19. ${}^{-}8 - 0$ 20. ${}^{-}5 - 4$

21. ${}^{-}8 - {}^{-}6$ 22. ${}^{-}2 - 11$ 23. $7 - {}^{-}5$

24. ${}^{-}12 - 15$ 25. $24 - 3$ 26. $19 - 64$

27. $28 - {}^{-}17$ 28. ${}^{-}17 - {}^{-}56$ 29. ${}^{-}29 - 18$

30. ${}^{-}42 - 26$ 31. ${}^{-}42 - {}^{-}72$ 32. ${}^{-}5 - 0$

33. $13 - {}^{-}54$ 34. ${}^{-}81 - {}^{-}2$ 35. ${}^{-}83 - {}^{-}92$

36. $24 - {}^{-}82$ 37. $48 - {}^{-}63$ 38. ${}^{-}56 - 32$

39. ${}^{-}40 - {}^{-}40$ 40. $35 - {}^{-}18$ 41. ${}^{-}51 - 40$

42. ${}^{-}80 - {}^{-}80$ 43. $100 - {}^{-}99$ 44. ${}^{-}7 - {}^{-}6$

Solve these problems.

45. On a winter day, Tony heard on the radio that the city temperature was ${}^{-}4°$ while the temperature in the country was ${}^{-}12°$. What was the difference in the temperature?

46. Compared to sea level, the elevation of Death Valley is ${}^{-}276$ feet. Mt. McKinley is 20,300 feet. How much higher is Mt. McKinley than Death Valley?

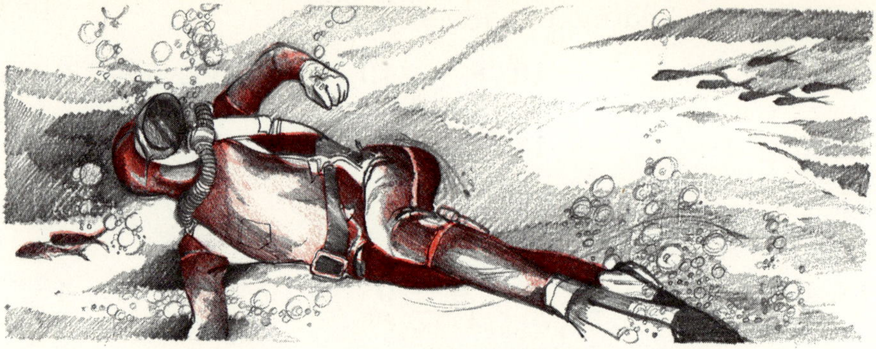

MULTIPLICATION OF INTEGERS

A scuba diver can dive to about 250 feet below sea level, or $^{-}250$ feet. A deep sea diver can dive 4 times that depth. What is $4 \times {}^{-}250$? Let's look at the product of a positive and a negative integer.

LET'S EXPLORE

1. Multiply. Look for a pattern.

$$2 \times 4$$
$$2 \times 3$$
$$2 \times 2$$
$$2 \times 1$$
$$2 \times 0$$

2. Continue the pattern of Item 1.

$$2 \times {}^{-}1$$
$$2 \times {}^{-}2$$
$$2 \times {}^{-}3$$

3. What do you think the product of a positive integer and a negative integer should be?

> The product of a positive integer and a negative integer is a negative integer.

4. How deep can a deep sea diver dive? Solve: $4 \times {}^{-}250 = n$.

5. Multiply.

Examples $3 \times {}^{-}4 = {}^{-}12$
$^{-}5 \cdot 6 = {}^{-}30$

a. $5 \times {}^{-}3$ b. $7 \cdot {}^{-}2$ c. $6 \cdot {}^{-}5$

d. $^{-}8 \times 0$ e. $^{-}9 \cdot 7$ f. $^{-}7 \times 10$

EXERCISES

Multiply.

1. $^{-}2 \cdot 5$
2. $^{-}7 \cdot 6$
3. $8 \times {}^{-}3$
4. $9 \times {}^{-}5$
5. $0 \cdot {}^{-}3$
6. $^{-}9 \times 8$
7. $11 \cdot {}^{-}9$
8. $^{-}6 \times 4$
9. $16 \cdot {}^{-}3$
10. $^{-}2 \cdot 24$
11. $^{-}34 \times 5$
12. $18 \times {}^{-}10$
13. $^{-}4 \times 21$
14. $^{-}10 \cdot 56$
15. $^{-}7 \times 20$
16. $16 \times {}^{-}8$
17. $28 \cdot {}^{-}13$
18. $42 \cdot {}^{-}29$
19. $101 \cdot {}^{-}17$
20. $^{-}56 \cdot 43$
21. $^{-}24 \times 66$
22. $^{-}21 \times 520$
23. $1,000 \cdot {}^{-}17$
24. $^{-}427 \cdot 0$

Let's Practice

Solve these problems.

1. Angelo charged \$1.50 for each car he washed. One weekend he washed 6 cars. How much did he earn?

2. Sylvia went on a diet. She was supposed to lose 3 pounds per week. She planned to stay on the diet for 5 weeks. How much should she lose?

MULTIPLYING TWO NEGATIVE INTEGERS

We know how to multiply a positive and negative integer. How can we multiply two negative integers?

LET'S EXPLORE

1. Multiply. Look for a pattern.

$$^{-}2 \times 6$$
$$^{-}2 \times 5$$
$$^{-}2 \times 4$$
$$^{-}2 \times 3$$
$$^{-}2 \times 2$$
$$^{-}2 \times 1$$
$$^{-}2 \times 0$$

2. Continue the pattern of Item 1.

$$^{-}2 \times {}^{-}1$$
$$^{-}2 \times {}^{-}2$$
$$^{-}2 \times {}^{-}3$$
$$^{-}2 \times {}^{-}4$$

3. What do you think the product of two negative integers should be?

> The product of two negative integers is a positive integer.

4. Multiply.

Examples $^{-}3 \times {}^{-}4 = 12$
$^{-}6 \cdot {}^{-}5 = 30$

a. $^{-}6 \times {}^{-}2$ b. $^{-}4 \cdot {}^{-}5$ c. $^{-}7 \times {}^{-}2$

d. $^{-}8 \times {}^{-}5$ e. $^{-}9 \times {}^{-}10$ f. $^{-}11 \cdot {}^{-}6$

5. Multiplication of integers is commutative. Thus, $^{-}5 \times 8 = 8 \times {}^{-}5$. Give two more examples of the commutative property of multiplication of integers.

6. Multiplication of integers is associative. Thus, $^{-}2 \times (5 \times {}^{-}3) = ({}^{-}2 \times 5) \times {}^{-}3$. Give two examples of the associative property of multiplication of integers.

> Multiplication of integers is commutative and associative.

EXERCISES

Multiply.

1. $^{-}9 \cdot {}^{-}3$	2. $^{-}5 \times {}^{-}6$	3. $^{-}8 \cdot {}^{-}9$
4. $^{-}5 \times {}^{-}4$	5. $0 \cdot {}^{-}7$	6. $^{-}7 \times {}^{-}2$
7. $^{-}10 \times {}^{-}6$	8. $^{-}1 \cdot {}^{-}6$	9. $^{-}2 \cdot {}^{-}9$
10. $^{-}3 \cdot 2$	11. $^{-}5 \times 7$	12. $4 \cdot {}^{-}6$
13. $^{-}9 \cdot {}^{-}4$	14. $^{-}3 \cdot 10$	15. $^{-}7 \times {}^{-}11$
16. $^{-}6 \times {}^{-}12$	17. $^{-}6 \times {}^{-}15$	18. $^{-}13 \cdot {}^{-}9$
19. $^{-}10 \times {}^{-}15$	20. $^{-}17 \cdot {}^{-}5$	21. $^{-}18 \times {}^{-}3$
22. $^{-}24 \cdot {}^{-}7$	23. $^{-}8 \times {}^{-}34$	24. $4 \cdot {}^{-}28$
25. $^{-}11 \cdot 90$	26. $^{-}11 \cdot {}^{-}34$	27. $^{-}23 \cdot 14$
28. 28×56	29. $^{-}11 \cdot {}^{-}25$	30. $^{-}16 \times {}^{-}23$
31. $^{-}21 \times 21$	32. $^{-}1 \cdot {}^{-}450$	33. $^{-}43 \cdot {}^{-}80$

UNEXPLORED TERRITORY

1. Multiply.

   a. 333667×1113

   b. 33336667×11133

2. Study the pattern. What comes next?

DIVISION OF INTEGERS

We know that the integer that makes the sentence $6 \div 3 = n$ true also makes the related sentence $3 \times n = 6$ true.

1. a. Write a related sentence for $12 \div {}^{-}3 = x$.

   b. Solve the related sentence.

   c. What is $12 \div {}^{-}3$?

2. a. Write a related sentence for ${}^{-}16 \div 4 = n$.

   b. Solve the related sentence.

   c. What is ${}^{-}16 \div 4$?

3. a. Write a related sentence for ${}^{-}20 \div {}^{-}5 = y$.

   b. Solve the related sentence.

   c. What is ${}^{-}20 \div {}^{-}5$?

To divide integers, we can write a related multiplication sentence. Then we solve the sentence.

4. Divide.

   Examples ${}^{-}24 \div 6 = {}^{-}4$ ${}^{-}35 \div {}^{-}7 = 5$

   a. ${}^{-}16 \div 2$ b. ${}^{-}18 \div {}^{-}9$ c. $24 \div {}^{-}6$

EXERCISES

Divide.

1. $36 \div 6$
2. $28 \div {}^{-}7$
3. $26 \div {}^{-}2$
4. ${}^{-}16 \div 8$
5. ${}^{-}11 \div {}^{-}1$
6. ${}^{-}42 \div 6$
7. ${}^{-}27 \div {}^{-}9$
8. ${}^{-}48 \div {}^{-}12$
9. $63 \div {}^{-}9$
10. ${}^{-}81 \div 3$
11. ${}^{-}144 \div {}^{-}12$
12. $26 \div {}^{-}13$
13. ${}^{-}108 \div {}^{-}9$
14. ${}^{-}72 \div 9$
15. ${}^{-}49 \div {}^{-}7$

DO YOU REMEMBER?

Add.

1. $2 + {}^{-}6$ **2.** ${}^{-}4 + {}^{-}10$ **3.** $26 + {}^{-}3$

4. $17 + {}^{-}4$ **5.** $0 + {}^{-}9$ **6.** ${}^{-}8 + {}^{-}5$

7. $25 + 7$ **8.** ${}^{-}16 + 8$ **9.** ${}^{-}6 + {}^{-}8$

10. $38 + {}^{-}17$ **11.** ${}^{-}43 + {}^{-}11$ **12.** ${}^{-}5 + 18$

Subtract.

13. $8 - {}^{-}2$ **14.** ${}^{-}3 - {}^{-}5$ **15.** $9 - 14$

16. $17 - 15$ **17.** ${}^{-}6 - 9$ **18.** ${}^{-}2 - 8$

19. $21 - 14$ **20.** ${}^{-}1 - {}^{-}9$ **21.** ${}^{-}3 - {}^{-}2$

22. $13 - 20$ **23.** ${}^{-}6 - {}^{-}18$ **24.** ${}^{-}19 - 23$

Multiply.

25. $5 \times {}^{-}2$ **26.** ${}^{-}3 \cdot 7$ **27.** $10 \times {}^{-}8$

28. $3 \cdot 8$ **29.** $9 \times {}^{-}5$ **30.** ${}^{-}4 \cdot {}^{-}6$

31. $14 \times {}^{-}2$ **32.** ${}^{-}10 \cdot 6$ **33.** ${}^{-}11 \cdot {}^{-}4$

34. ${}^{-}23 \times 3$ **35.** ${}^{-}14 \times {}^{-}5$ **36.** $6 \cdot {}^{-}12$

Divide.

37. ${}^{-}16 \div 4$ **38.** $27 \div {}^{-}9$ **39.** $42 \div {}^{-}6$

40. ${}^{-}28 \div {}^{-}2$ **41.** $56 \div {}^{-}7$ **42.** ${}^{-}18 \div {}^{-}3$

43. ${}^{-}64 \div {}^{-}8$ **44.** ${}^{-}72 \div 9$ **45.** ${}^{-}63 \div {}^{-}7$

46. ${}^{-}81 \div {}^{-}9$ **47.** ${}^{-}64 \div 4$ **48.** ${}^{-}28 \div {}^{-}7$

ABSOLUTE VALUE

1. a. Look at the number line. How far is 4 from 0?

   b. How far is $^{-}4$ from 0?

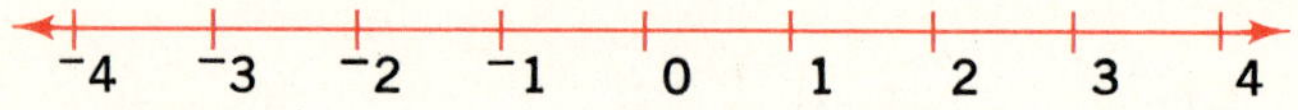

2. On a number line, the distance of a number from 0 is called the *absolute value* of the number.

   The absolute value of 5 is 5.
   The absolute value of $^{-}5$ is 5.

   a. What is the absolute value of $^{-}8$?

   b. What is the absolute value of 10?

3. a. On a number line, how far is 0 from 0?

   b. What is the absolute value of 0?

4. We use the symbol $|n|$ for absolute value. This is read "the absolute value of n." Read these.

   a. $|7|$ b. $|^{-}11|$ c. $|^{-}25|$ d. $|64|$

5. Here is a flow chart for finding the absolute value of an integer.

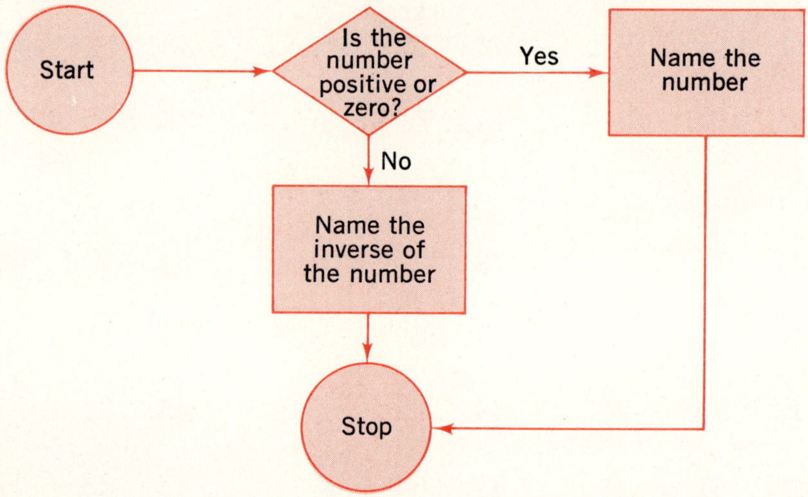

a. Suppose the number is 23. What is $|23|$?

b. Suppose the number is 0. What is $|0|$?

c. Suppose the number is $^{-}17$. What is $|^{-}17|$?

d. Suppose the number is $^{-}25$. What is $|^{-}25|$?

6. Find the absolute value of each.

a. $^{-}26$ b. 2 c. $^{-}12$ d. $^{-}6$

7. Which are true and which are false?

a. $|7| < |^{-}10|$ b. $|^{-}24| < |^{-}3|$ c. $|54| > |^{-}63|$

EXERCISES

Write simplest names.

1. $|^{-}8|$
2. $|16|$
3. $|0|$
4. $|^{-}42|$
5. $|7|$
6. $|^{-}63|$
7. $|^{-}124|$
8. $|37|$

Draw a number line for integers.

9. Circle all points whose absolute value is 3.

10. Draw a triangle around all points whose absolute value is 0.

11. Draw a square around all points whose absolute value is 9.

Which are true and which are false?

12. $|12| = |^{-}12|$
13. $|11| > |^{-}6|$
14. $|24| = |^{-}26|$
15. $|0| > |^{-}18|$
16. $|2| > |^{-}29|$
17. $|13| < |^{-}4|$
18. $|^{-}40| = |40|$
19. $|5| < |^{-}5|$
20. $|18| < |^{-}28|$
21. $|^{-}11| < |^{-}16|$
22. $|^{-}6| > |3|$
23. $|34| = |^{-}34|$

NUMBER LINE GRAPHS

The dots on the number line below show points for 7, $^-2$, and $^-5$.

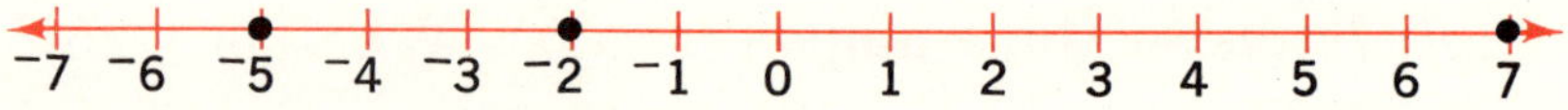

We say we have *graphed* 7, $^-2$, and $^-5$. Each point is the *graph* of a number.

1. Draw a different number line for parts a, b, and c.

   a. Graph $^-6$, 4, 0, and $^-3$.

   b. Graph all the integers between $^-5$ and 6.

   c. Graph all the odd numbers between 3 and 11.

2. When we graph the solution of the sentence $n + 5 = 8$, we say we have graphed the sentence.

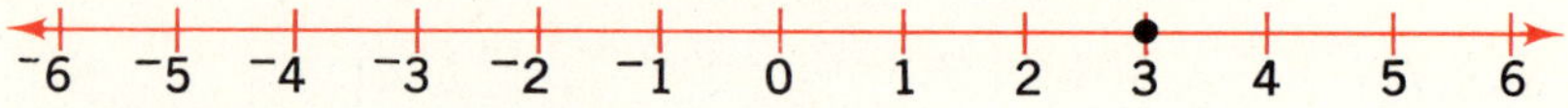

   Graph each sentence.

   a. $x + 5 = 7$ **b.** $11 - y = 8$ **c.** $5 \cdot n = {}^-20$

   d. $^-6 \cdot x = {}^-12$ **e.** $n = {}^-4 + 10$ **f.** $^-8 + y = {}^-3$

4. This is a graph of all the negative integers, x, that make $x > {}^-4$ a true sentence.

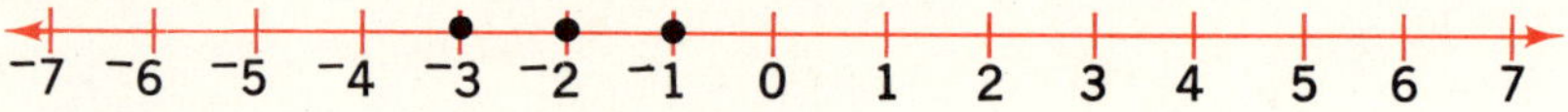

   a. Graph all the positive integers that make $x < 6$ true.

   b. Graph all the even integers that make $x > {}^-5$ true.

EXERCISES

Graph each sentence.

1. $n + {}^-6 = 4$ **2.** $9 \cdot y = 0$ **3.** $n - {}^-5 = {}^-4$

4. $^{-}5 + 9 = x$ 5. $x + 3 = 6$ 6. $\frac{24}{n} = {}^{-}3$

7. $6 \cdot y = {}^{-}12$ 8. $y - {}^{-}8 = 0$ 9. $\frac{x}{^{-}3} = 6$

Draw a different number line for each.

10. Graph all the integers between 6 and 8.

11. Graph all the even numbers between 5 and 16.

12. Graph all the positive integers that make $n < 11$ true.

13. Graph all the negative integers that make $x > {}^{-}8$ true.

Let's Practice

Solve these problems.

1. Timmy sold sodas at the Valentine dance. Each soda cost 15¢. How much did he take in if he sold 463 sodas?

2. Ernie earns $4.00 each Saturday delivering groceries. He spends $1.00 of it each week on desserts at school during lunch. How long will it take him to save $19.50 for a new tennis racket?

3. For selling hot dogs at the ball park, Henry received 10% of all he takes in. How much did he receive for selling 56 hot dogs?

4. Mr. Wild, the plumber, needed pieces of pipe measuring $2\frac{1}{2}$ feet, $3\frac{3}{4}$ feet, and $12\frac{1}{4}$ feet. What is the total length of pipe he needs?

GRAPHING INTEGER PAIRS

We can graph pairs of integers using two perpendicular number lines.

1. Look at the graph below.

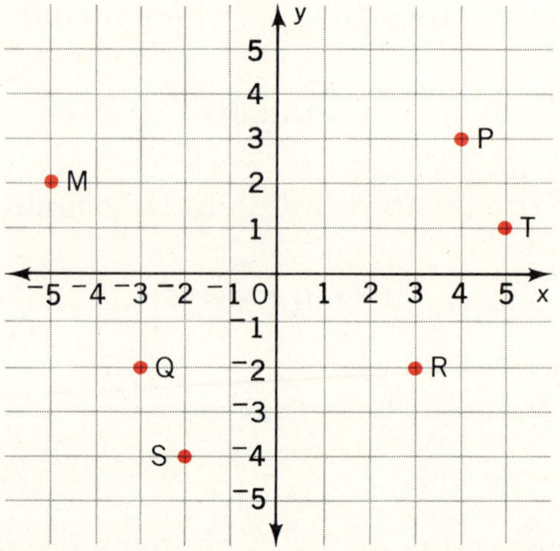

   a. Which direction is point *P* from the vertical number line?

   b. How many units from the vertical number line is it?

   c. Which direction is point *P* from the horizontal number line?

   d. How many units from the horizontal number line is it?

   e. An ordered pair for *P* is (4, 3). Give an ordered pair for point *T*.

2. We call the ordered pair ($^-5$, 2) *coordinates* of point *M*.

   a. What are the coordinates of point *R*? point *S*? point *Q*?

   b. Are ($^-8$, 6) and (6, $^-8$) coordinates for the same point? Why do we call these number pairs *ordered* pairs?

When graphing an ordered pair, the first coordinate tells the distance right or left of the vertical number line. The second coordinate tells the distance up or down from the horizontal number line.

3. Draw a horizontal and a vertical number line. Graph these ordered pairs.

a. $(2, 4)$ **b.** $(^-6, 4)$ **c.** $(0, ^-1)$ **d.** $(^-2, ^-4)$

e. $(0, ^-3)$ **f.** $(^-5, 6)$ **g.** $(4, 7)$ **h.** $(3, ^-6)$

EXERCISES

1. List the coordinates of each of the points *A* through *H*.

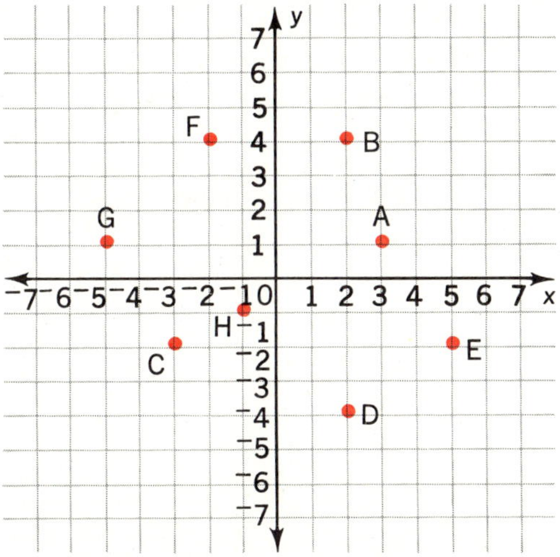

Graph these ordered pairs.

2. $(3, ^-2)$ **3.** $(0, ^-4)$ **4.** $(5, ^-4)$ **5.** $(^-8, ^-2)$

6. $(6, ^-3)$ **7.** $(^-2, 0)$ **8.** $(5, ^-3)$ **9.** $(^-3, 7)$

10. $(^-5, ^-5)$ **11.** $(9, ^-2)$ **12.** $(^-3, ^-3)$ **13.** $(2, ^-6)$

14. $(5, 0)$ **15.** $(^-4, ^-5)$ **16.** $(6, ^-1)$ **17.** $(^-2, 0)$

18. Tell why $(^-8, 3)$ and $(3, ^-8)$ are different ordered pairs.

GRAPHING SENTENCES

Let's graph a sentence like $x + y = {}^-4$.

LET'S EXPLORE

1. The chart shows some integer solutions to the sentence $x + y = {}^-4$. Copy and complete.

x	y
2	$^-6$
1	$^-5$
0	$^-4$
$^-1$	
$^-2$	
$^-3$	
	0

2. Below we have graphed the ordered pairs (x, y) from the chart. What kind of pattern does the graph form?

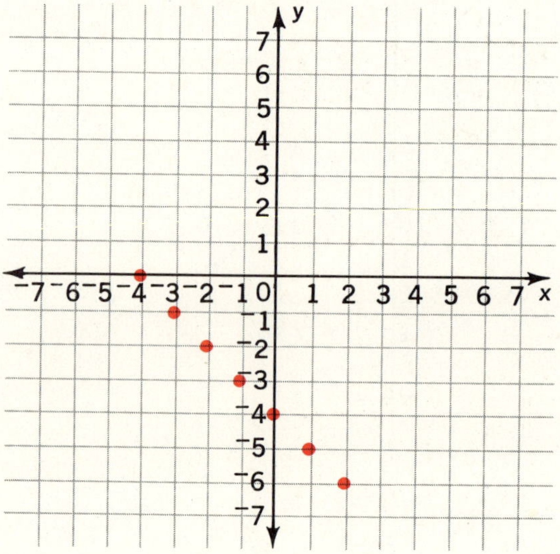

3. Graph eight solutions of the sentence $x = y$.

4. Graph eight solutions of the sentence $2 \cdot x = y$.

5. Graph eight solutions of the sentence $x - y = 2$.

6. Look at your graphs for Items 1-6. What pattern does each graph form?

7. Let's graph $y = x^2$. Copy and complete the chart below.

x	y
$^{-}4$	16
$^{-}3$	9
$^{-}2$	4
$^{-}1$	
0	
1	
2	
	9
	16

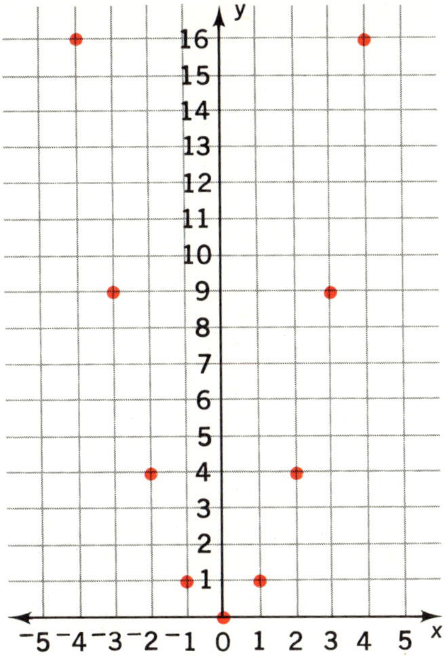

8. Next to the chart, we have graphed the ordered pairs (x, y) from the chart. What kind of pattern does the graph form?

EXERCISES

Graph these sentences.

1. $x + y = 2$
2. $x + y = 0$
3. $y - x = 3$
4. $3 \cdot x = y$
5. $x - y = 6$
6. $x - y = {}^{-}1$
7. $y - x = 6$
8. $y - x = 0$
9. ${}^{-}2 \cdot x = y$
10. $4 \cdot y = x$
11. $x + y = {}^{-}5$
12. $y - x = {}^{-}1$
13. $x = y^2$
14. $y = 2 \cdot x^2$
15. $y = {}^{-}1 \cdot x^2$

UNEXPLORED TERRITORY

Graph the sentence $y = \sqrt{x}$. *Hint* Use the table on page 409.

PROBLEM SOLVING

1. Mr. Cameron paid \$138 last year in school taxes. This year he paid \$162.84. What was the per cent of increase?

2. The temperatures last week were 8°, $^{-}2°$, 0°, 5°, $^{-}1°$, $^{-}3°$, 0°. What was the average temperature?

3. A barbecue sauce recipe to serve 8 people calls for $\frac{3}{4}$ cup of hot water. To make half a recipe, how much hot water should Mrs. Sanchez use?

4. Mike found 42 golf balls in the woods at the golf course. He sold them to some golfers for 35¢ each. How much money did he make?

5. Sara and Joan were selling brownies at the school May Mart for \$1.10 per pound. At the end of the Mart, they reduced the price by 20%. What was the new price?

6. The super market sold tomato soup at 5 cans for 99¢. Mrs. Cunningham bought only 2 cans. What did she pay?

7. Mr. Harmeyer had \$5,000 available for investing. He decided to buy 150 shares of stock at \$29.64 per share. He must also pay 5% of the total cost to the stock broker. How much will he spend?

8. One day in February, the temperature in Fairbanks, Alaska was $^{-}17°$. It was 49° in Dallas, Texas. What was the difference in temperature?

9. Dan earned the following amounts last week: \$1.50, \$3, \$2.75, \$5.50, \$4.25, and \$10. He made purchases in the following amounts: \$1.75, \$2, \$.85, \$5.90, and \$1.25. How much money did he have left after his purchases?

CHAPTER REVIEW

Which are true and which are false?

1. $^{-}2 > 2$ **2.** $0 > {}^{-}2$ **3.** $0 < {}^{-}5$ **4.** $^{-}7 < 2$

Describe each with an integer. Name the inverse of each.

5. five seconds after lift off **6.** gain of 8 yards

7. bank withdrawal of $50 **8.** 5° below zero

Add.

9. $^{-}3 + {}^{-}7$ **10.** $5 + 8$ **11.** $0 + {}^{-}2$

12. $7 + {}^{-}9$ **13.** $^{-}3 + 10$ **14.** $3 + {}^{-}3$

15. $7 + {}^{-}8$ **16.** $^{-}10 + {}^{-}20$ **17.** $30 + {}^{-}12$

Subtract.

18. $^{-}3 - 5$ **19.** $7 - 18$ **20.** $^{-}2 - {}^{-}3$

21. $0 - 4$ **22.** $0 - {}^{-}17$ **23.** $25 - 37$

24. $^{-}20 - 15$ **25.** $^{-}25 - {}^{-}35$ **26.** $100 - {}^{-}225$

Multiply.

27. 8×15 **28.** $^{-}3 \cdot 42$ **29.** $13 \times {}^{-}5$

30. $^{-}7 \times {}^{-}9$ **31.** $0 \times {}^{-}18$ **32.** $15 \cdot {}^{-}30$

33. $^{-}3 \cdot {}^{-}4 \cdot 5$ **34.** $^{-}2 \times 15 \times 7$ **35.** $^{-}5 \cdot {}^{-}2 \cdot {}^{-}8$

Divide.

36. $25 \div 5$ **37.** $75 \div {}^{-}3$ **38.** $^{-}100 \div 4$

39. $^{-}36 \div 9$ **40.** $^{-}24 \div {}^{-}8$ **41.** $^{-}121 \div 11$

42. $150 \div {}^{-}15$ **43.** $^{-}210 \div 70$ **44.** $^{-}81 \div 27$

Write the simplest name.

45. $|^-7|$ 46. $|5|$ 47. $|0|$ 48. $|^-12|$

Which are true and which are false?

49. $|22| > |^-22|$ 50. $|0| < |^-5|$ 51. $|^-3| = |3|$

52. $|32| < |^-18|$ 53. $|7| < |^-8|$ 54. $|^-100| > |^-81|$

Graph each sentence.

55. $x > {}^-5$ 56. $2 \cdot y = 0$ 57. $m < 2$

58. $c + 3 = 7$ 59. $\frac{12}{b} = {}^-3$ 60. $^-6 + 3 = a$

Graph these ordered pairs.

61. $(1,^-1)$ 62. $(0,2)$ 63. $(^-3,0)$ 64. $(4,3)$

65. $(^-2,^-1)$ 66. $(^-3,2)$ 67. $(1,^-3)$ 68. $(1,3)$

69. $(2,^-2)$ 70. $(4,0)$ 71. $(0,^-4)$ 72. $(0,0)$

Graph these sentences.

73. $x = y$ 74. $2 \cdot x = y$ 75. $x + y = 4$

76. $x = {}^-1 \cdot y$ 77. $x - y = 2$ 78. $y = x^2$

Solve this problem.

79. At one point, the depth of the ocean is $^-190$ feet. At another point nearby, it is $^-320$ feet. How much deeper is the second ocean depth?

Terms You Should Know

integer	inverse	number line
positive integer	absolute value	coordinates
negative integer	graph	ordered pair

CHAPTER TEST

Which are true and which are false?

1. $^{-}5 < 5$ **2.** $^{-}3 < 0$ **3.** $^{-}2 < {}^{-}4$

Describe each with an integer. Name the inverse of each.

4. debt of $15 **5.** 10 seconds before splash down

Add.

6. $^{-}7 + 5$ **7.** $34 + 25$ **8.** $^{-}18 + {}^{-}25$

Subtract.

9. $25 - 39$ **10.** $0 - {}^{-}23$ **11.** $^{-}17 - {}^{-}13$

Multiply.

12. $32 \times {}^{-}5$ **13.** $^{-}12 \cdot {}^{-}4$ **14.** $^{-}20 \times 0$

Divide.

15. $28 \div {}^{-}7$ **16.** $^{-}39 \div {}^{-}3$ **17.** $^{-}36 \div 12$

Write the simplest name.

18. $|{}^{-}14|$ **19.** $|20|$ **20.** $|0|$

Which are true and which are false?

21. $|{}^{-}42| > |{}^{-}50|$ **22.** $|12| < |{}^{-}20|$

Graph these ordered pairs.

23. $(2,{}^{-}2)$ **24.** $({}^{-}2,{}^{-}2)$ **25.** $({}^{-}2,2)$

Graph these sentences.

26. $x < {}^{-}2$ **27.** $3 \times a = 9$ **28.** $x - y = 0$

Solve this problem.

29. The temperature one afternoon was 92° F. After a sudden rainstorm, the temperature dropped to 76° F. Write an integer for the change in temperatures.

CHAPTER SEVEN

Exponential Notation

EXPONENTIAL NOTATION

Two names for 125 are $5 \cdot 5 \cdot 5$ and 5^3. We read 5^3 as "the third power of 5" or "five to the third power."

1. Read these.

   a. 6^3 b. 7^4 c. 20^5 d. 17^8

A symbol like 5^4 is called **exponential notation**. The 5 is called the **base** and the 4 is called the **exponent**. The exponent tells how many times the base is used as a factor.

2. Name the base and the exponent.

   a. 4^2 b. 8^5 c. 9^{12} d. 102^6

3. The standard numeral for 2^3 is 8. Write standard numerals.

   a. 3^3 b. 9^2 c. 12^2 d. 5^4

4. Exponential notation for 125 is 5^3. Write exponential notation.

   a. 64 b. 121 c. 16 d. 1,000

EXERCISES

Write exponential notation.

1. $6 \cdot 6 \cdot 6 \cdot 6$
2. $5 \cdot 5 \cdot 5 \cdot 5 \cdot 5 \cdot 5$
3. $2 \cdot 2 \cdot 2 \cdot 2 \cdot 2 \cdot 2 \cdot 2 \cdot 2$
4. $50 \cdot 50 \cdot 50 \cdot 50 \cdot 50$
5. 36
6. 169
7. 196
8. 144
9. 625
10. 10,000
11. 225
12. 125

Write standard numerals.

13. 2^5
14. 7^2
15. 4^2
16. 5^3
17. 6^3
18. 8^3
19. 9^4
20. 20^3
21. 10^4
22. 4^5

0 AND 1 AS EXPONENTS

We agreed that 10^1 means 10, and 10^0 means 1. What about 1 and 0 as exponents with bases other than 10?

LET'S EXPLORE

1. Copy and complete.

2^6	2^5	2^4	2^3	2^2
64				

   a. What pattern do you see?

   b. Extend the pattern for 2^1 and 2^0.

2. Copy and complete.

3^6	3^5	3^4	3^3	3^2
729				

   a. What pattern do you see?

   b. Extend the pattern for 3^1 and 3^0.

3. What will any number to the power of 1 always mean?

4. What will any number (except 0) to the power of 0 always mean?

For all numbers n, $n^1 = n$.
For all numbers n (except 0), $n^0 = 1$.

5. Give the meaning of each.

   a. 7^1 b. 6^0 c. 5^0

   d. $(8^1) \cdot (16^0)$ e. $(24) + (4^0)$ f. $(48^0) \div (56^0)$

EXERCISES

Give the meaning of each.

1. 19^1
2. 56^0
3. 893^0
4. 147^1
5. 43^0
6. 1^0
7. $(19^1) \times (7^0)$
8. $(42^1) + (18^0)$
9. $(482^1) \div (2^0)$
10. $(97^1) \div (17^0)$
11. $(785^1) - (987^0)$
12. $(563^1) + (4^1)$
13. $(5{,}900^0) + (87^0)$
14. $(4{,}987^1) - (4{,}000^1)$
15. $(50^1) \times (7^0)_{50}$
16. $(562^1) \div (562^0)$
17. $(283^1) \times (5{,}836^0)$
18. $(3{,}951^1) \div (3^0)$

Let's Practice

An expanded numeral for 74,082 is $(7 \times 10^4) + (4 \times 10^3) + (0 \times 10^2) + (8 \times 10^1) + (2 \times 10^0)$.

Write expanded numerals.

1. 5,286
2. 19,434
3. 97,804
4. 624,083
5. 1,403,562
6. 89,979,657
7. 300,609
8. 10,098

UNEXPLORED TERRITORY

Every natural number can be named in terms of powers of 3. Sums and differences may be used. In any numeral for a natural number, no power of 3 may be used more than once.

Examples $2 = 3^1 - 3^0$ $4 = 3^1 + 3^0$

$7 = (3^2 - 3^1) + 3^0$ $10 = 3^2 + 3^0$

Name these natural numbers in terms of powers of 3.

1. 8
2. 3
3. 9
4. 12
5. 14
6. 30
7. 21
8. 15

MULTIPLYING POWERS

How can we multiply using exponential notation? Let's look for some patterns.

LET'S EXPLORE

1. We know that 2^5 means $2 \cdot 2 \cdot 2 \cdot 2 \cdot 2$.

   a. What does 2^3 mean?

   b. What does $2^5 \cdot 2^3$ mean?

   c. Use a single exponent to write another numeral for the product $2^5 \cdot 2^3$.

2. When we factor 5^4, we write $5 \cdot 5 \cdot 5 \cdot 5$.

   a. Factor 5^6.

   b. Factor $5^4 \times 5^6$.

   c. Use a single exponent to write another numeral for the product $5^4 \times 5^6$.

3. Write exponential notation for the product $5^5 \cdot 5^8$. (*Hint* Factor first, and then write exponential notation.)

4. Write exponential notation for the product $10^{20} \cdot 10^5$.

5. Write exponential notation for $3^7 \cdot 3^2 \cdot 3^5$.

When the bases are alike, we can find exponential notation for a product like $3^6 \cdot 3^8$ by adding the exponents.

$$3^6 \cdot 3^8 = 3^{6+8} = 3^{14}$$

6. Write exponential notation.

   Examples $5^6 \cdot 5^3 = 5^{6+3} = 5^9$

   $3^4 \cdot 9^4 \cdot 3^8 = 3^{4+8} \cdot 9^4 = 3^{12} + 9^4$

   a. $10^3 \cdot 10^{12}$ b. $8^0 \times 8^3$ c. $19^{10} \cdot 10^4 \cdot 19^6$

EXERCISES

Write exponential notation.

1. $3^2 \cdot 3^3$
2. $5^4 \cdot 5^2$
3. $4^8 \times 4^3$
4. $10^3 \times 10^5$
5. $2^3 \cdot 2^4 \cdot 2^3$
6. $6^2 \cdot 6^4$
7. $3^5 \times 3$
8. $4^7 \cdot 4^1$
9. $10^6 \cdot 10^2 \cdot 10^0$
10. $7^5 \times 7^1 \times 7^0$
11. $12^1 \times 12^3 \times 12^5$
12. $2^7 \times 2^0 \times 2^6 \times 2^4$
13. $10^3 \times 10^0 \times 10^5$
14. $8^5 \times 8^{13} \times 8^0 \times 8^4$
15. $19^4 \times 19^5 \times 19^0$
16. $4^8 \times 4^{13} \times 4^{11}$
17. $32^4 \times 32^{11}$
18. $20^6 \times 20^{20}$
19. $5^0 \times 5^{15} \times 5^1$
20. $14^6 \times 14^0 \times 14^8$
21. $100^9 \times 100^{15} \times 100^6$
22. $53^0 \times 53^{10}$
23. $4^6 \times 8^{10} \times 8^9 \times 4^4$
24. $15^4 \times 9^6 \times 15^{11} \times 9^4$
25. $3^6 \cdot 7^4 \cdot 3^2 \cdot 7^2$
26. $7^0 \times 5^4 \times 6^0$
27. $9^0 \times 5^1 \times 5^8 \times 9^{10}$
28. $18^4 \cdot 16^0 \cdot 18^8 \cdot 16^1$

Let's Practice

Solve these problems.

1. Mr. Jacobs received a state tax refund of \$220.16 and a federal tax refund of \$434.92. What was his total refund?
2. Frankie said that bus fare had increased from 30¢ to 35¢. What was the per cent of increase?

DIVIDING USING EXPONENTIAL NOTATION

Let's find an easy way to divide, using exponential notation.

LET'S EXPLORE

1. Consider the division $\frac{2^4}{2^2}$.

   a. Factor the numerator and denominator.

   b. Simplify and then write exponential notation.

2. a. Factor the numerator and denominator of $\frac{3^5}{3^2}$.

   b. Simplify and then write exponential notation.

3. Write exponential notation for the quotient $\frac{4^6}{4^3}$.

4. Write exponential notation for $\frac{6^4}{6^3}$.

When the bases are alike, we can find exponential notation for a quotient like $\frac{7^5}{7^3}$ by subtracting the exponents.

$$\frac{7^5}{7^3} = 7^{5-3} = 7^2$$

Examples

$$\frac{8^5}{8^3} = 8^{5-3} = 8^2$$

$$\frac{10^7}{10^6} = 10^{7-6} = 10^1, \text{ or } 10$$

$$\frac{7^6 \times 7^3}{7^2 \times 7^4} = \frac{7^{6+3}}{7^{2+4}} = \frac{7^9}{7^6}$$

$$= 7^{9-6} = 7^3$$

5. Write exponential notation.

   a. $\frac{17^8}{17^5}$ b. $\frac{3^7}{3^2}$ c. $\frac{11^9}{11^8}$ d. $\frac{10^{15}}{10^8}$

   e. $\frac{12^5}{12^0}$ f. $\frac{5^6 \cdot 5^3}{5^2}$ g. $\frac{123^5 \cdot 123^8}{123^9}$ h. $\frac{11^7 \times 11^3}{11^2 \times 11^5}$

EXERCISES

Write exponential notation.

1. $\frac{2^5}{2^3}$ 2. $\frac{6^8}{6^4}$ 3. $\frac{15^{25}}{15^{20}}$ 4. $\frac{18^6}{18^6}$

5. $\frac{12^4}{12^0}$ 6. $\frac{4^0}{4^0}$ 7. $\frac{9^{12}}{9^6}$ 8. $\frac{12^3}{12^2}$

9. $\frac{5^6}{5^1}$ 10. $\frac{7^9}{7^3}$ 11. $\frac{24^{10}}{24^6}$ 12. $\frac{4^{13}}{4^5}$

13. $\frac{19^4}{19^3}$ 14. $\frac{81^{31}}{81^{15}}$ 15. $\frac{16^{19}}{16^{19}}$ 16. $\frac{144^{13}}{144^{11}}$

17. $\frac{16^8}{16^3}$ 18. $\frac{16^{25}}{16^{19}}$ 19. $\frac{11^{11}}{11^4}$ 20. $\frac{90^5}{90^2}$

21. $\frac{24^9}{24^5}$ 22. $\frac{73^{14}}{73^{13}}$ 23. $\frac{29^{50}}{29^8}$ 24. $\frac{23^{16}}{23^{12}}$

Write exponential notation.

25. $\frac{7^3 \times 7^2}{7^4}$ 26. $\frac{4^7 \times 4^5}{4^4}$ 27. $\frac{8^2 \times 8^{10}}{8^3 \times 8^5}$

28. $\frac{16^4 \cdot 16^5}{16^8}$ 29. $\frac{21^5 \cdot 21^9}{21^2 \cdot 21^{12}}$ 30. $\frac{100^4 \cdot 100^6}{100^8}$

31. $\frac{11^8 \times 11^6}{11^2}$ 32. $\frac{19^4 \times 19^{12}}{19^{10}}$ 19^6 33. $\frac{16^4 \times 16^8}{16^5 \times 16^6}$

34. $\frac{16^0 \times 16^{11}}{16^2}$ 35. $\frac{21^4 \times 21^6}{21^8}$ 21^2 36. $\frac{21^4 \times 21^9 \times 21^3}{21^8 \times 21^7}$

37. $\frac{23^4 \times 23^8 \times 23^5}{23^5 \times 23^9}$ 38. $\frac{15^6 \times 15^8 \times 15^{10}}{15^4}$ 39. $\frac{72^8 \times 72^9}{72^0 \times 72^3 \times 72^5}$

Let's Practice

Subtract.

1. $3 - 5$ 2. $7 - 8$ 3. $0 - 4$ 4. $2 - 6$

5. $6 - 11$ 6. $4 - 6$ 7. $^{-}8 - {}^{-}3$ 8. $10 - {}^{-}5$

9. $^{-}12 - {}^{-}7$ 10. $24 - {}^{-}8$ 11. $16 - 23$ 12. $16 - 25$

NEGATIVE INTEGERS AS EXPONENTS

We know how to use positive integers and zero as exponents. How might we use negative integers as exponents?

LET'S EXPLORE

1. Copy and complete.

		Factor	*Simplify*
a.	$\frac{5^3}{5^5}$	$\frac{5 \cdot 5 \cdot 5 \cdot 1}{5 \cdot 5 \cdot 5 \cdot 5 \cdot 5}$	$\frac{1}{5^2}$
b.	$\frac{2^5}{2^8}$	$\frac{2 \cdot 2 \cdot 2 \cdot 2 \cdot 2 \cdot 1}{2 \cdot 2 \cdot 2 \cdot 2 \cdot 2 \cdot 2 \cdot 2 \cdot 2}$	$\frac{1}{2^3}$
c.	$\frac{3^2}{3^5}$	$\frac{3 \cdot 3 \cdot 1}{3 \cdot 3 \cdot 3 \cdot 3 \cdot 3}$	$\frac{1}{3^3}$
d.	$\frac{4}{4^4}$		
e.	$\frac{10^2}{10^6}$		
f.	$\frac{8^6}{8^{11}}$		
g.	$\frac{7^4}{7^8}$		

2. Remember that exponential notation for $\frac{6^4}{6^2}$ is 6^{4-2}, or 6^2. Suppose we try subtracting exponents for $\frac{5^3}{5^5}$. We get 5^{3-5}, or 5^{-2}. Copy and complete the table on page 185. Use the idea of subtracting exponents.

		Subtract Exponents	Simplify
a.	$\frac{5^3}{5^5}$	5^{3-5}	5^{-2}
b.	$\frac{2^5}{2^8}$	2^{5-8}	2^{-3}
c.	$\frac{3^2}{3^5}$	3^{2-5}	3^{-3}
d.	$\frac{4}{4^4}$	4^{1-4}	
e.	$\frac{10^2}{10^6}$		
f.	$\frac{8^6}{8^{11}}$		
g.	$\frac{7^4}{7^8}$		

3. Compare the tables in Items 1 and 2. Are the columns on the left alike? Are the columns on the right alike?

4. How should we agree to use negative integers as exponents?

We shall agree that a symbol like 5^{-3} means $\frac{1}{5^3}$.

5. Write exponential notation.

Examples $\frac{1}{4^2} = 4^{-2}$ $\frac{1}{8^6} = 8^{-6}$

a. $\frac{1}{6^2}$ b. $\frac{1}{10^5}$ c. $\frac{1}{7^6}$ d. $\frac{1}{2^{15}}$

6. Write fractional numerals.

Examples $3^{-5} = \frac{1}{3^5}$ $2^{-6} = \frac{1}{2^6}$

a. 8^{-4} b. 7^{-6} c. 9^{-3} d. 11^{-5}

7. Which are true and which are false?

a. $5^{-3} = \frac{1}{5^3}$ b. $8^{-3} = \frac{1}{3 \times 8}$ c. $\frac{1}{6^2} = 6^2$

d. $3^{-2} = \frac{1}{3^2}$ e. $\frac{1}{9^3} = 9^5$ f. $10^{-4} = \frac{1}{4 \times 10}$

EXERCISES

Write exponential notation.

1. $\frac{1}{8^3}$ 2. $\frac{1}{5^7}$ 3. $\frac{1}{6^3}$ 4. $\frac{1}{4^8}$ 5. $\frac{1}{9^2}$

6. $\frac{1}{12^8}$ 7. $\frac{1}{20^6}$ 8. $\frac{1}{11^{17}}$ 9. $\frac{1}{10^6}$ 10. $\frac{1}{100^4}$

11. $\frac{1}{25^7}$ 12. $\frac{1}{37^{11}}$ 13. $\frac{1}{15^3}$ 14. $\frac{1}{9^{10}}$ 15. $\frac{1}{64^3}$

16. $\frac{1}{32^4}$ 17. $\frac{1}{81^3}$ 18. $\frac{1}{15^6}$ 19. $\frac{1}{92^{20}}$ 20. $\frac{1}{100^5}$

Write fractional numerals.

21. 2^{-8} 22. 5^{-7} 23. 8^{-7} 24. 10^{-4} 25. 17^{-6}

26. 20^{-9} 27. 42^{-6} 28. 28^{-23} 29. 19^{-6} 30. 54^{-3}

31. 11^{-20} 32. 14^{-21} 33. 19^{-5} 34. 21^{-8} 35. 29^{-11}

36. 93^{-5} 37. 56^{-11} 38. 62^{-9} 39. 56^{-4} 40. 31^{-8}

Which are true and which are false?

41. $8^{-9} = \frac{1}{8^{-9}}$ 42. $\frac{1}{4^3} = 4^{-3}$ 43. $16^{-3} = \frac{1}{16}$ 44. $9^{-8} = \frac{1}{9^8}$

45. $11^{-8} = \frac{1}{8^{11}}$ 46. $29^9 = \frac{1}{29^9}$ 47. $42^{-5} = \frac{1}{5^{42}}$ 48. $28^{-2} = \frac{1}{28^2}$

Let's Practice

Solve these problems.

1. The diameter of the barrel of a 12-gauge shotgun is .729 in. The diameter of the barrel of a 20-gauge shotgun is .615 in. How much larger is the 12-gauge shotgun barrel?

2. Mr. Allen claimed the following deductions on his income tax: \$134.23 for medical expenses; \$1,435.64 for taxes; \$216 for contributions; and \$2,162.89 for interest. Find his total deduction.

PROBLEM SOLVING

1. During a price war, gasoline prices dropped 30%. The usual price was 34.9¢ per gallon. What was the new price?

2. Mr. Shultz bought motor oil on sale at 95¢ for 5 quarts. What was the saving?

3. Dick took a test with 30 problems. He got $\frac{5}{6}$ of the problems correct. How many problems were correct?

4. Mr. Browning said that about 78% of his math students passed the final exam. He taught 134 students. About how many passed the exam?

5. The Birds won 91 out of 130 games. What per cent of their games did they win?

6. In math class, the ratio of boys to girls was 2 to 3. How many boys were in the class?

7. During a rainy period, the weather bureau recorded the following amounts of rain: 1.26 inches; .72 inches; and 2.16 inches. How much rain fell?

8. In the first six months of the year, the Neumayer Architect Agency recorded the following profits and losses: \$1,600; \$2,300; \$1,523.65; $^{-}\$234$; \$965 and $^{-}\$1,254.68$. Find the overall income.

POWERS OF POWERS

The third power of 5 is 5^3. The fifth power of the third power of 5 is $(5^3)^5$. How can we find a simple name for $(5^3)^5$?

LET'S EXPLORE

1. Copy and complete.

a.	$(2^2)^3$	$2^2 \cdot 2^2 \cdot 2^2 = 2^{2+2+2}$	2^6
b.	$(5^2)^4$	$5^2 \cdot 5^2 \cdot 5^2 \cdot 5^2 = 5^{2+2+2+2}$	5^8
c.	$(7^3)^5$		
d.	$(11^6)^3$		
e.	$(9^4)^6$		

2. Look at the first and last columns. What shortcut do you see?

To find a power of a power like $(5^3)^5$, we multiply the exponents.

$$(5^3)^5 = 5^{3\times5} = 5^{15}$$

3. Find a simple name for each.

Example $(6^2)^7 = 6^{2\times7} = 6^{14}$

a. $(10^3)^4$ b. $(5^4)^3$ c. $(7^6)^3$ d. $(4^3)^5$

EXERCISES

Simplify.

1. $(5^6)^2$ 2. $(2^2)^{10}$ 3. $(6^2)^1$ 4. $(3^4)^4$

5. $(9^4)^2$ 6. $(4^1)^3$ 7. $(5^0)^3$ 8. $(7^2)^0$

9. $(4^3)^8$ 10. $(19^3)^9$ 11. $(6^7)^0$ 12. $(5^8)^7$

13. $(5^5)^9$ 14. $(16^2)^{21}$ 15. $(42^8)^9$ 16. $(56^{11})^1$

EXPANDED NUMERALS

How will negative exponents work in expanded numerals?

1. Let's write an expanded numeral for 724.563. Keep the usual pattern of decreasing exponents.

   $(7 \cdot 10^2) + (2 \cdot \_\_\_) + (4 \cdot \_\_\_)$
   $+ (5 \cdot 10^{-1}) + (6 \cdot \_\_\_) + (3 \cdot \_\_\_)$

2. $5 \cdot 10^{-1}$ means $5 \cdot \frac{1}{10}$, or .5. Copy and complete.

   a. $6 \cdot 10^{-2}$ means $6 \cdot$ ___, or ___.

   b. $4 \cdot 10^{-3}$ means $4 \cdot$ ___, or ___.

3. Write expanded numerals.

Example $275.403 = (2 \cdot 10^2) + (7 \cdot 10^1) + (5 \cdot 10^0) + (4 \cdot 10^{-1}) + (0 \cdot 10^{-2}) + (3 \cdot 10^{-3})$

a. 12.73 **b.** 56.125 **c.** 183.4275

EXERCISES

Write expanded numerals.

1. 362.143 **2.** 324.3427 **3.** 3.02145

4. 35.10023 **5.** 105.00352 **6.** 9,724.10342

7. 9,071.6028 **8.** .006742 **9.** 8.9301

10. 15.129 **11.** .682 **12.** .05431

13. 8.0051 **14.** 8.2134 **15.** 40.9701

Let's Practice

Multiply.

1. $^-2 \cdot 5$ **2.** $8 \times {}^-3$ **3.** $4 \cdot {}^-6$ **4.** $^-7 \cdot {}^-8$

5. $^-8 \times {}^-4$ **6.** $^-3 \cdot 11$ **7.** $15 \times {}^-4$ **8.** $5 \cdot {}^-6$

POWERS OF POWERS: NEGATIVE EXPONENTS

Perhaps we can find a way to find the power of a power when some of the exponents are negative.

LET'S EXPLORE

1. Copy and complete.

a. $(3^{-2})^3$	$\frac{1}{3^2} \cdot \frac{1}{3^2} \cdot \frac{1}{3^2}$	$\frac{1}{3^6}$	3^{-6}
b. $(2^{-4})^2$	$\frac{1}{2^4} \cdot \frac{1}{2^4}$	$\frac{1}{2^8}$	2^{-8}
c. $(3^{-5})^5$			
d. $(7^{-3})^4$			
e. $(5^{-4})^6$			

2. Look at the first and last columns. What shortcut do you see?

3. Multiply exponents to simplify.

Example $(8^{-4})^2 = 8^{-4\times2} = 8^{-8}$

a. $(4^{-3})^4$ **b.** $(6^{-2})^3$ **c.** $(7^{-5})^4$ **d.** $(9^{-6})^5$

4. Remember that $(2^6)^{-3} = \frac{1}{(2^6)^3}$. Copy and complete.

a. $(3^2)^{-3}$	$\frac{1}{(3^2)^3}$	$\frac{1}{3^{2\times3}} = \frac{1}{3^6}$	3^{-6}
b. $(2^4)^{-2}$	$\frac{1}{(2^4)^2}$	$\frac{1}{2^{4\times2}} = \frac{1}{2^8}$	2^{-8}
c. $(8^3)^{-3}$	$\frac{1}{(8^3)^3}$		
d. $(3^5)^{-4}$			
e. $(7^3)^{-5}$			

5. Look at the first and last columns. What shortcut do you see?

6. Multiply exponents to simplify.

Example $(5^3)^{-4} = 5^{3 \times -4} = 5^{-12}$

a. $(7^3)^{-2}$ b. $(5^4)^{-6}$ c. $(8^4)^{-3}$ d. $(3^5)^{-3}$

To find a power of a power like $(5^{-2})^5$ or $(6^3)^{-2}$, we multiply the exponents.

$$(5^{-2})^5 = 5^{-2 \times 5} = 5^{-10}$$

$$(6^3)^{-2} = 6^{3 \times -2} = 6^{-6}$$

7. Simplify.

a. $(2^4)^{-5}$ b. $(5^{-3})^3$ c. $(8^{-6})^4$ d. $(10^9)^{-6}$

EXERCISES

Simplify.

1. $(3^{-4})^2$
2. $(5^3)^{-4}$
3. $(7^4)^{-7}$
4. $(6^{-7})^6$
5. $(4^{-5})^2$
6. $(2^3)^6$
7. $(3^2)^{-5}$
8. $(9^3)^{-6}$
9. $(21^7)^3$
10. $(4^{-2})^8$
11. $(15^{-6})^3$
12. $(13^{10})^{-4}$
13. $(56^2)^{-8}$
14. $(4^9)^{20}$
15. $(18^{-6})^{11}$
16. $(11^{-6})^{10}$
17. $(19^{-4})^5$
18. $(24^7)^{-11}$
19. $(15^6)^{12}$
20. $(19^{12})^{-11}$
21. $(30^{-6})^5$
22. $(29^8)^{-7}$
23. $(56^{-4})^9$
24. $(129^3)^5$

UNEXPLORED TERRITORY

1. We have not yet studied a case like $(2^{-3})^{-4}$. Try to find a rule for simplifying a symbol like $(2^{-3})^{-4}$.

2. Simplify.

a. $(5^{-4})^{-2}$ b. $(6^{-3})^{-5}$ c. $(9^{-2})^{-8}$ d. $(4^{-6})^{-7}$

e. $(10^{-6})^{-4}$ f. $(9^{-12})^{-3}$ g. $(2^{-4})^{-4}$ h. $(7^{-6})^{-6}$

DO YOU REMEMBER?

Write exponential notation.

1. $8^5 \cdot 8^7$
2. $9^2 \cdot 9^0$
3. $10^3 \cdot 10^{11}$
4. $6^5 \cdot 6^3$
5. $4^7 \cdot 4^0$
6. $7^5 \cdot 7^3 \cdot 7^0$
7. $3^8 \cdot 3^4 \cdot 3^5$
8. $10^6 \cdot 10^1 \cdot 10^9$
9. $4^7 \cdot 4^8$

Write exponential notation.

10. $\frac{12^5}{12^3}$
11. $\frac{4^{11}}{4^6}$
12. $\frac{9^8}{9^2}$
13. $\frac{10^{11}}{10^7}$
14. $\frac{15^6}{15^0}$
15. $\frac{24^{15}}{24^{12}}$
16. $\frac{16^4 \cdot 16^2}{16^3}$
17. $\frac{19^4 \cdot 19^{11}}{19^9}$
18. $\frac{20^5 \cdot 20^{13}}{20^9}$

Write exponential notation.

Example $\frac{1}{5^4} = 5^{-4}$

19. $\frac{1}{6^2}$
20. $\frac{1}{8^{14}}$
21. $\frac{1}{5^6}$
22. $\frac{1}{4^7}$
23. $\frac{1}{12^4}$
24. $\frac{1}{9^7}$
25. $\frac{1}{11^6}$
26. $\frac{1}{2^{16}}$

Write expanded numerals.

27. 264.379
28. 418.1034
29. 206.0038

Simplify.

30. $(5^4)^3$
31. $(2^3)^8$
32. $(12^6)^{-4}$
33. $(56^4)^{-2}$
34. $(9^{-2})^5$
35. $(10^8)^5$
36. $(24^{-2})^6$
37. $(32^8)^9$
38. $(16^5)^{12}$
39. $(20^{-1})^9$
40. $(11^5)^{-7}$
41. $(54^{-6})^9$
42. $(124^{-6})^8$
43. $(98^0)^{-4}$
44. $(15^{-2})^{20}$
45. $(12^{-3})^8$
46. $(56^1)^{-8}$
47. $(93^5)^0$
48. $(28^{-7})^{12}$
49. $(158^{-6})^6$

POWERS OF TEN

Let's think about exponential notation for powers of ten and see what pattern we can find.

1. Multiply.

   a. 563×100 b. 56.3×100 c. 5.63×100 d. $.563 \times 100$ e. $.0563 \times 100$

2. Remember that $100 = 10^2$. Multiply.

   a. 563×10^2 b. 56.3×10^2 c. 5.63×10^2 d. $.563 \times 10^2$ e. $.0563 \times 10^2$

3. Write standard numerals for these products.

   a. 246×10^4 b. 24.6×10^3 c. $.246 \times 10^2$ d. $.246 \times 10^5$

4. Did you find a pattern? How would you use it to multiply 3.246 by 10^3?

> We can multiply by a power of ten like 10^3 by moving the decimal point three places to the right.

5. Write standard numerals.

   Examples $2.463 \times 10^2 = 246.3$
   $42.16 \times 10^4 = 421{,}600$

   a. 16.249×10^3 b. 82.42×10^3 c. 294.642×10^5

   d. 3.962×10^2 e. 395.48×10^1 f. 1.6952×10^4

> What happens when we multiply by a power of 10, where the exponent is negative?

6. Consider 500×10^{-2}.

$$500 \times 10^{-2} = 500 \times \frac{1}{10^2} = 500 \times \frac{1}{100} = 5$$

   What is 500×10^{-2}?

7. Consider $23{,}000 \times 10^{-3}$.

$$23{,}000 \times 10^{-3} = 23{,}000 \times \frac{1}{10^3}$$
$$= 23{,}000 \times \frac{1}{1{,}000}$$
$$= 23$$

8. In what direction could you have moved the decimal point in 23,000 to get the same answer?

We can multiply by a power of ten like 10^{-4} by moving the decimal point four places to the left.

9. Write standard numerals.

Examples $423.689 \times 10^{-2} = 4.23689$

$2.65 \times 10^{-3} = .00265$

a. 143.25×10^{-2} b. 8.3×10^{-3} c. 19.56×10^{-4}

EXERCISES

Write standard numerals.

1. 3.468×10^2
2. 327×10^2
3. 24.63×10^3
4. 24.65×10^3
5. 562.3×10^{-4}
6. 89.2×10^5
7. $.0345 \times 10^3$
8. $.243 \times 10^{-2}$
9. 4563×10^{-2}
10. 1.00056×10^4
11. 267×10^{-3}
12. $.346 \times 10^{-6}$
13. 326.5×10^3
14. 342.78×10^{-1}
15. 8.274×10^{-5}
16. 4.68×10^{-1}
17. $.2634 \times 10^3$
18. 1.23×10^{18}
19. 8.4×10^{20}
20. 6.342×10^{15}
21. 268.14×10^{-2}
22. $.56 \times 10^8$
23. 17.823×10^{-9}
24. 9.3×10^{15}
25. 14.9×10^{-3}
26. $.0568 \times 10^{-5}$
27. 144.32×10^{10}

SCIENTIFIC NOTATION

The distance from the earth to the North Star is about 2,400,000,000,000,000 miles. The mass of a hydrogen atom is about $\frac{1}{273,171,800,000,000,000,000,000,000}$ of a pound. As you can see, scientists must work with very large and very small numbers. To make their work easier, they use *scientific notation.*

1. Use exponential notation to name the missing factor.

Examples $726 = 7.26 \times 10^2$ $\quad .439 = 4.39 \times 10^{-1}$

a. $.524 = 5.24 \times n$ **b.** $27,000 = 2.7 \times n$

c. $6,260,000 = 6.26 \times n$ **d.** $.052394 = 5.2394 \times n$

Scientific notation for a number consists of:
(a) a decimal numeral for a number from 1 to 10
(b) exponential notation for a power of ten.

2. The speed of light is about 186,000 miles per second.

a. $186,000 = 1.86 \times$ ____.

b. What is scientific notation for 186,000?

3. An oxygen atom is about .000000001 of the size of a drop of water.

a. $.000000001 = 1 \times$ ____.

b. What is scientific notation for .000000001?

4. Write scientific notation.

a. 147,289 b. 17,342,000

c. .0561 d. .0000864

5. Write standard numerals.

Examples $5.6 \times 10^{4} = 56{,}000$
$2.8 \times 10^{-6} = .0000028$

a. 2.9×10^{8} b. 3.2×10^{-5} c. 1.8×10^{9}

d. 8.6×10^{-8} e. $.0065 \times 10^{6}$ f. $.0087 \times 10^{-5}$

EXERCISES

Write scientific notation.

1. 32,000,000
2. 6,503,000,000
3. 37,400,000,000
4. 895,200,000,000,000
5. .015
6. .00241
7. .000004138
8. .002103
9. .34567
10. 1,000
11. .0000005821
12. 818,000,000,000,000,000

Write standard numerals.

13. 3.6×10^{7}
14. 1.23×10^{9}
15. 4.1×10^{-6}
16. 2.83×10^{-8}
17. 5.34×10^{12}
18. 1.29×10^{-5}
19. 2.936×10^{6}
20. 7.862×10^{-4}
21. 56.7×10^{-3}
22. 98.65×10^{2}
23. 87.9×10^{4}
24. 9.546×10^{-5}

PROBLEM SOLVING

1. The sun is about 93,000,000 miles from the earth. Write scientific notation for this distance.

2. The closest star to our sun is Alpha Centauri. It is about 25,800,000,000,000 miles away. Write scientific notation for this distance.

3. The diameter of the giant red star Betelgeuse is about 3.72×10^8 miles. Write standard notation for this.

4. The volume of the sun is only 1.4×10^{-8} that of Betelgeuse. Write standard notation for this.

5. The weight of the earth is about 6×10^{21} tons. How many pounds is this?

6. Light travels at a speed of six trillion miles per year. It takes light 400 years to reach the earth from the North Star.

   a. How many miles from the earth is the North Star? Write standard notation for this number.

   b. Write scientific notation for this distance.

7. The Great Nebula is the nearest galaxy to our own. It is about 4,200,000,000,000,000,000 miles from the earth. Write scientific notation for this distance.

8. The mass of a hydrogen atom is about

$$\frac{1}{273{,}171{,}800{,}000{,}000{,}000{,}000{,}000{,}000}$$

   of a pound. Write scientific notation for this number.

9. Scientific notation for the mass of the sun in tons is 2×10^{27}. Write standard notation for this.

10. A googol is a number named by a 1 followed by 100 zeros. Write a name for a googol using scientific notation.

DO YOU REMEMBER?

An early system of writing numerals was that of the Egyptians. Here are their basic symbols.

Our Numeral	Egyptian Numeral	Object Represented
1		vertical staff
10		heel bone
100		coiled rope
1,000		lotus flower
10,000		bent reed
100,000		burbot fish
1,000,000		astonished man

The Egyptians combined basic symbols by writing them together. A numeral for 3 was . A numeral for 123 was  Here are some other numerals for 123.

Notice that each basic numeral means the same thing, no matter where it is placed.

1. Tell what these numerals mean.

   a.  b. 

   c.  d. 

2. Write Egyptian numerals.

   a. 233 b. 100,053

   c. 10,316 d. 1,203,410

3. What advantage does our numeration system have over the Egyptian system?

CHAPTER REVIEW

Write exponential notation.

1. $3 \cdot 3 \cdot 3 \cdot 3 \cdot 3$

2. $2 \cdot 2 \cdot 2 \cdot 2 \cdot 2 \cdot 2$

3. $100 \times 100 \times 100$

4. 121

Write standard numerals.

5. 5^4 **6.** 2^5 **7.** 100^3 **8.** 10^{10} **9.** 20^4

Write exponential notation.

10. $\frac{1}{3^5}$ **11.** $\frac{1}{2^4}$ **12.** $\frac{1}{10^3}$ **13.** $\frac{1}{6^{-2}}$

Write exponential notation.

14. $2^3 \cdot 2^4$

15. $3^8 \cdot 3^{12}$

16. $5^3 \cdot 5^2 \cdot 5^4$

17. $7^0 \cdot 7^1 \cdot 7^3$

18. $3^0 \cdot 4^0 \cdot 5^0 \cdot 7^2$

19. $10^5 \cdot 10^7 \cdot 10^{-15}$

20. $2^6 \cdot 5^3 \cdot 3^{-2} \cdot 2^{-4} \cdot 5^0$

21. $7^3 \cdot 2^5 \cdot 7^{-4} \cdot 2^0 \cdot 7^2$

22. $\frac{3^4}{3^2}$

23. $\frac{4^8}{4^5}$

24. $\frac{2^3}{2^5}$

25. $\frac{5^7 \cdot 5^{-4}}{5^6}$

26. $\frac{2^0 \cdot 3^5 \cdot 5^3}{3^3 \cdot 5^3}$

27. $\frac{2^6 \cdot 3^4 \cdot 5^7 \cdot 2^{-3} \cdot 5^{-2}}{3^2 \cdot 5^4 \cdot 2^5 \cdot 3^3}$

28. $(3^2)^5$

29. $(5^3)^{-2}$

30. $(2^{-2})^3 \cdot (2^5)^3$

31. $(10^3)^{-2} \cdot (10^2)^3 \cdot (10^4)^7$

Write expanded notation.

32. 32.48 **33.** 150.372 **34.** .5693

35. 2.83702 **36.** 314.0073 **37.** .00345

Write standard numerals.

38. 3.243×10^3

39. 2.102×10^5

40. $.003 \times 10^5$

41. $.0578 \times 10^6$

42. 12.3×10^4

43. 7.34×10^{-2}

44. $.004 \times 10^{-2}$

45. 11.3×10^{-4}

46. 254×10^{-2}

47. $1{,}362 \times 10^{-4}$

Write scientific notation.

48. 214

49. .56

50. 1,324

51. 283,000

52. 7,289,000

53. .00034

Write standard numerals.

54. 1.34×10^3

55. 1.567×10^6

56. 7.24×10^{-6}

57. 3.4×10^{10}

58. 1.523×10^7

59. 2.5×10^{-15}

Solve these problems.

60. The distance from the earth to the moon is about 239,000 miles. Write scientific notation for this distance.

61. The Murray County budget for the fiscal year was $\$2.4 \times 10^6$. Write standard notation for this.

Terms You Should Know

exponential notation	power
exponent	negative exponent
base	scientific notation

CHAPTER TEST

Write exponential notation.

1. $5 \cdot 5 \cdot 5 \cdot 5 \cdot 5 \cdot 5 \cdot 5$
2. 49

Write standard numerals.

3. 10^4 *10,000*
4. 2^6

Write exponential notation.

5. $\frac{1}{5^3}$
6. $\frac{1}{3^{-2}}$

Write exponential notation.

7. $3^4 \cdot 3^6$
8. $7^0 \cdot 7^5 \cdot 7^1$
9. $5^{-2} \cdot 5^0 \cdot 5^4$
10. $2^3 \cdot 3^4 \cdot 2^{-5} \cdot 3^2$
11. $\frac{10^7}{10^3}$
12. $\frac{2^5 \cdot 3^2 \cdot 2^{-2}}{2^4 \cdot 3^2}$
13. $(5^2)^3$
14. $(10^{-8})^{-2}$

Write expanded notation.

15. 12.38
16. .3407

Write standard numerals.

17. 5.103×10^7
18. 3.45×10^3

Write scientific notation.

19. 39,000,000
20. .0000017

Write standard numerals.

21. 121.3×10^{-2}
22. 1.05×10^{-8}

Solve this problem.

23. The polar circumference of the earth is 24,857 miles. Write scientific notation for this.

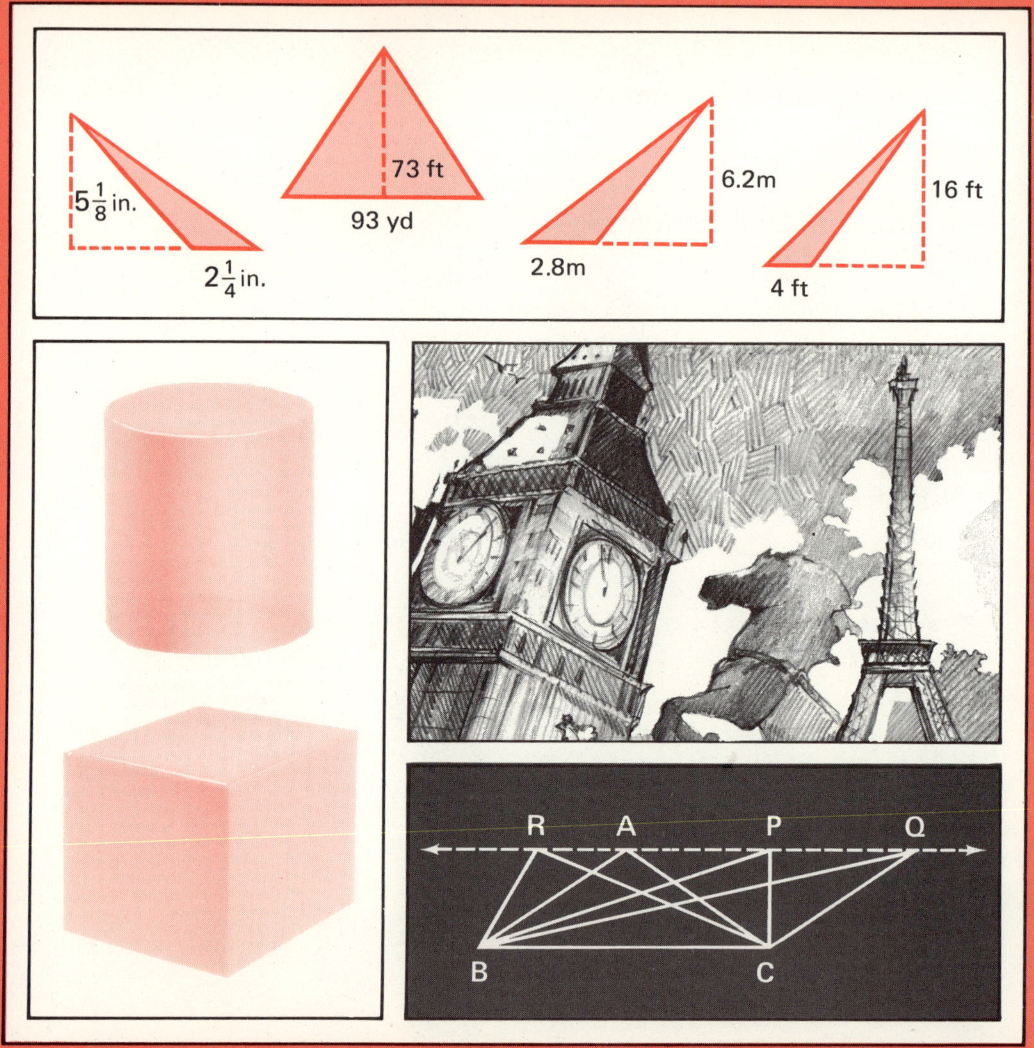

CHAPTER EIGHT

Areas and Volumes

CLASSIFICATION OF QUADRILATERALS

We can draw a diagram to show the relationships between types of quadrilaterals.

1. A *quadrilateral* is a polygon with four sides. Draw a diagram like this to represent all quadrilaterals.

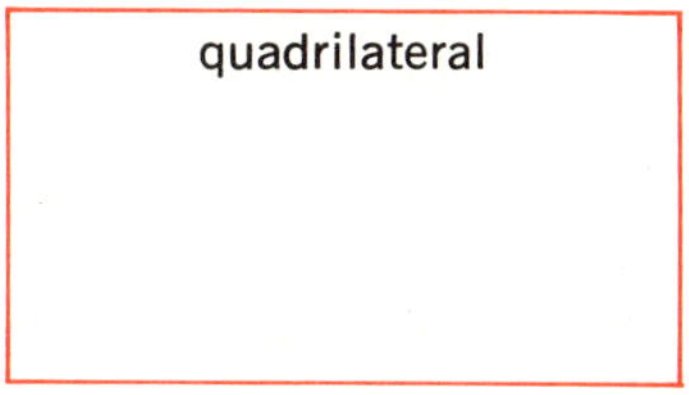

2. A *trapezoid* is a quadrilateral with at least one pair of opposite sides parallel. Add a figure like this to your diagram to represent all trapezoids.

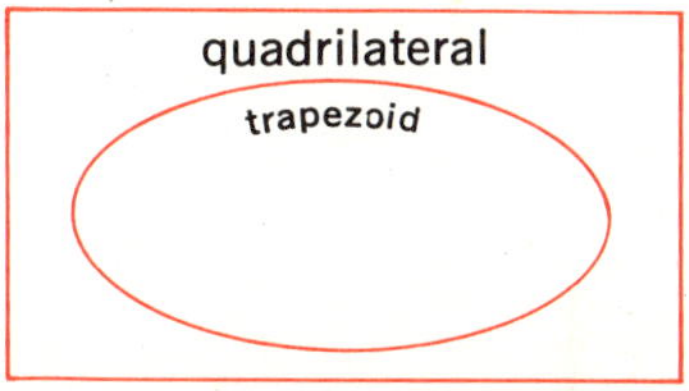

3. A *parallelogram* is a trapezoid with both pairs of opposite sides parallel. Draw another figure to represent all parallelograms.

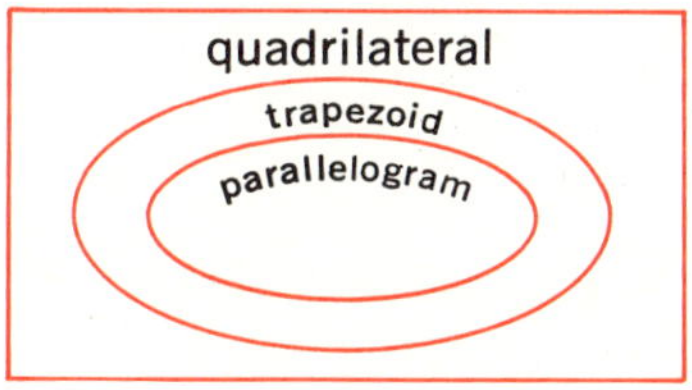

4. A *rectangle* is a parallelogram with four right angles. Add another figure to represent all rectangles.

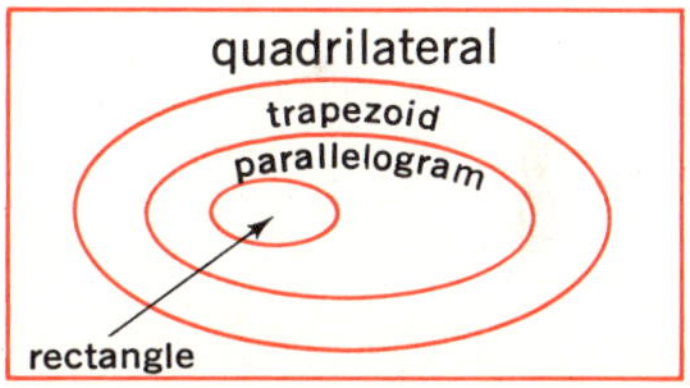

5. A *rhombus* is a parallelogram with all four sides congruent. Represent all rhombi (or rhombuses) this way.

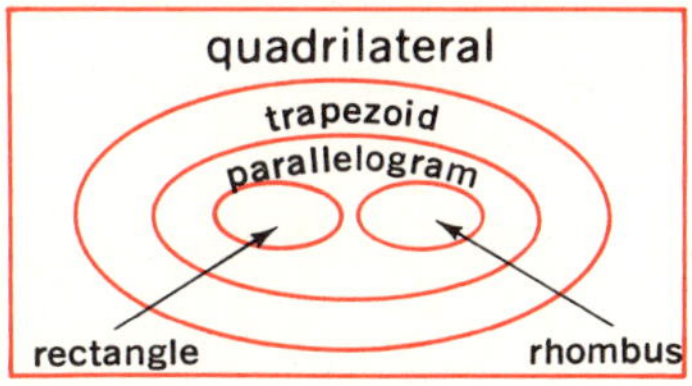

6. A *square* is a rectangle with all sides congruent. So, a square is both a rectangle and a rhombus. Represent all squares by overlapping like this.

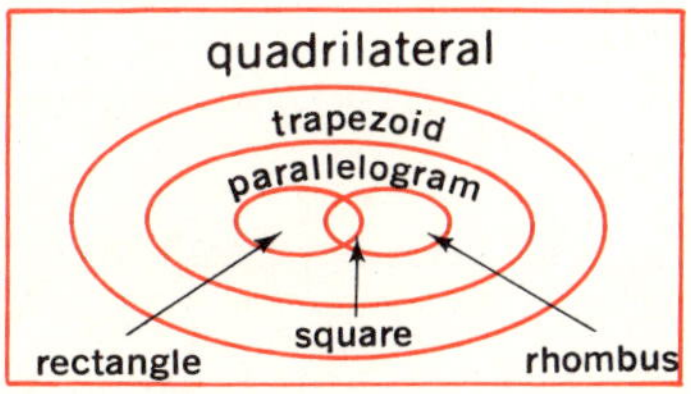

7. Look at the diagram on page 203. Are these true or false?

   a. All trapezoids are quadrilaterals.

   b. Some parallelograms are squares.

   c. All rectangles are trapezoids.

   d. All rhombi are squares.

EXERCISES

Give as many names as you can for each figure.

1. 

2. 

3. 

4. 

5. 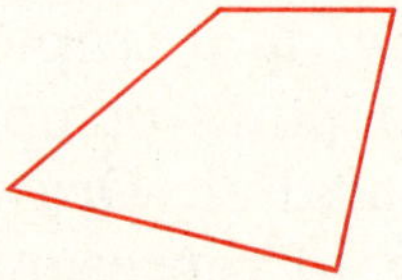

6. 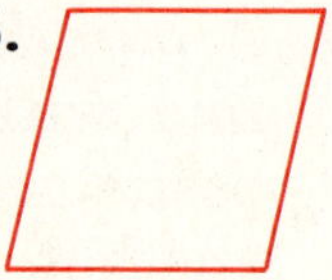

Which are true and which are false?

7. All squares are parallelograms.

8. All rectangles are trapezoids.

9. All rectangles are rhombi.

10. All trapezoids are parallelograms.

11. Some rhombi are rectangles.

12. All squares are rhombi.

13. All trapezoids are quadrilaterals.

14. All rectangles are parallelograms.

15. All parallelograms are rectangles.

DO YOU REMEMBER?

The perimeter of a polygon is the sum of the measures of its sides. Find the perimeter of each.

1.
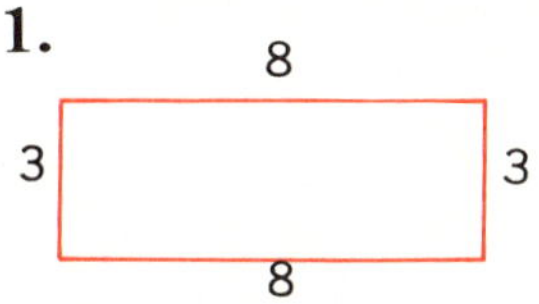

2.
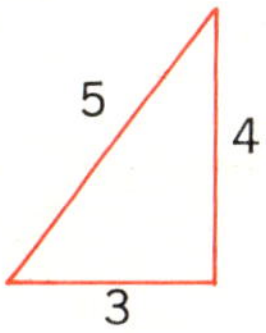

3.
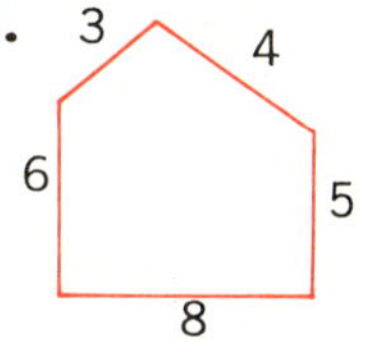

4.
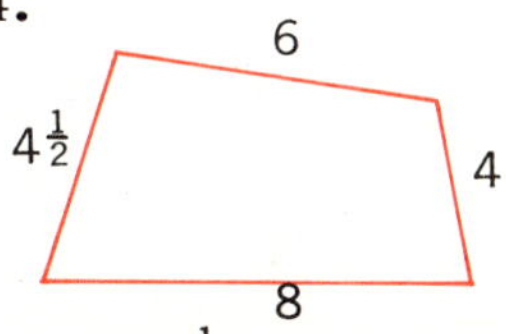

5.
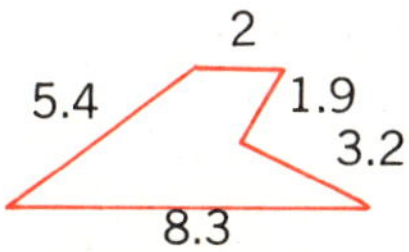

6.
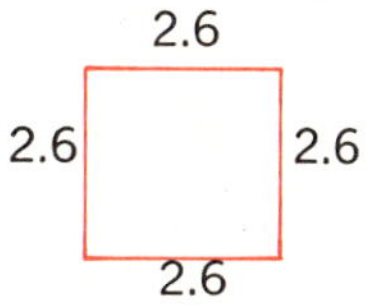

7.
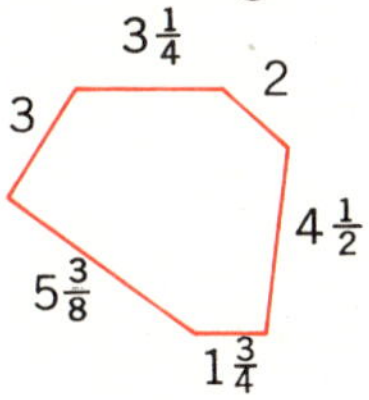

8.
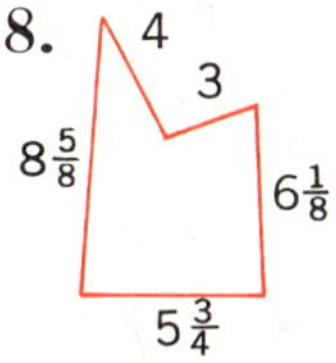

9.
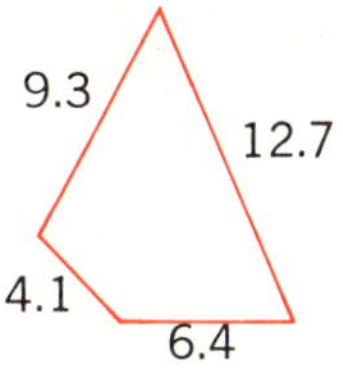

10. Draw a square whose perimeter is 20 inches.

11. Draw a rectangle whose perimeter is 10 inches.

Solve these problems.

12. Frank was building a rectangular dog run. The run measured 12 feet by 7 feet. How many feet of wire fence did Frank need to enclose the run? Allow $1\frac{1}{2}$ extra feet for waste.

13. Jack is manager of the football team. To hold back the crowd, he needs a rope fence around the field. The fence will be a rectangle about 70 yards by 140 yards.

    a. How much rope should he buy?

    b. Rope comes in 200 foot packages for $1.49 per package. What will the rope cost?

AREA

Any closed figure on a plane and its interior is a two-dimensional figure. Such figures are called *regions*. The measure of a region is its *area*.

1. Here are pictures of several regions. Draw two more regions.

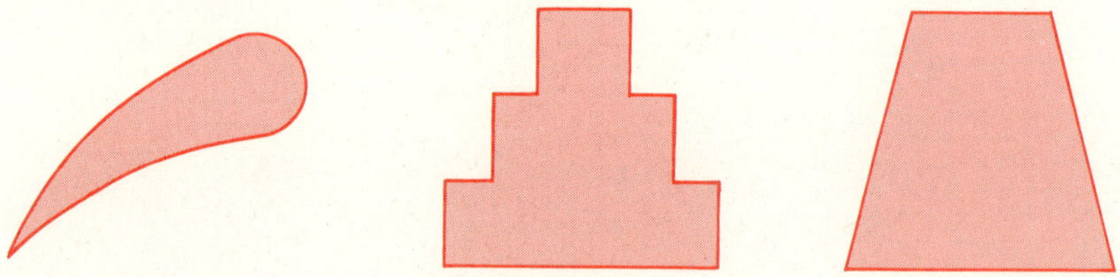

2. In the English system, we use the square inch, square foot, square yard, etc., to measure areas. Copy and complete.

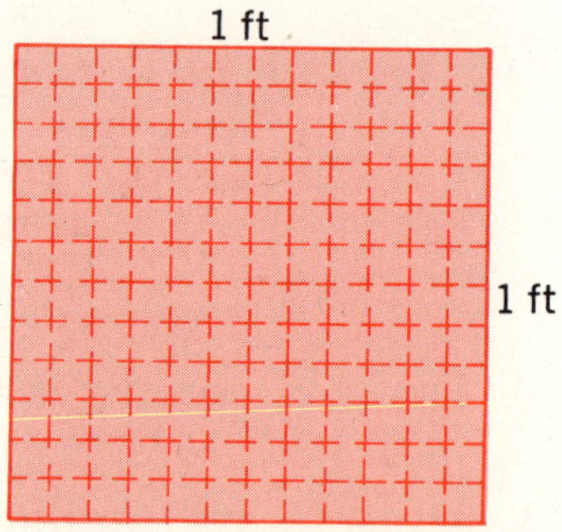

   a. 1 square foot = ____ square inches.

   b. 1 square yard = ____ square feet.

3. The metric system also has standard area units. The square centimeter, square meter, etc., are used. Copy and complete.

   a. 1 square centimeter = ____ square millimeters.

   b. 1 square meter = ____ square centimeters.

4. We can write 8 in.^{2} to mean 8 square inches. Write another name for each.

   a. 20 square inches
   b. 5 sq ft
   c. 10 sq yd
   d. 24 sq cm
   e. 16 yd^2
   f. 54 mm^2
   g. 109 ft^2
   h. 11 sq cm

5. Copy and complete.

Examples $3 \text{ ft}^2 = (3 \times 144) \text{ in.}^2 = 432 \text{ in.}^2$
$3 \text{ cm}^2 = (3 \times 100) \text{ mm}^2 = 300 \text{ mm}^2$

a. $2 \text{ ft}^2 =$ ____ in.^2
b. $6 \text{ ft}^2 =$ ____ in.^2
c. $4 \text{ yd}^2 =$ ____ ft^2
d. $4 \text{ cm}^2 =$ ____ mm^2
e. $2 \text{ m}^2 =$ ____ cm^2
f. $3 \text{ yd}^2 =$ ____ ft^2

6. Copy and complete.

Examples $288 \text{ in.}^2 = (288 \div 144) \text{ ft}^2 = 2 \text{ ft}^2$
$700 \text{ mm}^2 = (700 \div 100) \text{ cm}^2 = 7 \text{ cm}^2$

a. $576 \text{ in.}^2 =$ ____ ft^2
b. $54 \text{ ft}^2 =$ ____ yd^2
c. $900 \text{ mm}^2 =$ ____ cm^2
d. $27 \text{ ft}^2 =$ ____ yd^2

EXERCISES

1. Draw several two-dimensional regions.

Copy and complete.

2. $5 \text{ yd}^2 =$ ____ ft^2
3. $14 \text{ ft}^2 =$ ____ in.^2
4. $6 \text{ cm}^2 =$ ____ mm^2
5. $3 \text{ m}^2 =$ ____ cm^2
6. $21 \text{ ft}^2 =$ ____ in.^2
7. $9 \text{ yd}^2 =$ ____ ft^2
8. $10 \text{ cm}^2 =$ ____ mm^2
9. $4 \text{ m}^2 =$ ____ cm^2

Copy and complete.

10. $63 \text{ ft}^2 =$ ____ yd^2
11. $720 \text{ in.}^2 =$ ____ ft^2
12. $1{,}000 \text{ mm}^2 =$ ____ cm^2
13. $60{,}000 \text{ cm}^2 =$ ____ m^2
14. $25{,}000 \text{ cm}^2 =$ ____ m^2
15. $81 \text{ ft}^2 =$ ____ yd^2
16. $37{,}000 \text{ cm}^2 =$ ____ m^2
17. $108 \text{ ft}^2 =$ ____ yd^2
18. $1{,}008 \text{ in.}^2 =$ ____ ft^2
19. $3{,}000 \text{ mm}^2 =$ ____ cm^2

AREAS OF RECTANGULAR REGIONS

We can find the area of a rectangular region by multiplying the length by the width.

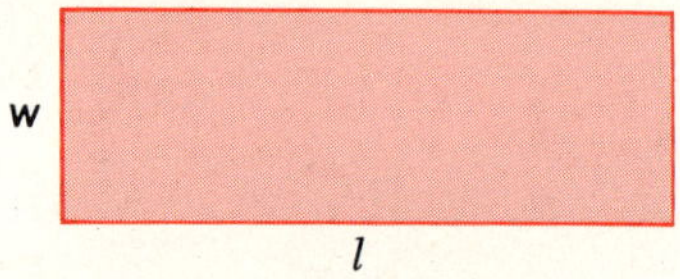

$$A = l \cdot w$$

1. Find the areas.

a. 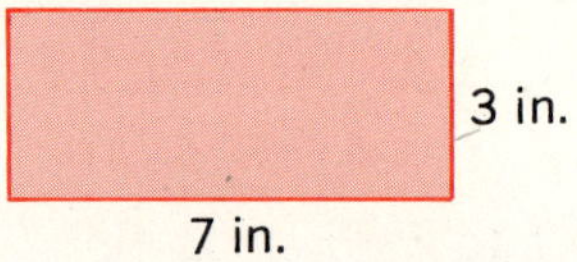

b. 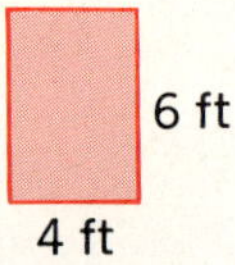

c. 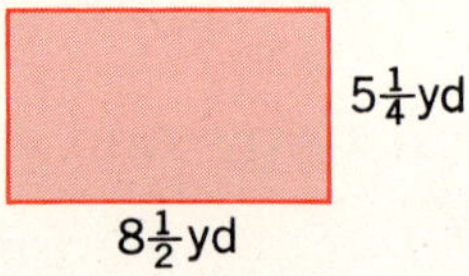

2. A square is a rectangle with all sides having the same length s. In the formula $A = l \cdot w$, replace l and w with s. You now have a formula for the area of a square.

3. Find the areas.

a. 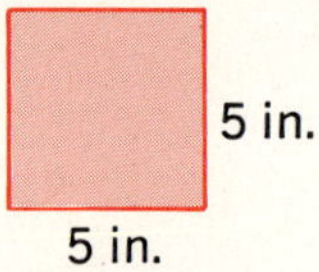

b. 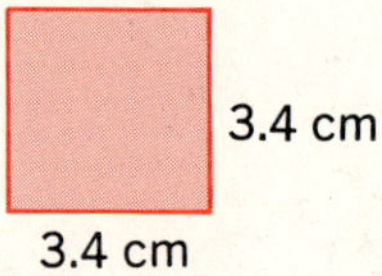

c. 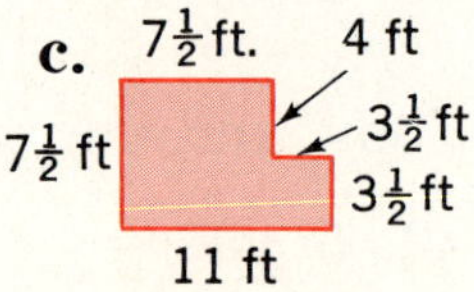

EXERCISES

Find the areas.

1. 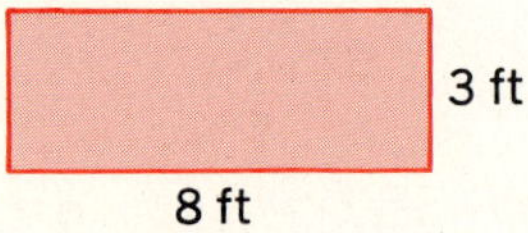

2. 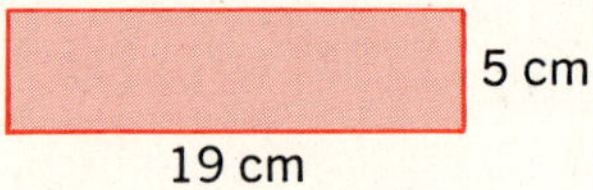

3. 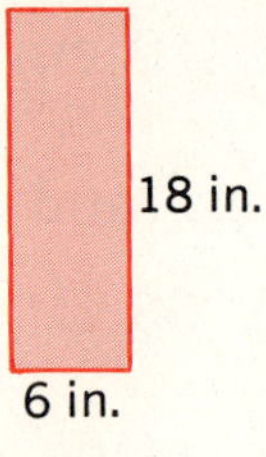

4. 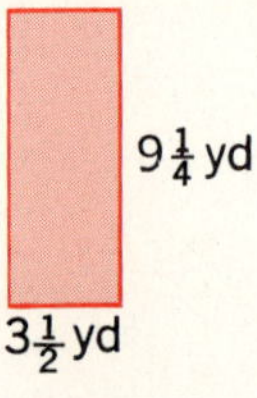

5. 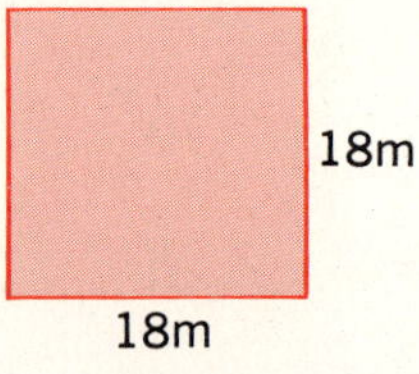

6. 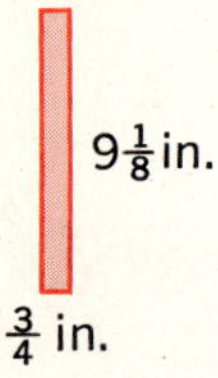

7. 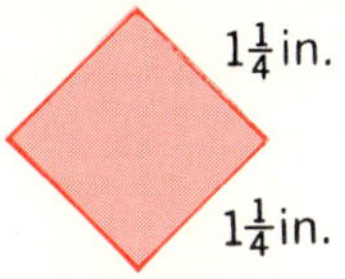

8. 

9. 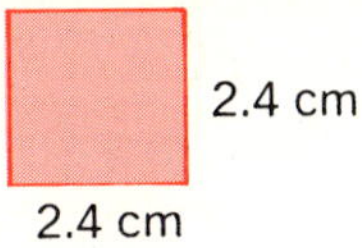

10. 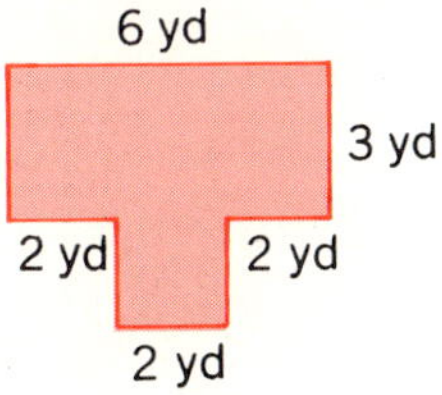

11. 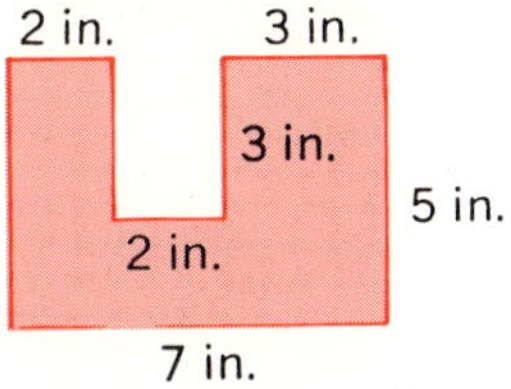

12. 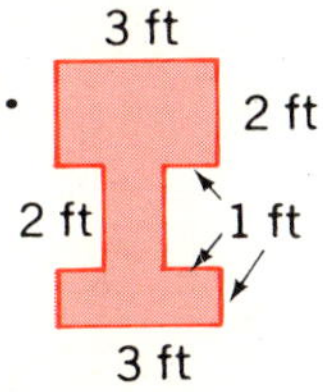

Solve these problems.

13. Mr. Watson planted grass on a rectangular plot which was 42 feet by 128 feet. He can buy lawn fertilizer in bags for 1,000 square feet at $4.98. How much will he spend for fertilizer.

14. Here is a plan that Mr. Wood drew for a new blacktop driveway. Blacktop cost 38¢ per square foot. What will the driveway cost?

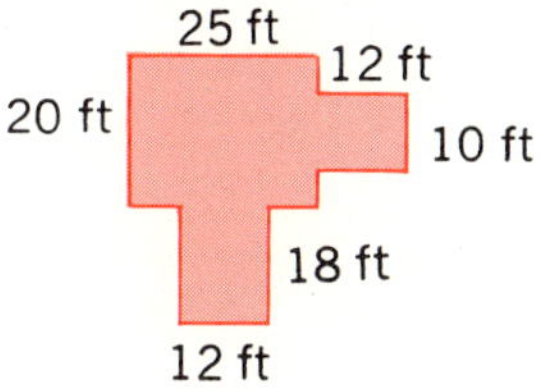

Let's Practice

Recall that the distance around a circle is called the *circumference*. We can find the circumference by multiplying the length of a diameter by π.

Find the circumferences. Use $\frac{22}{7}$ for π.

1. 

2. 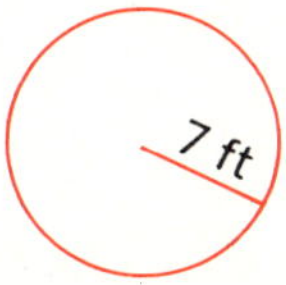

3. 

4. 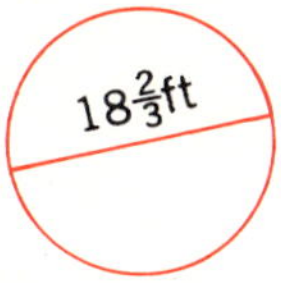

Find the circumferences. Use 3.14 for π.

5. 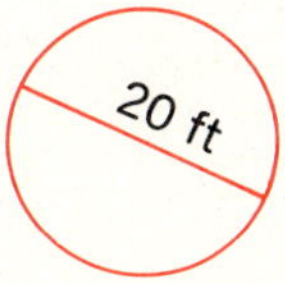

6. 

7. 

8. 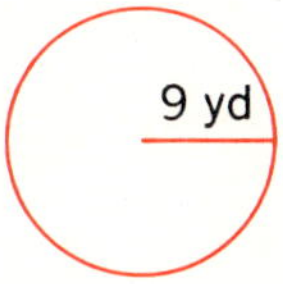

AREAS OF PARALLELOGRAMS

We can find the area of a parallelogram by multiplying the length of a base by the height.

$$A = b \cdot h$$

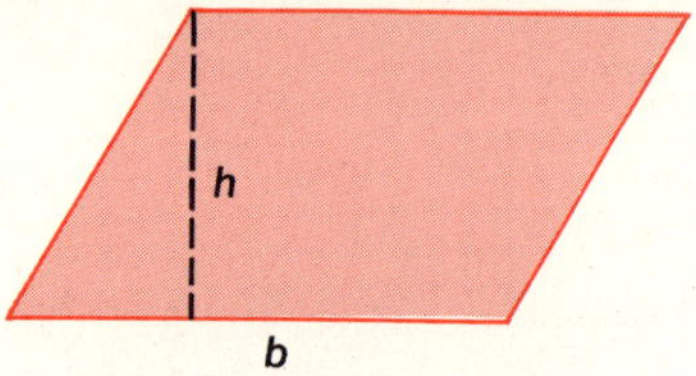

1. What is the height of each parallelogram?

a.
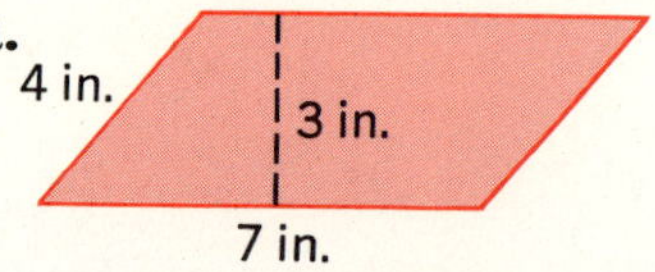

b.
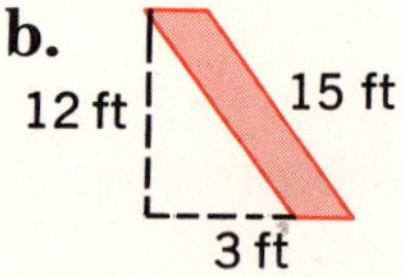

c.
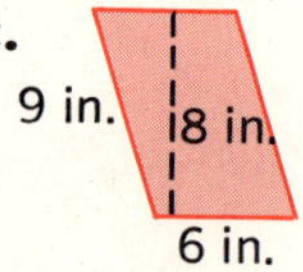

2. What is the area of each parallelogram above?

3. Find the area of each parallelogram.

a.
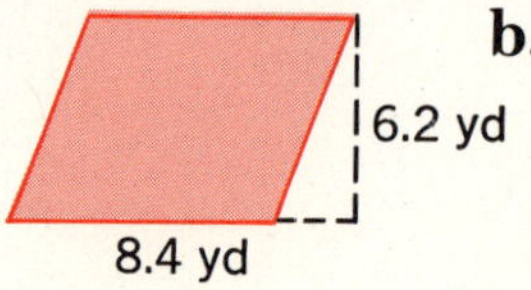

b.
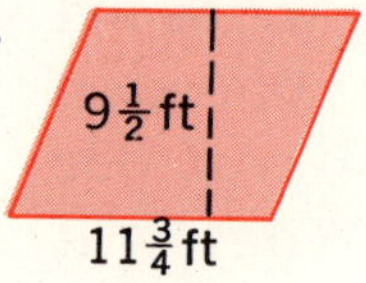

c.
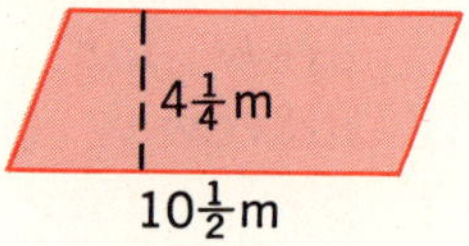

EXERCISES

Find the area of each parallelogram.

1.
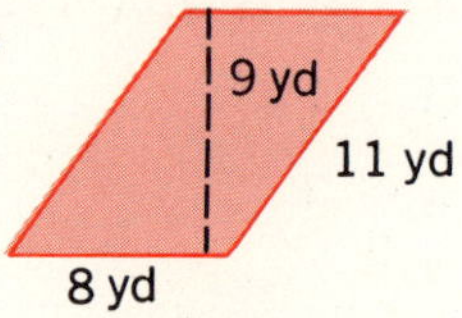

2.
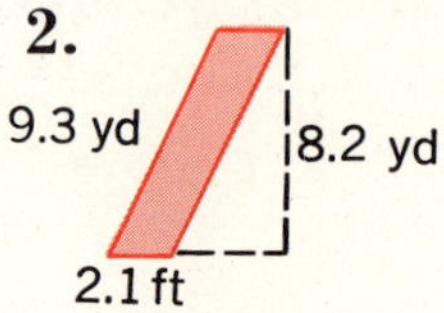

3.
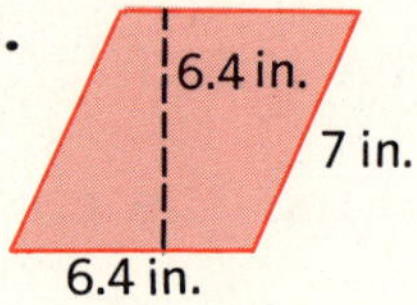

4.
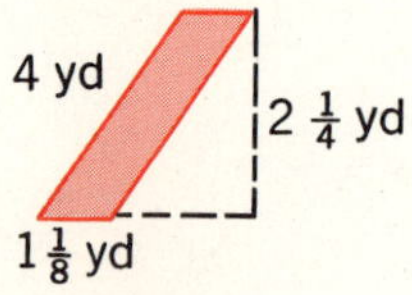

5.
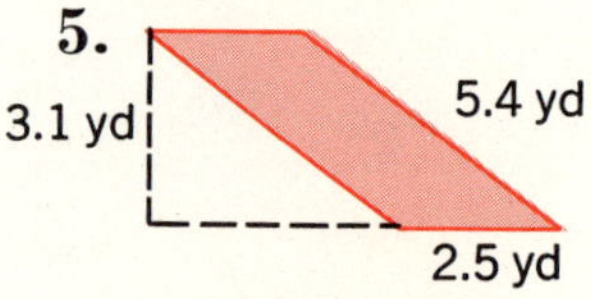

6.
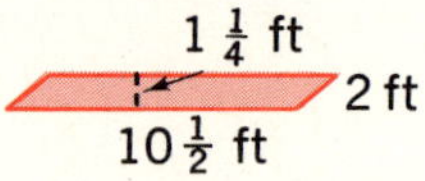

7.
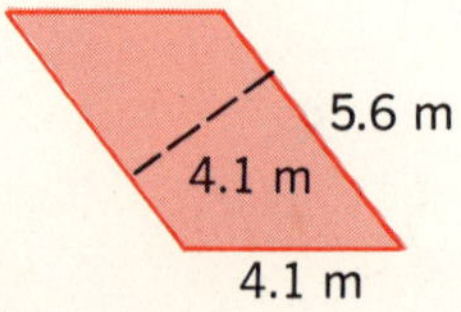

8.
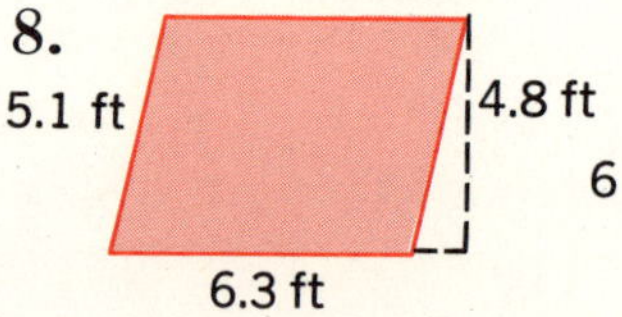

9.
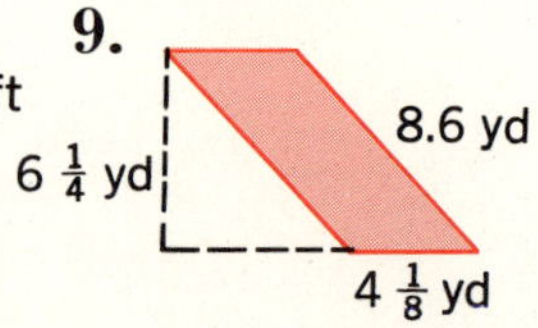

PROBLEM SOLVING

1. Mrs. Bernstein plans to put wall-to-wall carpeting in her living room. The room is $16\frac{1}{2}$ feet wide and 24 feet long. How much should Mrs. Bernstein plan to spend?

2. Mr. Chenoweth, a builder, needs bricks for a new house at the rate of 7 bricks per square foot.

   a. About how many bricks will he need for the following:

      2 walls each 44 feet × 11 feet

      2 walls each 29 feet × 11 feet

   b. Bricks cost $67 per 1,000. What will be the total cost of the bricks for the four walls?

3. Mr. Wade plans to put peg board on a section of his garage wall. Peg board comes in 4 ft by 8 ft sheets which cost $4.80 per sheet.

   a. How many sheets will be needed for a section which is 10 ft high and 24 ft long?

   b. What will they cost?

4. Mr. Fritz wants to remove dandelions from his lawn. He estimates that his lawn covers 24,000 square feet. Weed killer costing $8.95 comes in bags which cover 10,000 square feet. How much should he expect to spend?

AREAS OF TRIANGLES

The area of any triangular region is half the length of the base times the height.

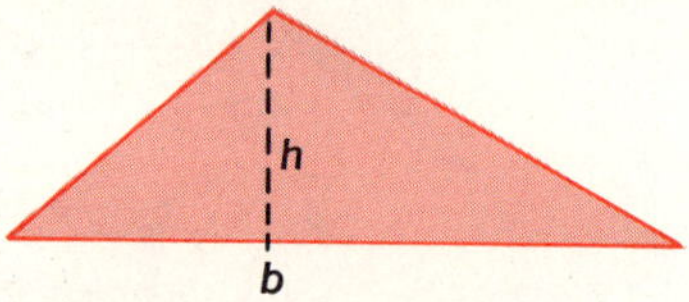

$$A = \frac{1}{2} \cdot b \cdot h$$

1. What is the height of each triangle?

a.

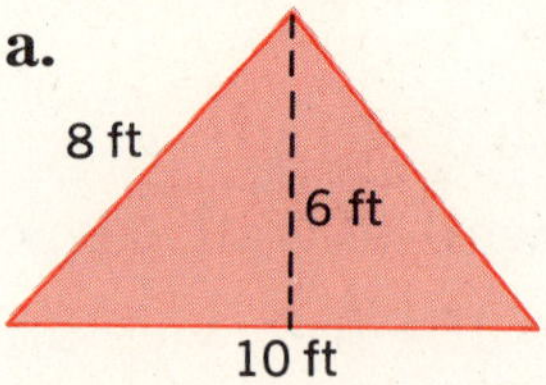

b.

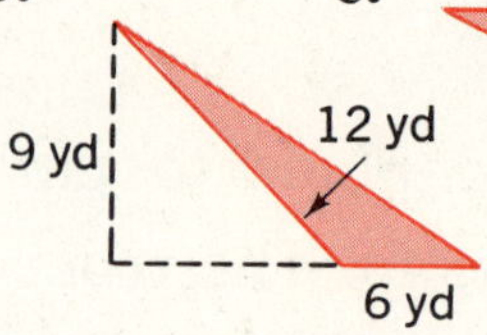

c.

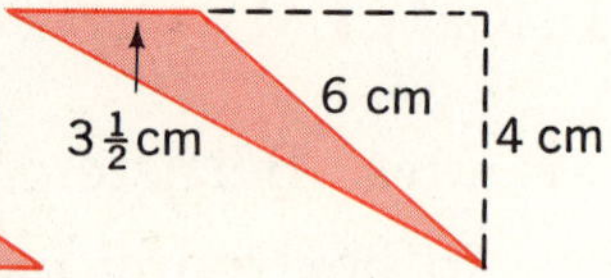

2. What is the area of each triangle above?

3. Find the area of each triangle.

a.

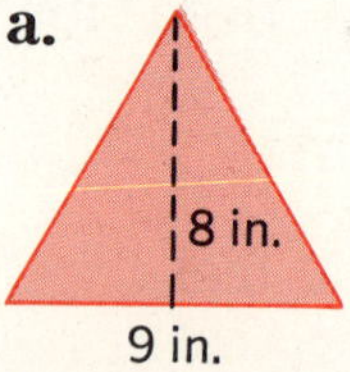

b.

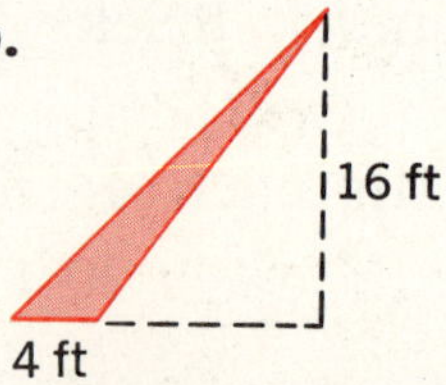

c.

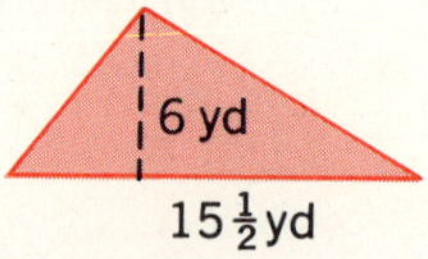

EXERCISES

Find the area of each triangle.

1.

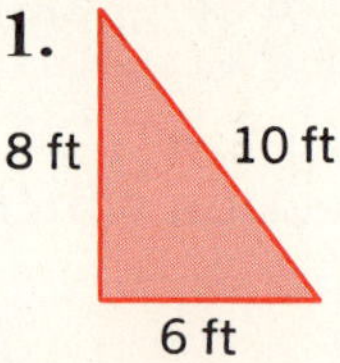

2.

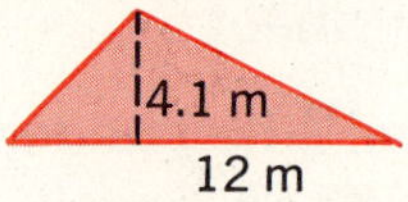

3.

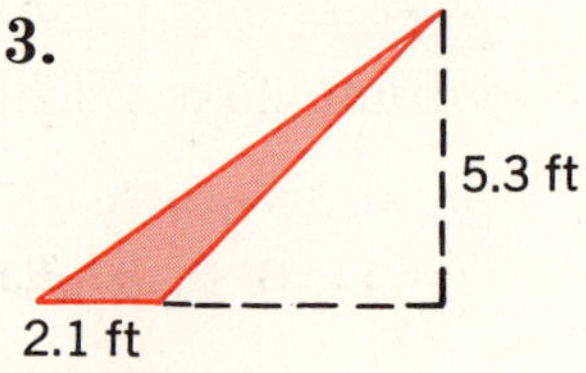

4.

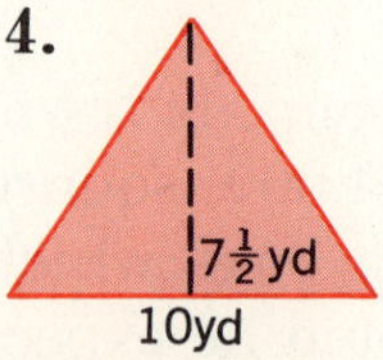

5.

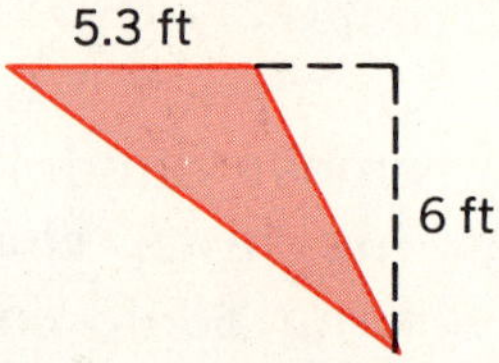

6.

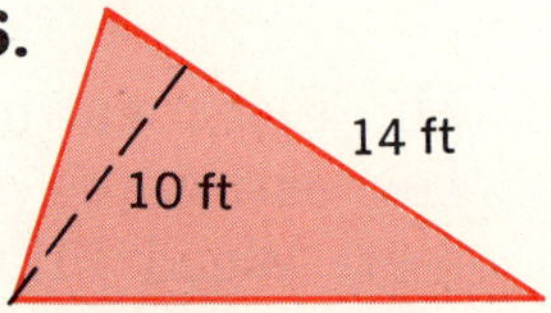

7.

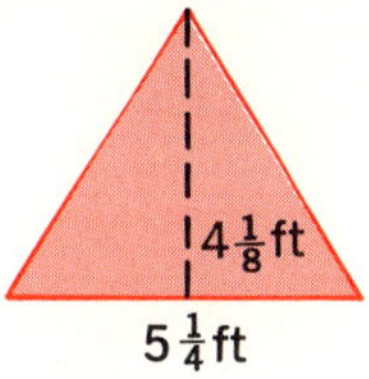

8.

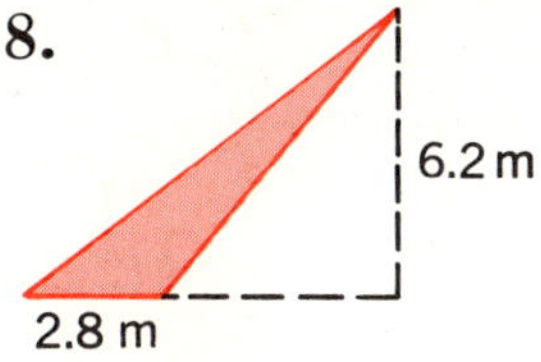

9.

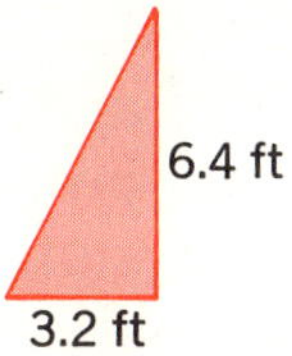

10.

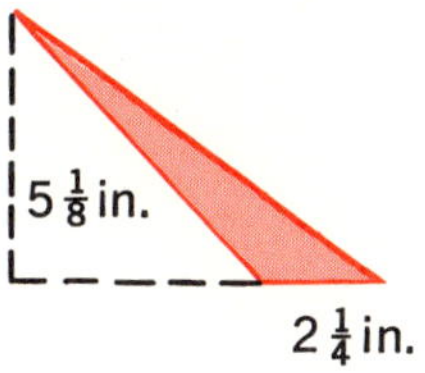

11.

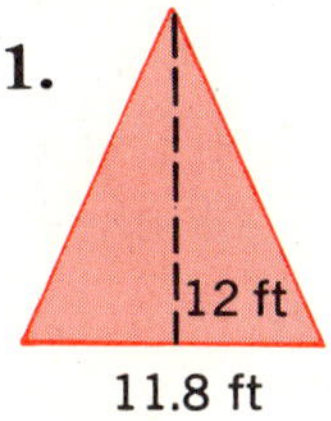

12.

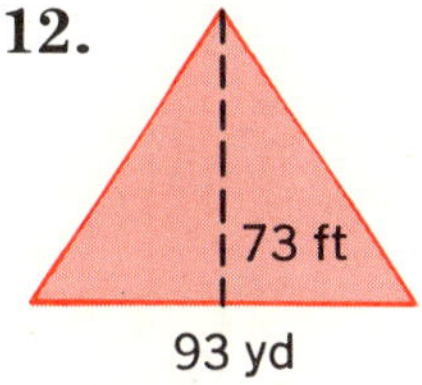

Solve these problems.

13. Roofing shingles are ordered by the *square*. A square covers 100 square feet. How many squares will be needed for the roof shown? (Don't forget the hidden sides.)

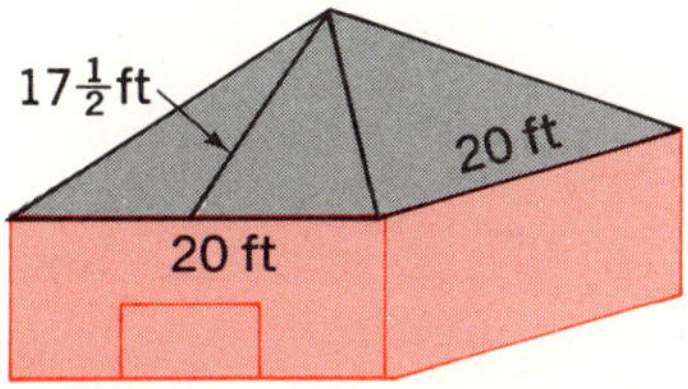

14. About 2 pounds of nails are needed for each square of roof surface. How many pounds of nails should be ordered for the roof in Exercise 13?

15. Janet is making triangular-shaped scarves for the school play. About how much material will she need for 28 scarves?

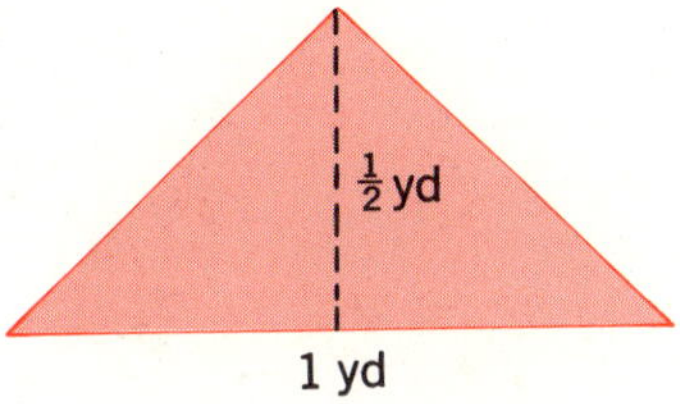

UNEXPLORED TERRITORY

Consider $\triangle ABC$, $\triangle RBC$, $\triangle PBC$, and $\triangle QBC$. How do their areas compare? Explain your answer.

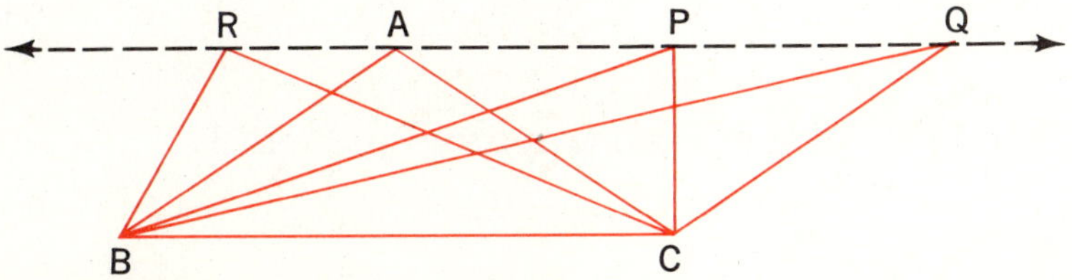

AREAS OF TRAPEZOIDS

The area of any trapezoid is half the product of the height and the sum of the lengths of the bases.

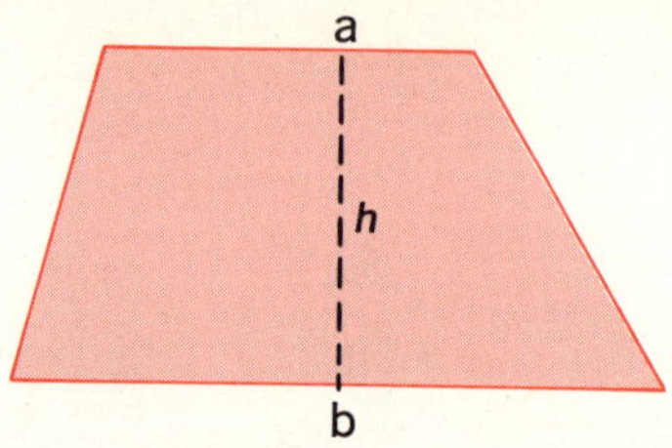

$$A = \frac{1}{2} \cdot h \cdot (a + b)$$

1. What is the height of each trapezoid?

a.

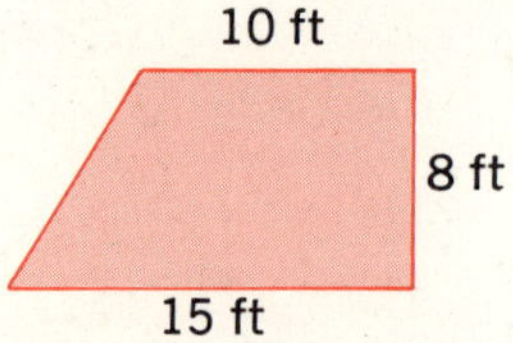

b.

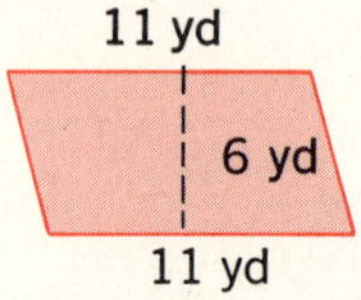

c.

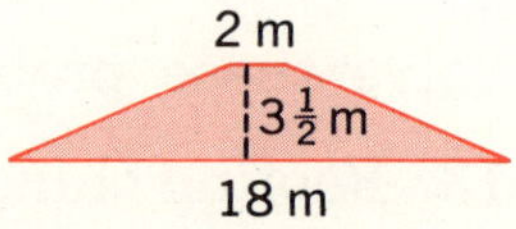

2. What are the lengths of the bases of each trapezoid above?

3. Find the area of each trapezoid.

Example

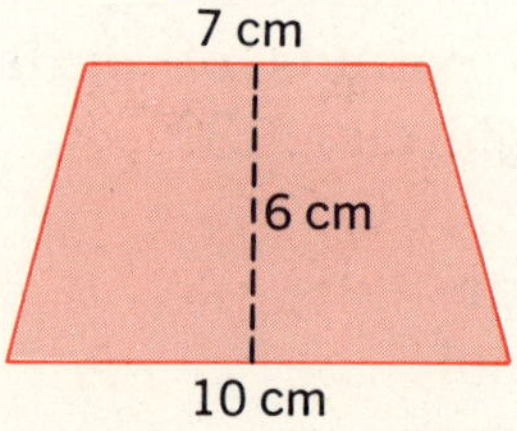

$$\begin{aligned} A &= \frac{1}{2} \cdot h \cdot (a + b) \\ &= \frac{1}{2} \cdot 6 \cdot (7 + 10) \\ &= \frac{1}{2} \cdot 6 \cdot 17 \\ &= 3 \cdot 17 \\ &= 51 \end{aligned}$$

The area is 51 cm^2.

a.

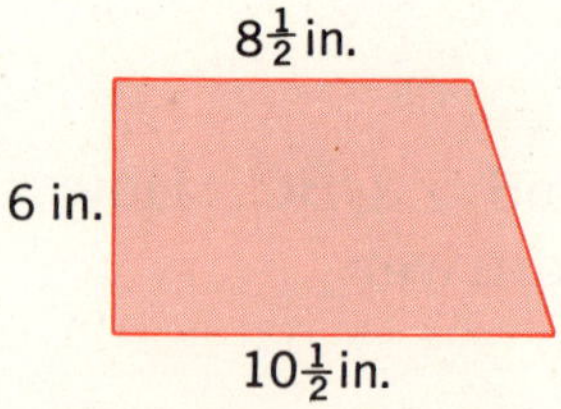

b.

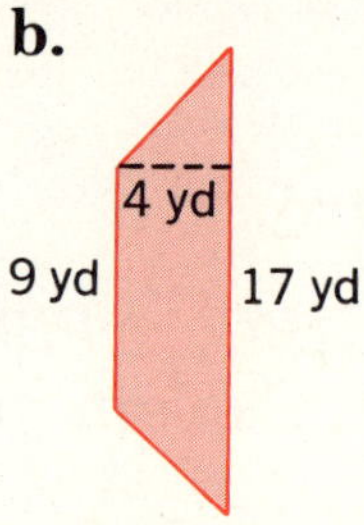

c.

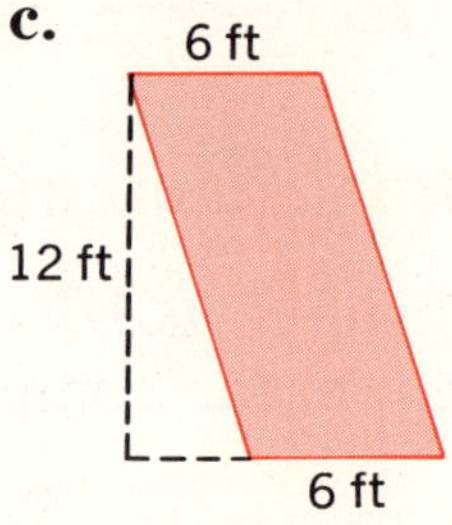

4. Find the area of each trapezoid in Item 1.

EXERCISES

Find the area of each trapezoid.

1. 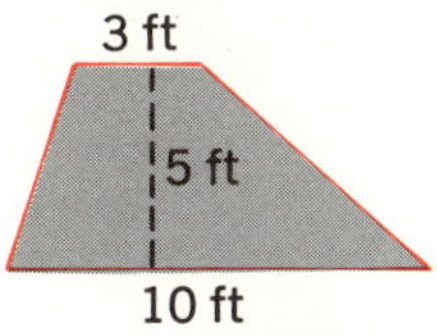

2. 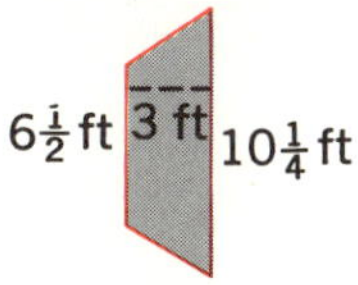

3. 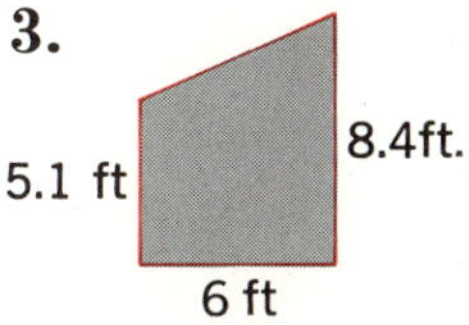

4. 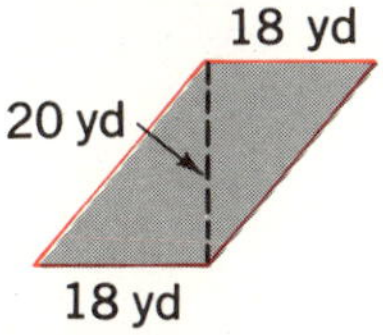

5. 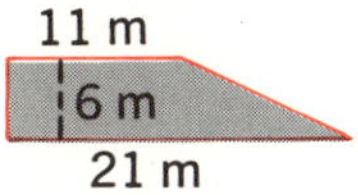

6. 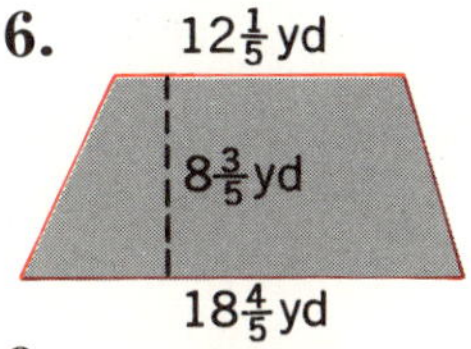

7. 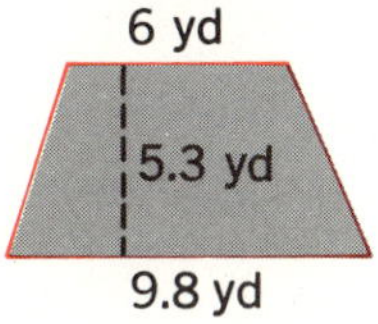

8. 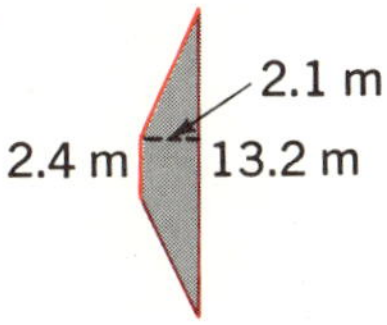

9. 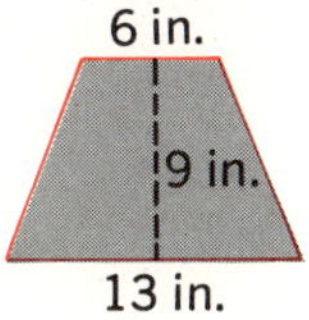

Solve these problems.

10. To make vent covers, Mac had to cut trapezoids out of sheet metal. The dimensions are shown on the right. About how many covers can be cut from a sheet of metal 48 in. by 36 in. Allow 100 sq. in. for waste.

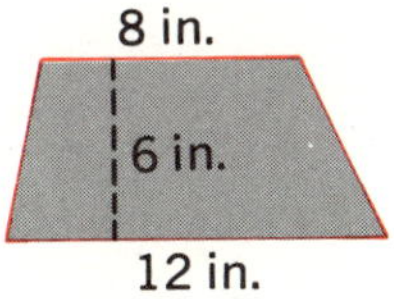

11. The school custodian has to put new tile flooring on a trapezoidal-shaped platform in the music room. The bases are 16 feet and 24 feet long, and the height is 14 feet. How many square feet of flooring will he need? Allow 10 square feet for waste.

UNEXPLORED TERRITORY

Every parallelogram is a trapezoid. Explain how the formula for finding the area of a trapezoid will also work for a parallelogram. (*Hint* Remember that the bases of a parallelogram are the same length.)

AREAS OF CIRCULAR REGIONS

The area of any circular region with radius r is the product of π and r^2.

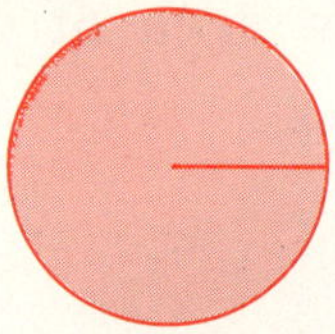

$$A = \pi \cdot r^2$$

1. Find the area of each circular region. Use $\frac{22}{7}$ for π.

a.

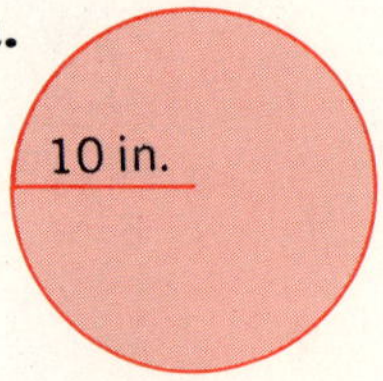

b.

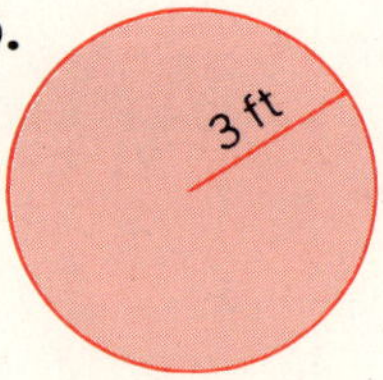

c.

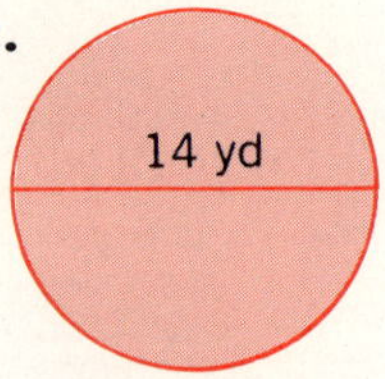

2. Find the area of each circular region. Use 3.14 for π.

a.

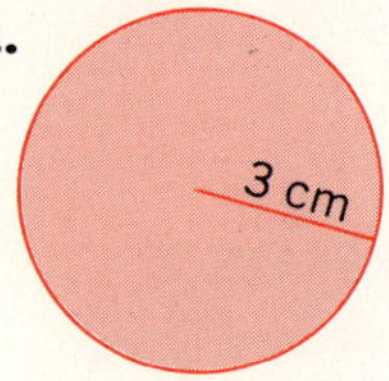

b.

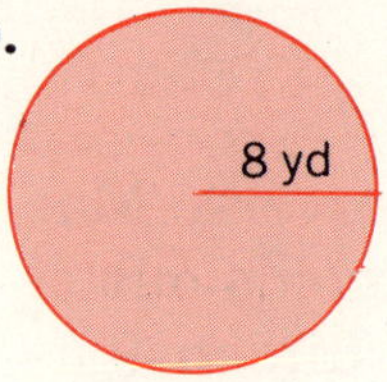

c.

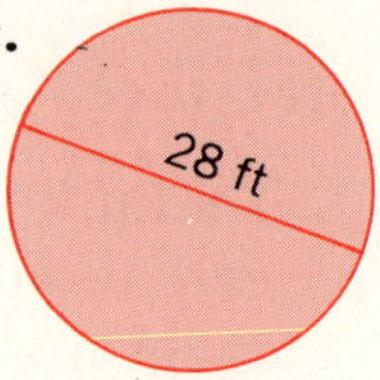

EXERCISES

Find the area of each circular region. Use $\frac{22}{7}$ for π.

1.

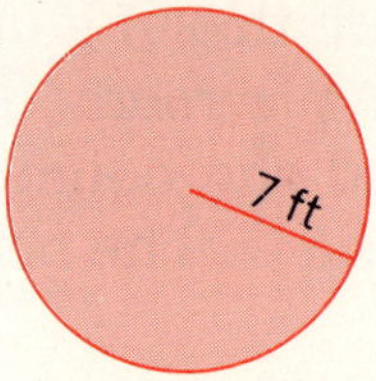

2.

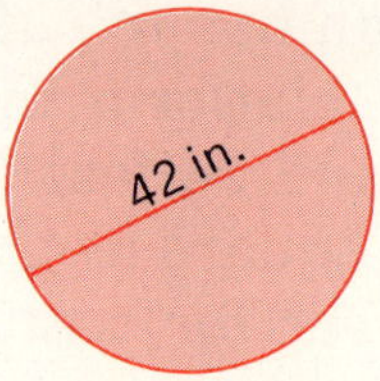

3.

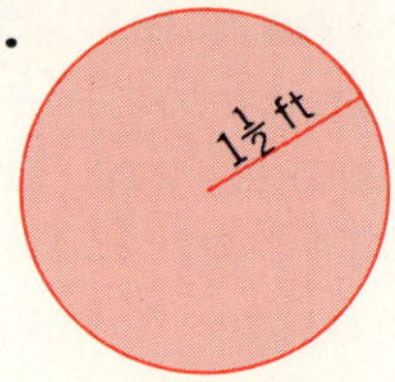

4.

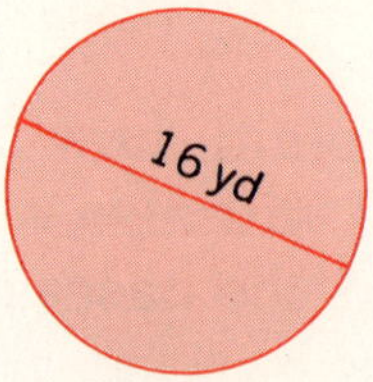

5.

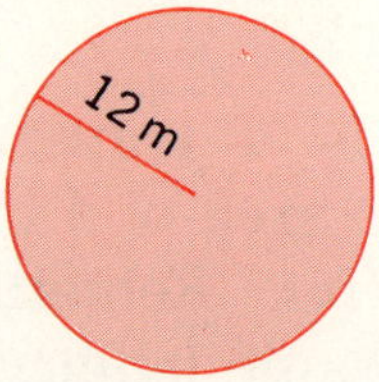

6.

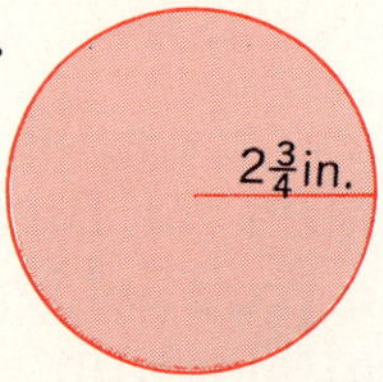

Find the area of each circular region. Use 3.14 for π.

7.

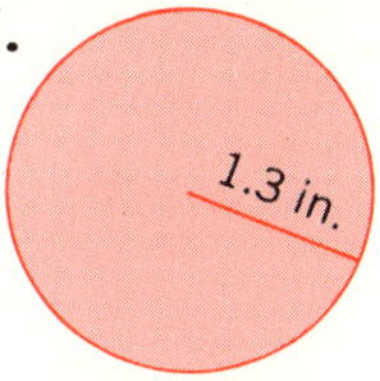

8.

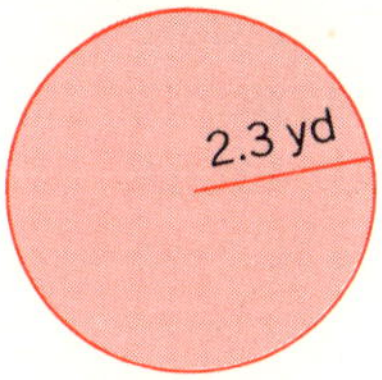

9.

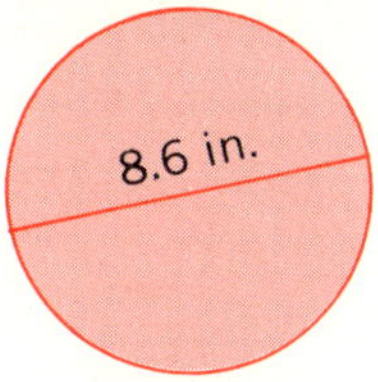

10.

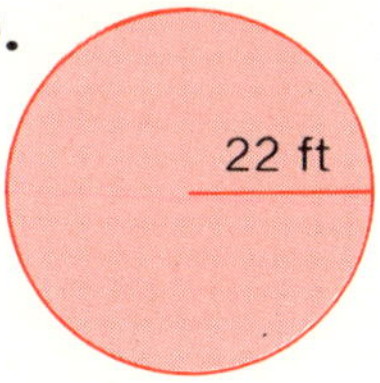

11.

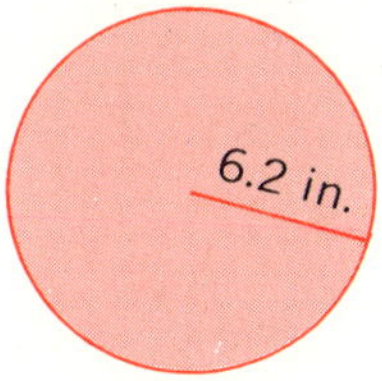

12.

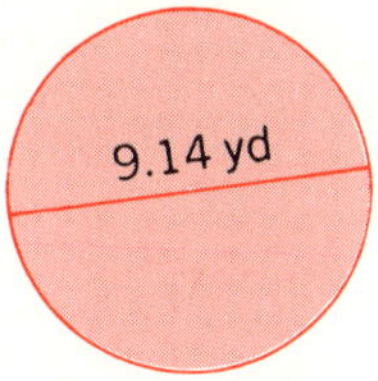

Find the area of each shaded region. Use 3.14 for π.

13.

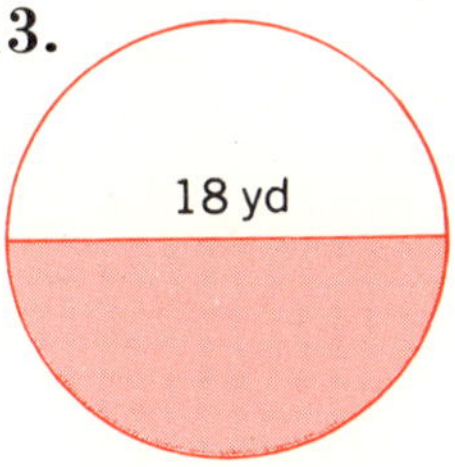

14.

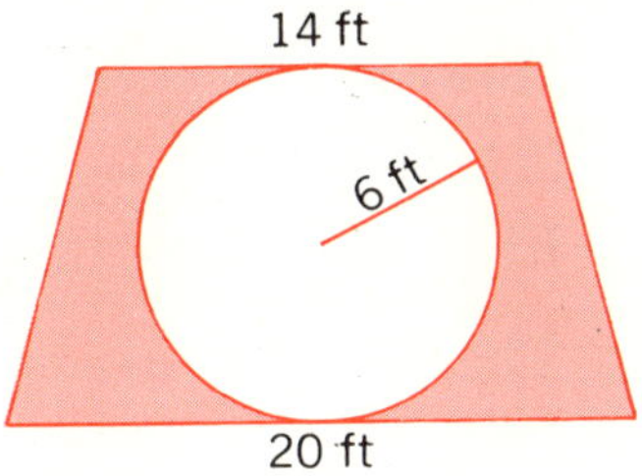

Solve these problems.

15. The radius of a circular island in a highway intersection is 80 feet. How much fertilizer will be needed for grass on the island at the rate of 10 pounds per 5,000 square feet?

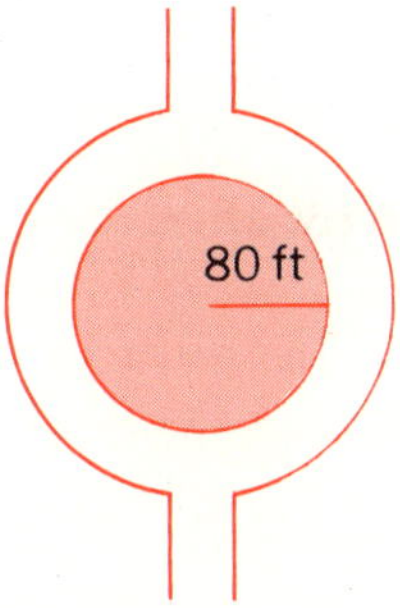

16. The diameter of a piston is 3.4 inches. Find the area of the top of the piston.

17. Mr. Marchione's lawn sprinkler covers a circular region 28 feet in diameter. How many square feet of lawn are covered at one time?

PRISMS AND SURFACE AREA

1. A *right prism* has two congruent parallel polygonal bases. All of its other faces, called *lateral faces*, are rectangles. Look at this picture of a prism.

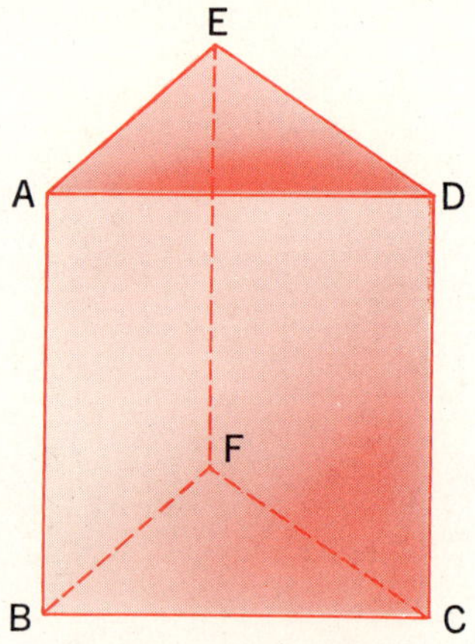

   a. *ADE* is a base of the prism. Name the other base.

   b. *ABCD* is a lateral face of the prism. Name the two other faces.

2. Here is a rectangular prism.

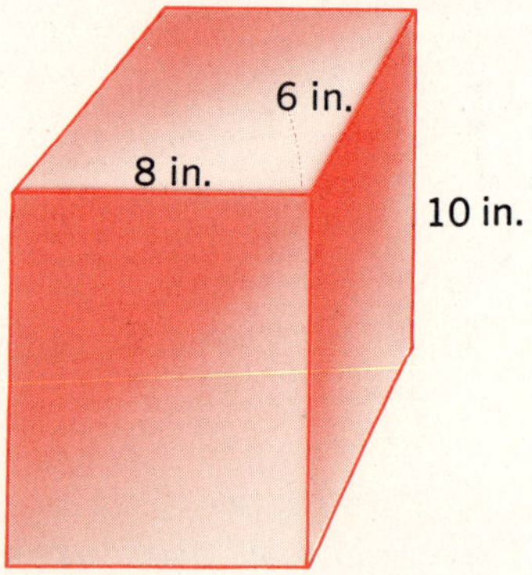

   a. How many lateral faces are there?

   b. Find the area of each.

   c. The *lateral area* is the sum of the areas of the lateral faces. Find the lateral area of the prism.

   d. How many bases are there?

   e. What is the area of each?

   f. The *surface area* of a prism is the sum of its lateral area and the areas of its bases. Find the surface area of the prism.

3. Find the lateral area and surface area of each.

a.

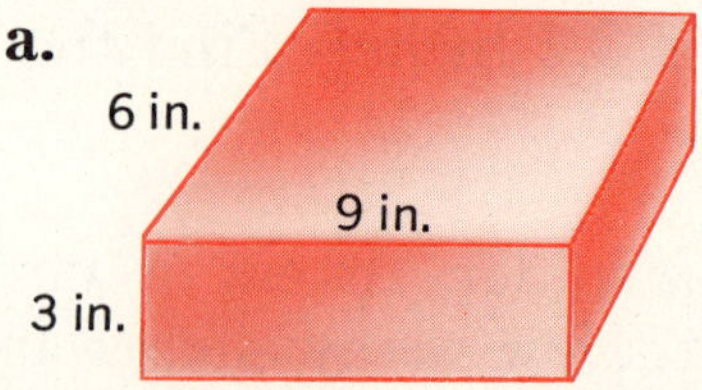

b.

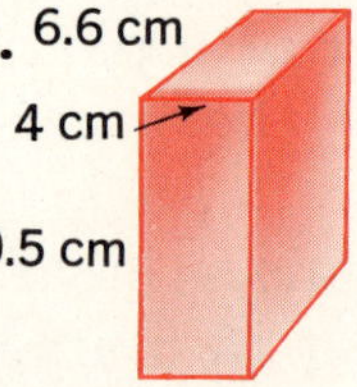

c.

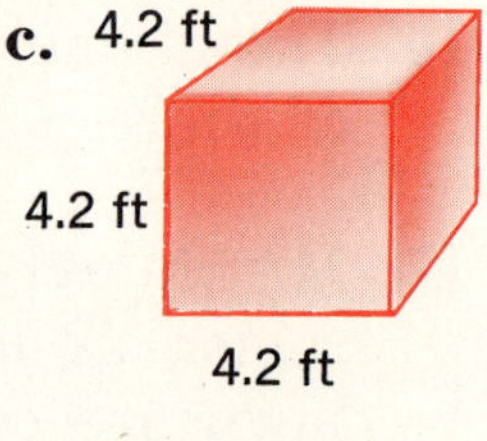

EXERCISES

Find the lateral area and surface area of each.

1. 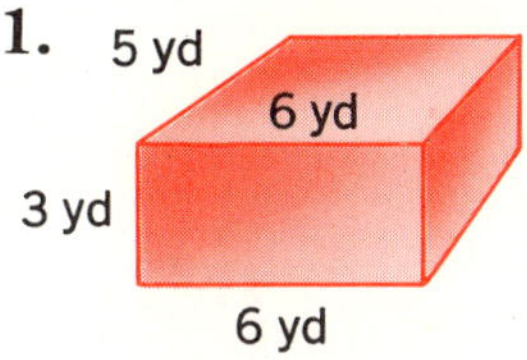

2. 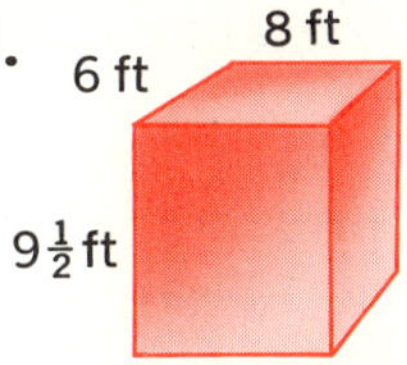

3. 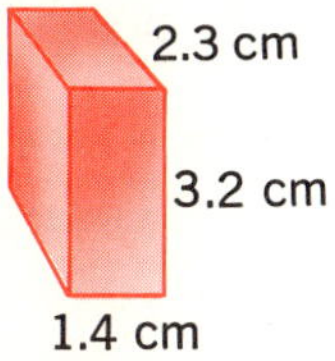

4. 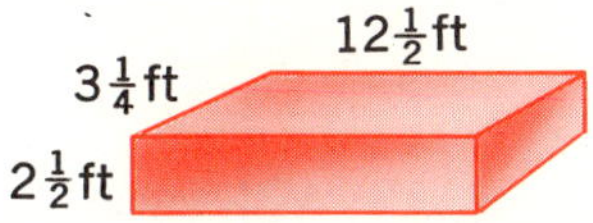

5. 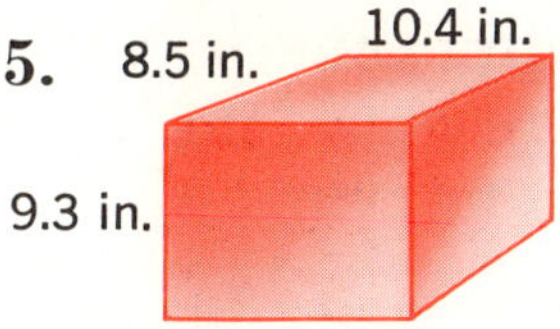

6. 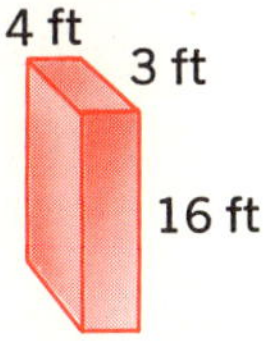

Solve these problems.

7. Write a formula for the surface area of a rectangular prism. Use *l*, *w*, and *h* for the lengths of the edges.

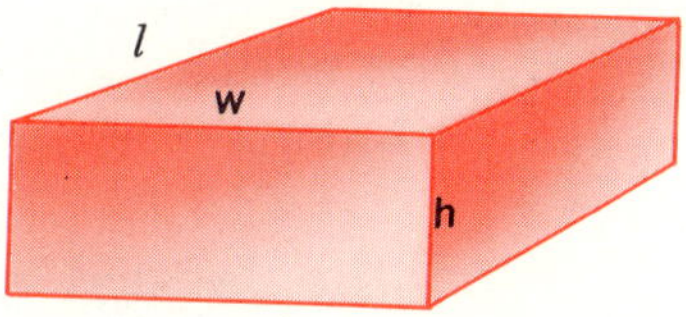

8. Find the surface area of a rectangular prism whose length is 10 in., width is 8 in., and height is 7 in.

UNEXPLORED TERRITORY

Mr. Spiropolis is estimating the cost of painting the walls and ceilings of a house. He needs the lateral area of the rooms plus the area of the ceilings. Find the total area. (Do not consider windows or doors.)

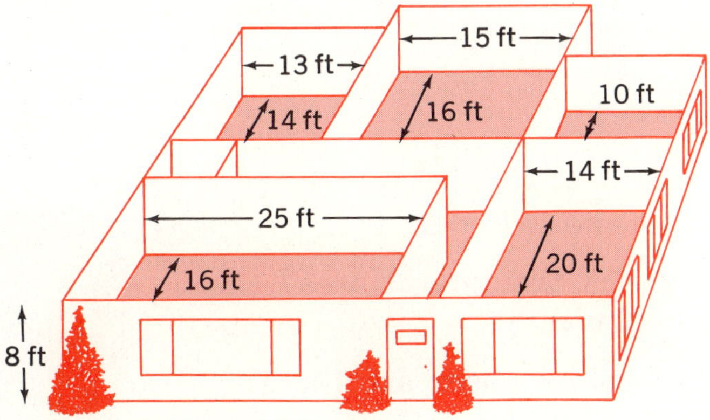

SURFACE AREAS OF CYLINDERS

1. A right circular cylinder also has two congruent bases. They are circles.

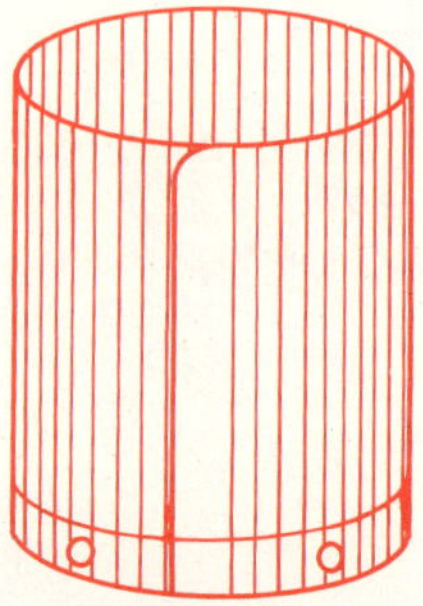

   a. Take a piece of notebook paper. Bring the top and bottom of it together so that the edges form a circle. You have formed a cylinder with your paper. The paper represents the *lateral surface* of the cylinder.

   b. Now lay the paper flat on your desk. What shape is the paper?

   c. How would you find the area of the paper?

   d. The width of the paper is the height of the cylinder. What part of the cylinder is the length of the paper?

2. Find the lateral area of each.

Example

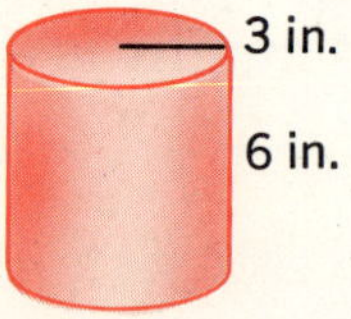

lateral area = circumference × height
$= (2 \cdot \pi \cdot r) \cdot h$
$\approx (2 \cdot 3.14 \cdot 3) \cdot 6$
$\approx 18.84 \cdot 6$
≈ 113.04

The lateral area is approximately 113.04 $in.^2$.

a. 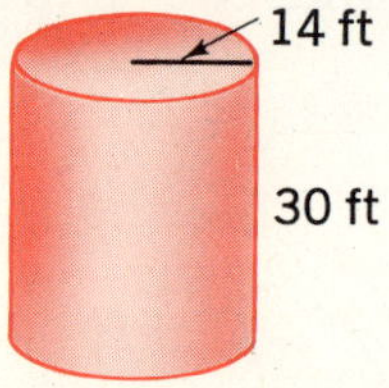

b. 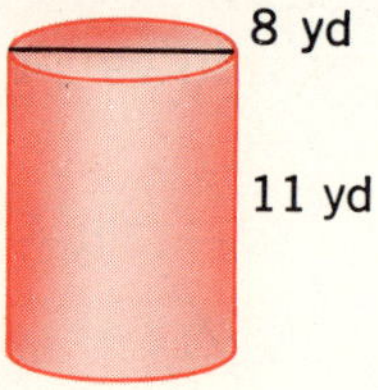

c. 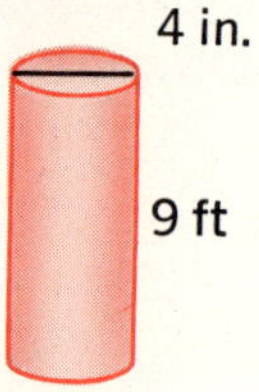

3. Look at the cylinder.

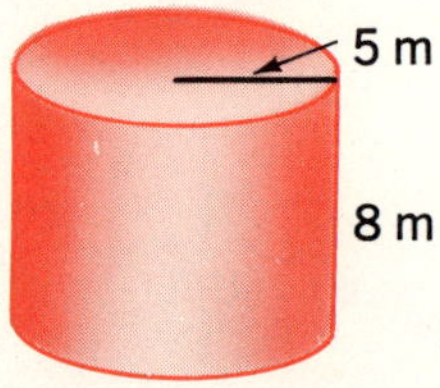

   a. What is the lateral area? Use 3.14 for π.

   b. How many bases are there?

c. What is the area of each base?

d. The *surface area* of a cylinder is the sum of its lateral area and the areas of its bases. Find the surface area of the cylinder.

4. Find the lateral area and surface area of each. Use 3.14 for π.

a. 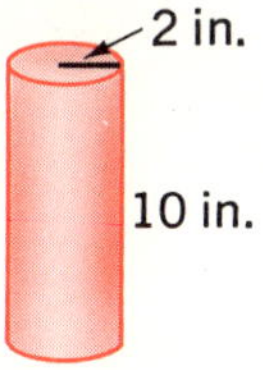

b. 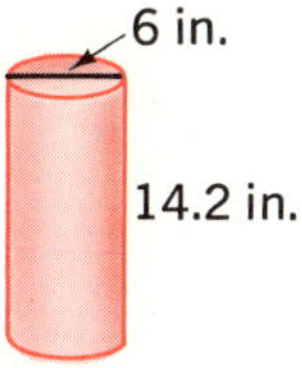

c. 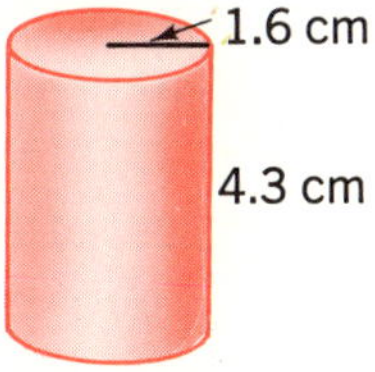

EXERCISES

Find the lateral area and surface area of each. Use 3.14 for π.

1. 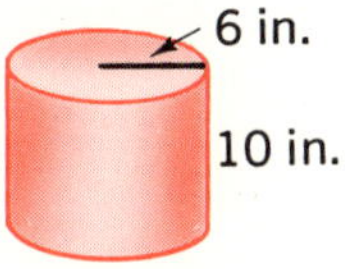

2. 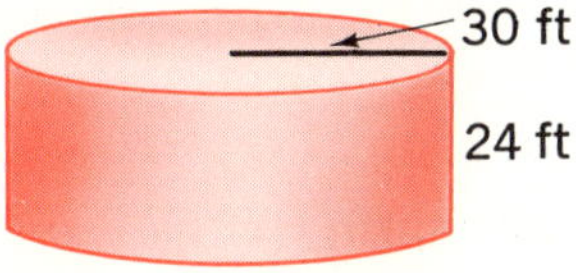

3. 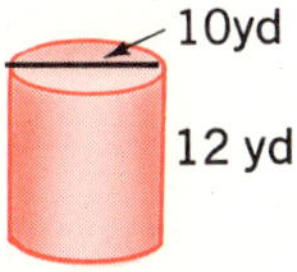

4. 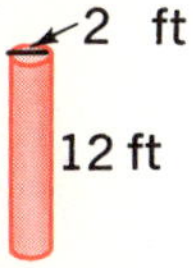

5. 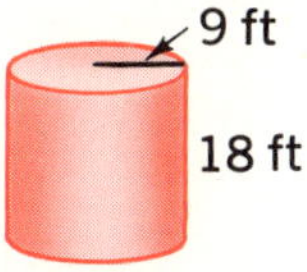

6. 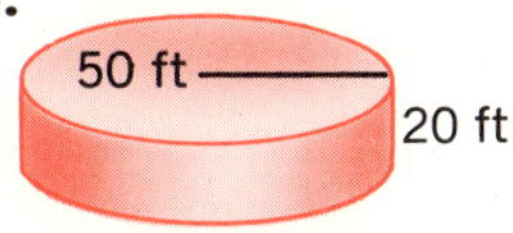

7. 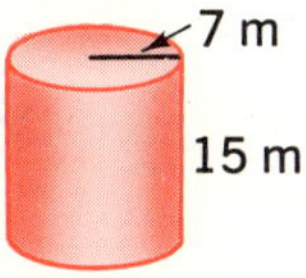

8. 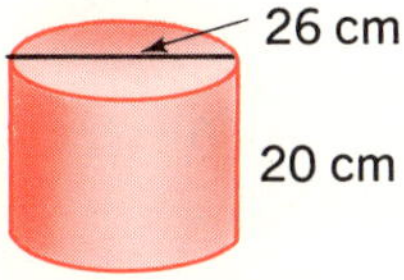

9. 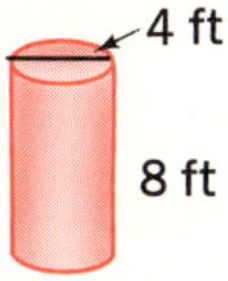

Solve these problems.

10. Mr. Roberts wanted to cover a cold water pipe with insulation. The pipe has a 2-inch radius and is 8 feet long. About how many square inches of insulation should he order?

11. How many square feet of asbestos are needed to cover a furnace pipe that is 9 feet long with a diameter of 1 foot?

VOLUMES OF PRISMS AND CYLINDERS

The measure of a three-dimensional figure is called its *volume.* It is the number of cubic units contained in the figure.

1. **a.** How many cubic inches does the box hold?

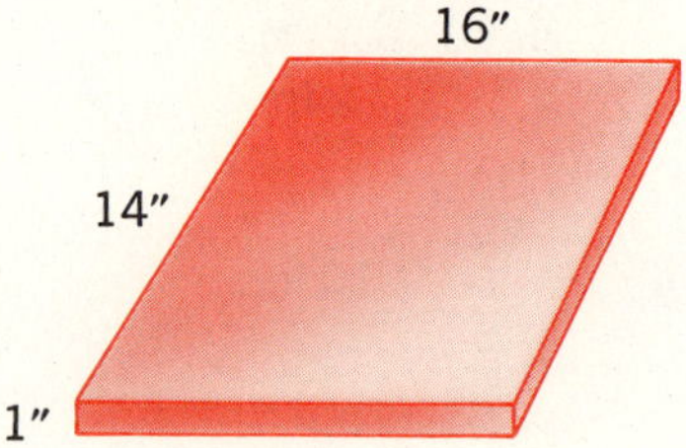

   b. Suppose the height of the box were 3 inches. What would be the volume of the box?

   c. What would be the volume of the box if the height were 5 inches?

> A formula for the volume of a rectangular prism is
> $$V = l \cdot w \cdot h$$

2. Find the volume of each.

   a.

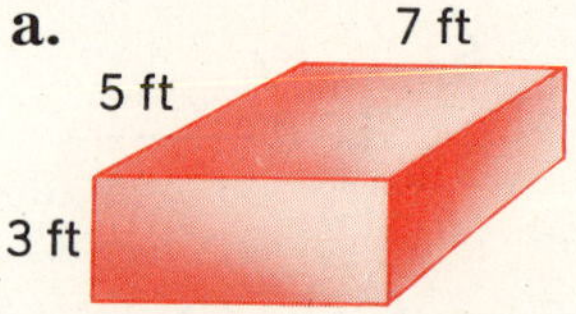

   b.

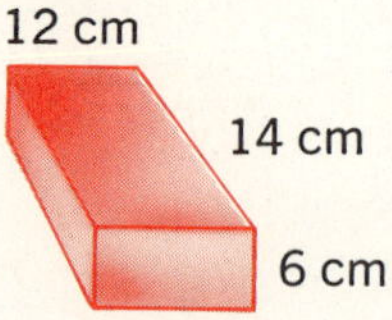

   c.

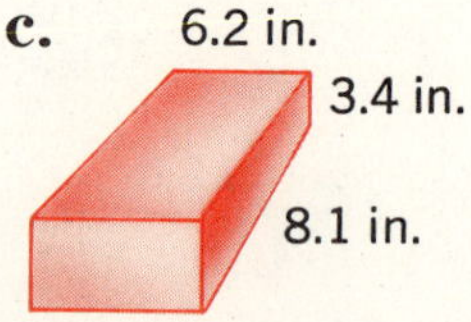

3. We can write "10 in.^{3}" to mean 10 cubic inches. Write another name for each.

   a. 25 in.^{3} **b.** 34 cubic cm

   c. 53 cubic ft **d.** 62 m^3

   e. 293 cubic yards **f.** 473.2 cm^3

4. We can find another formula for the volume of a prism.

   a. Write a formula for the area of a base with length l and width w.

   b. Replace "$l \cdot w$" with B in the volume formula, $V = l \cdot w \cdot h$.

5. We can use the formula $V = B \cdot h$ to find the volume of a cylinder.

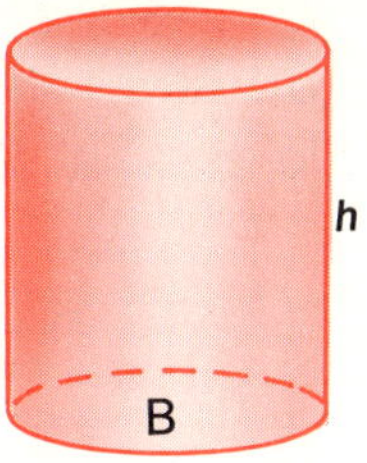

a. Find B if the radius of the base is 5 feet. Use 3.14 for π.

b. Find the volume if the height is 10 feet.

> A formula for the volume of a prism or a cylinder is
>
> $$V = B \cdot h$$
>
> B is the area of the base and h is the height.

6. Find the volume of each. Use 3.14 for π.

a. 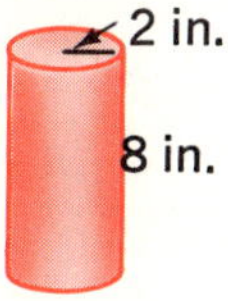

b. 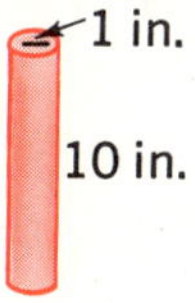

c. 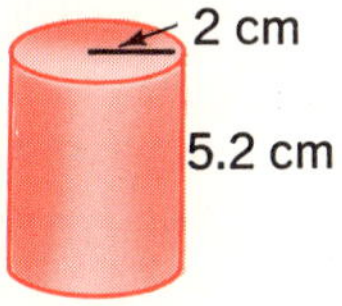

EXERCISES

Find the volume of each. Use 3.14 for π.

1. 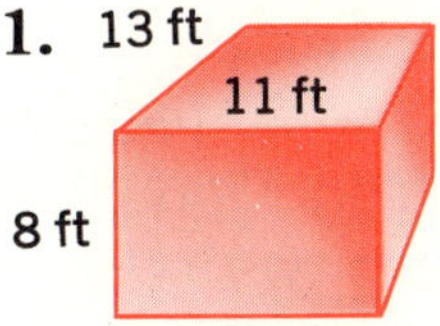

2. 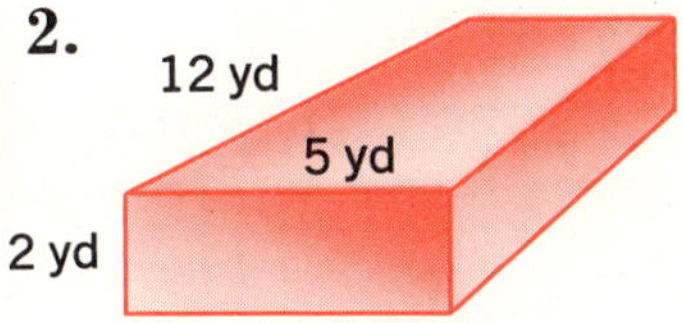

3. 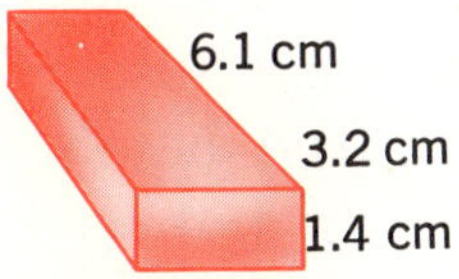

4. 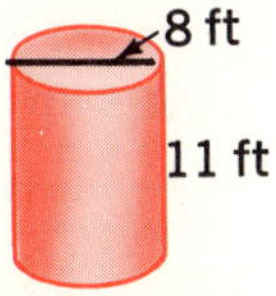

5. 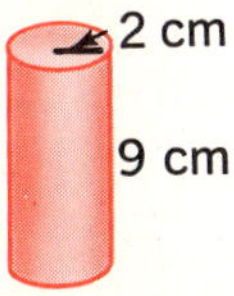

6. 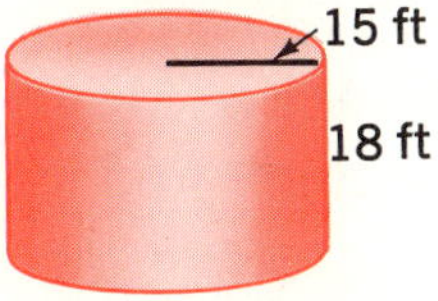

7. 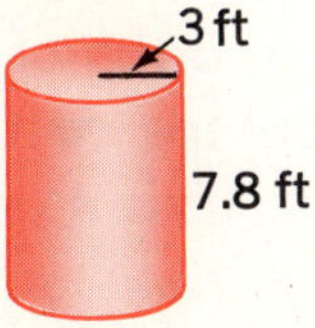

8. 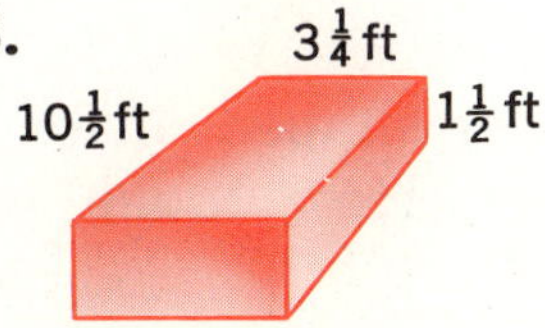

9. 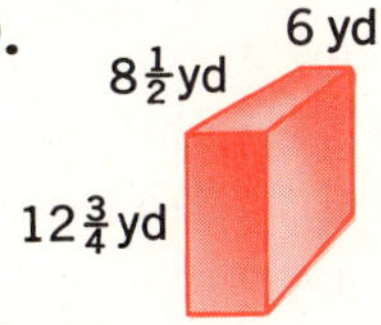

Solve these problems.

10. A 50-foot high petroleum tank has a 20-foot radius. How many cubic feet of petroleum will it hold?

11. Mr. Dunn is building a concrete slab patio. The slab is to be 8 inches thick, 24 feet long, and 12 feet wide. How many cubic yards of concrete should he order to make the slab.

12. The silo on Mr. Burn's farm is 38 feet high and 14 feet in diameter. How many cubic feet of ensilage will it hold?

13. The circular swimming pool in Kim's yard is 16 feet in diameter and 5 feet deep.

    a. About how many cubic feet of water will it hold?

    b. A cubic foot of water is about 7.5 gallons. How many gallons of water will the pool hold?

14. Don's ski house is in the shape of a cylinder. There are 3 floors. The diameter of each floor is 20 feet and the height of each of the walls is 8 feet. Don plans to put wooden panels on all the interior walls. How many square feet will he cover with paneling?

15. Atlas Moving Company's van is 30 feet long, 9 feet high and 8 feet wide. What is its volume?

16. In Mrs. Sherman's new home, the living room dimensions are 30 feet by 20 feet. In the middle of the room is a cylinder-shaped brick fireplace with a diameter of 4 feet. How much of the living room floor is open area?

DIMENSIONS IN SPORTS

Find the perimeter and area of each of the following playing fields.

1. BASEBALL DIAMOND

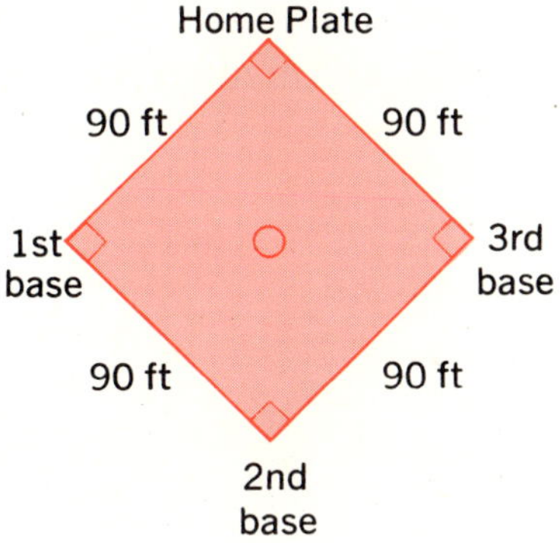

2. FOOTBALL FIELD

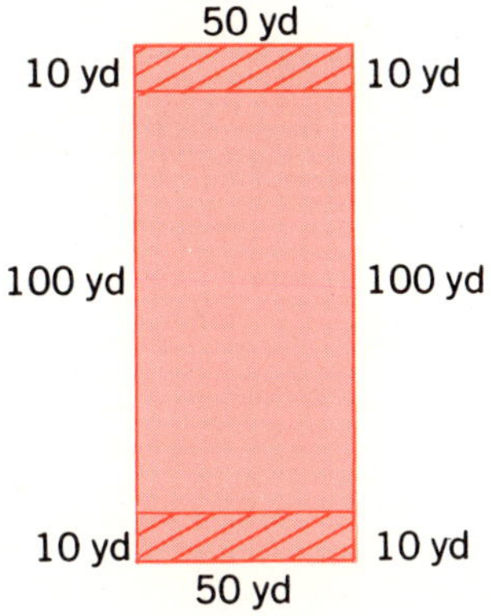

3. TENNIS COURT—DOUBLES

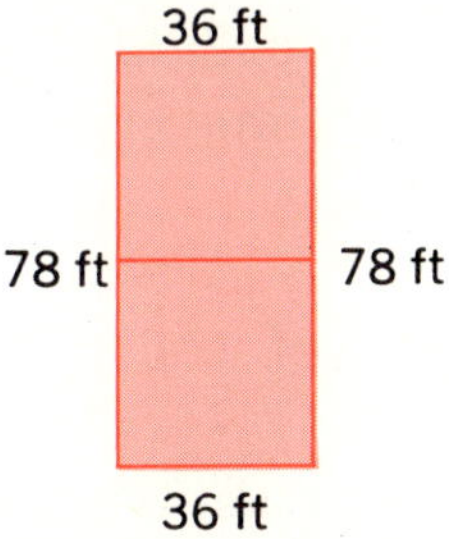

4. SOFTBALL DIAMOND

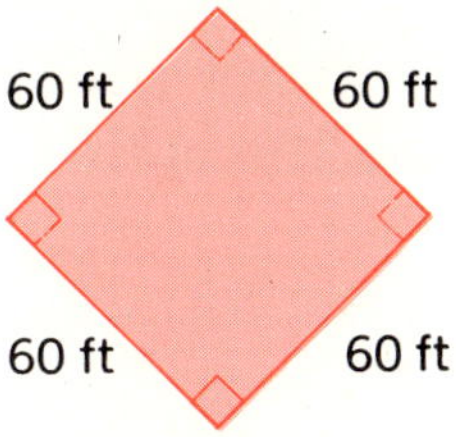

5. VOLLEYBALL COURT

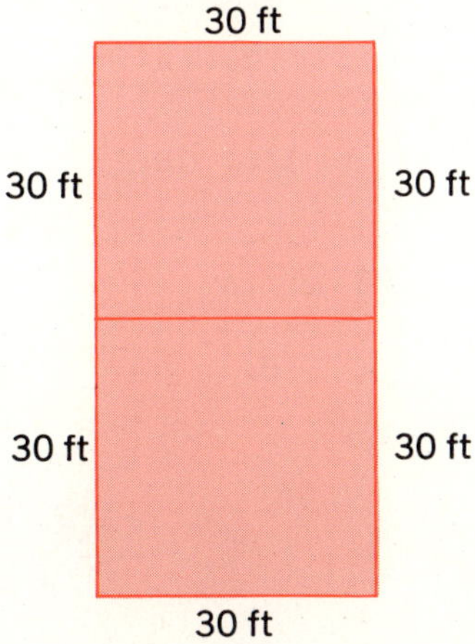

6. ICE HOCKEY RINK

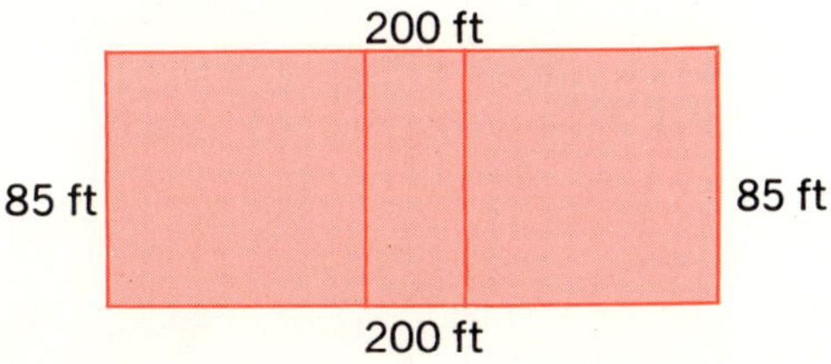

VOLUMES OF CONES

We will develop a formula for the volume of a cone experimentally.

LET'S EXPLORE

1. Make models of a cone and a cylinder out of cardboard. They must have congruent bases and the same height.

2. a. Fill the cone with sand.

   b. Pour this sand into the cylinder.

   c. Repeat this procedure until the cylinder is full. How many times did you fill the cone and transfer before the cylinder was full?

3. What is a formula for the volume of the cylinder?

4. What is a formula for the volume of the cone?

A formula for the volume of a cone is

$$V = \frac{1}{3} \cdot B \cdot h$$

B is the area of the base and h is the height.

5. Find the volume of each. Use 3.14 for π.

a.

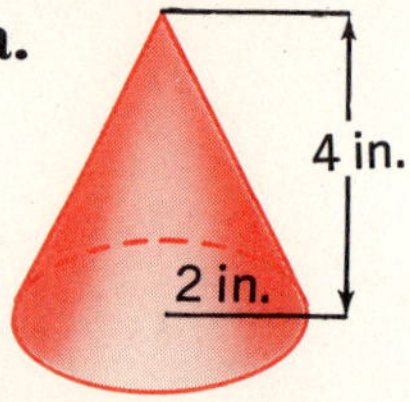

b.

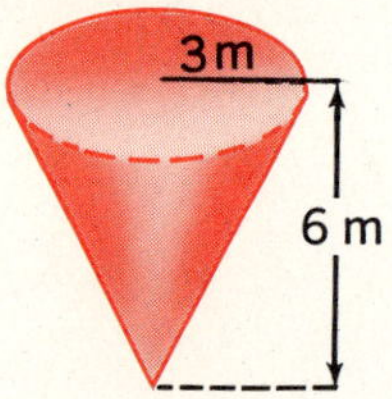

c.

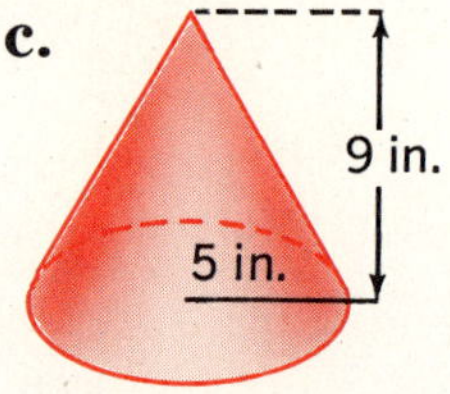

EXERCISES

Find the volume of each.

1.
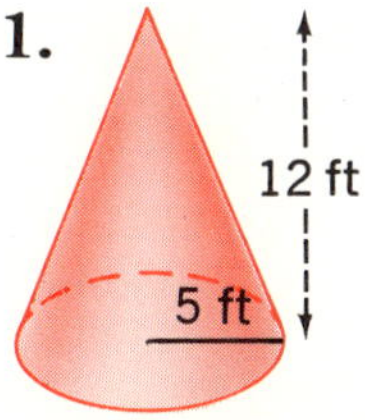

2.
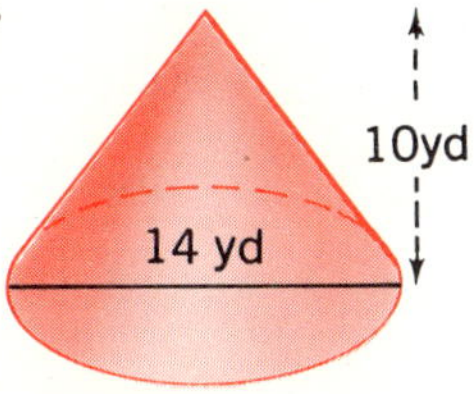

3.
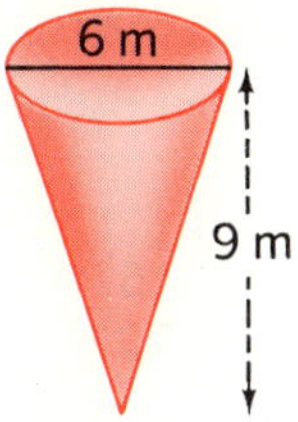

4.
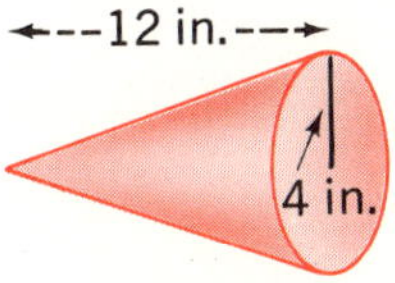

5.
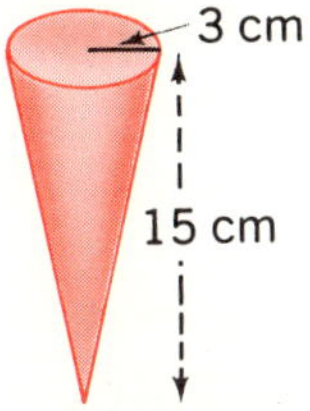

6.
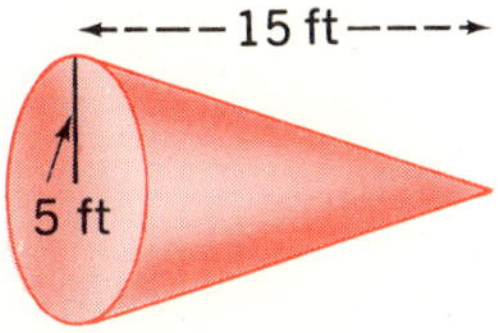

7.
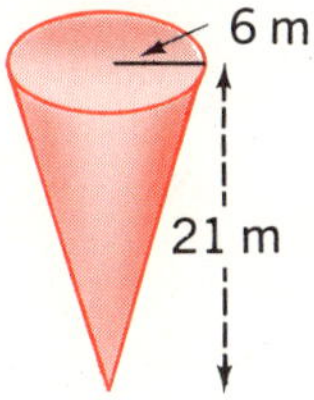

8.
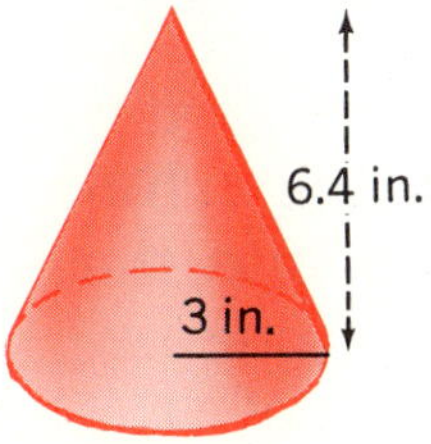

9.
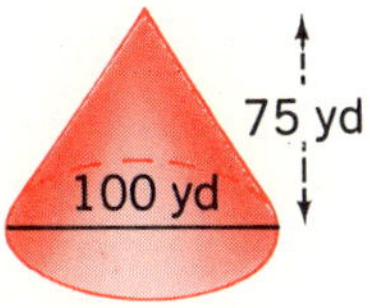

10. The radius of the base of a cone is 3 inches and its height is 9 inches. Find the volume.

11. For transplanting seedlings, Debbie made conical-shaped cups out of cardboard. The cones were 4 inches in diameter and 6 inches high. How many cubic inches of potting soil will each cone hold?

UNEXPLORED TERRITORY

1. What happens to the volume of a cone if you double its height?

2. What happens to the volume of a cone if you double its radius?

VOLUMES OF SPHERES

Nancy, Mike, and Fred did an experiment to find a formula for the volume of a sphere. They used a cone, a cylinder, and a sphere with these measures.

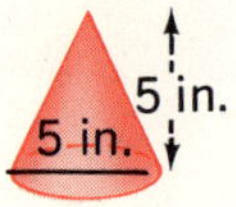

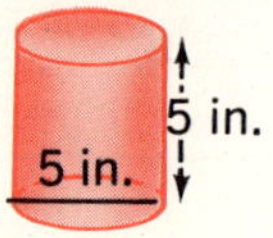

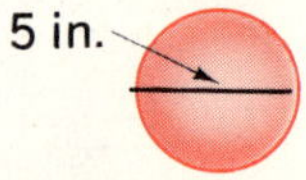

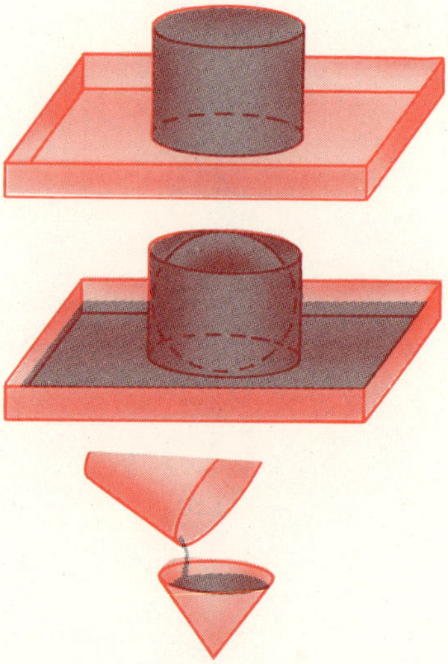

1. Nancy filled the cylinder with water and put it in a pan.

   Then she placed the sphere into the cylinder and let the water overflow.

   She then took the water that remained in the cylinder and poured it into the cone. It filled the cone exactly. What part of the original water was this?

2. We know that $\frac{1}{3}$ the volume of water in the cylinder did not overflow. How much did overflow?

3. a. A formula for the volume of the cylinder is $V = B \cdot h$. This means that a formula for the volume of a sphere is $V = \_\_\_\_ \cdot B \cdot h$.

   b. The B in the formula means the area of the base of the cylinder. Replace B in the formula with "$\pi \cdot r^2$."

   $$V = \frac{2}{3} \cdot \_\_\_\_ \cdot h$$

   c. The h is the height of the cylinder, which was the same as the diameter of the sphere. Replace h in the formula with "$2 \cdot r$."

   $$V = \frac{2}{3} \cdot \pi \cdot r^2 \cdot \_\_\_\_$$

d. Simplify this formula.

$$V = \frac{2}{3} \cdot \pi \cdot r^2 \cdot 2 \cdot r$$

A formula for the volume of a sphere is

$$V = \frac{4}{3} \cdot \pi \cdot r^3$$

r is the length of a radius.

4. Find the volume of each sphere. Use 3.14 for π.

Example

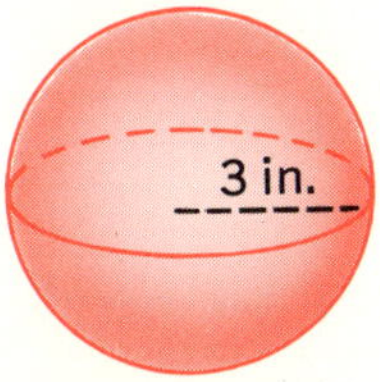

$$V = \frac{4}{3} \cdot \pi \cdot r^3$$
$$\approx \frac{4}{3} \cdot 3.14 \cdot 3^3$$
$$\approx \frac{4}{3} \cdot 3.14 \cdot 27$$
$$\approx 113.04$$

The volume is approximately 113.04 $in.^3$.

a.

b.

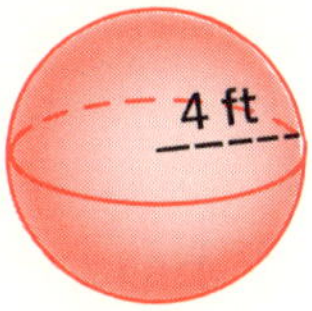

c.

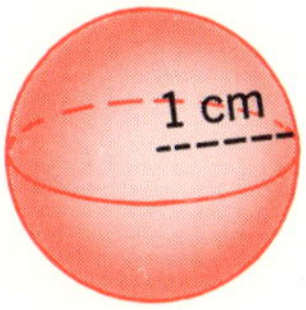

EXERCISES

Find the volume of each sphere. Use 3.14 for π.

1.

2.

3.

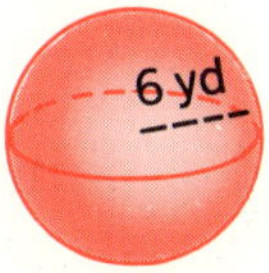

4.

5.

6.

Solve this problem.

7. The earth is a sphere with a radius of about 4,000 miles. What is its volume?

PROBLEM SOLVING

1. The inside radius of a pipe is 8 inches, while the outside radius is 9 inches. The pipe is 12 feet long. What volume of metal is in the pipe?

2. For a new fireplace, Mr. and Mrs. King ordered oversized bricks. The supplier told them they could estimate about 5 bricks per square foot. The fireplace is shaped as shown at the right.

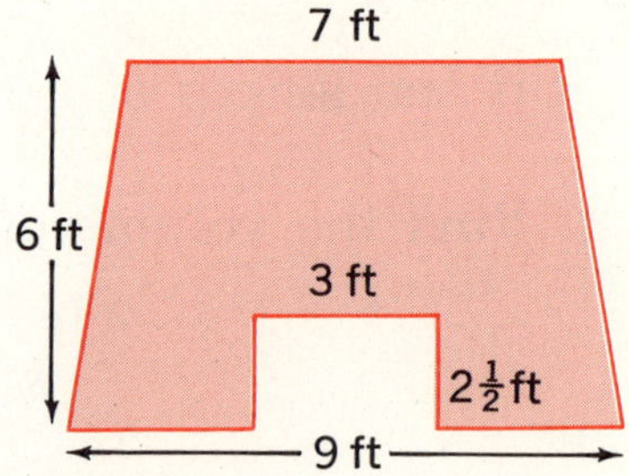

   a. About how many bricks will be needed?

   b. The bricks cost $82 per 1,000. How much can the Kings expect to spend?

3. Angie and Irv were hired to paint the outside and bottom of 18 trash cans at the city park. The cans are $3\frac{1}{4}$ feet high, and have bases with diameters of 2 feet.

   a. About how many square feet will they have to paint?

   b. A gallon of paint covers about 450 square feet. How many gallons will they need?

4. Mr. Jameson is building an enclosed water tank at his shop. The tank is in the shape of a cylinder 20 feet high and 20 feet in diameter.

   a. How many gallons will the tank hold? (Recall that 1 ft^3 holds about 7.5 gallons.)

   b. How many square feet of metal will be needed to construct it? Allow 50 ft^2 for waste.

5. When ordering hardwood flooring, it is necessary to order 34% more than the actual area. Find the number of square feet of flooring necessary for a room 28 feet long and 18 feet wide.

CHAPTER REVIEW

Which are true and which are false?

1. All squares are parallelograms.

2. All trapezoids are quadrilaterals.

3. All parallelograms are trapezoids.

4. All quadrilaterals are trapezoids.

Copy and complete.

5. 2 yd^2 = ____ ft^2

6. 2 m^2 = ____ cm^2

7. 36 ft^2 = ____ yd^2

8. 10 cm^2 = ____ mm^2

Find the perimeter and area of each.

9.
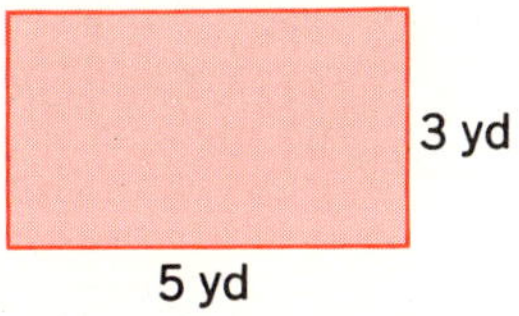

10.
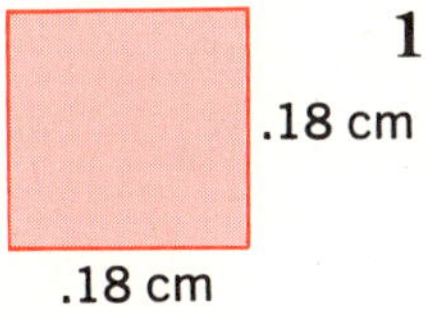

11.
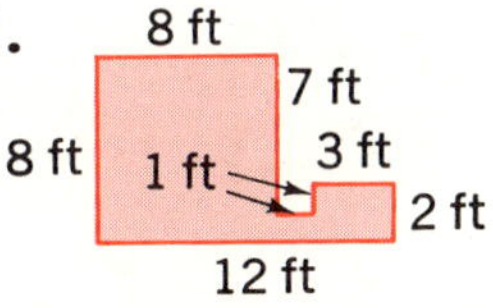

Find the area of each.

12.
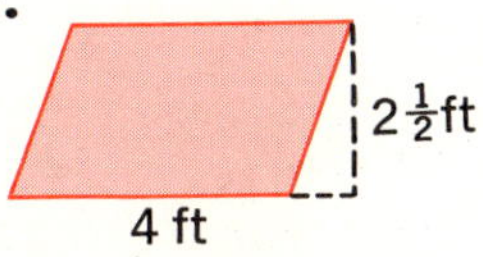

13.
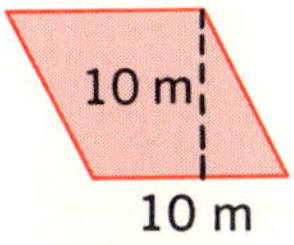

14.
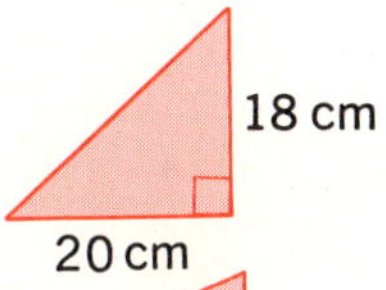

15.
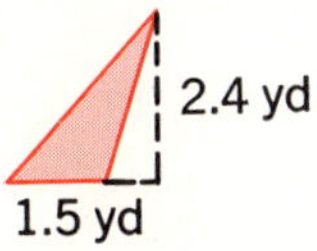

16.
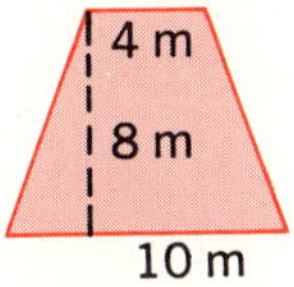

17.
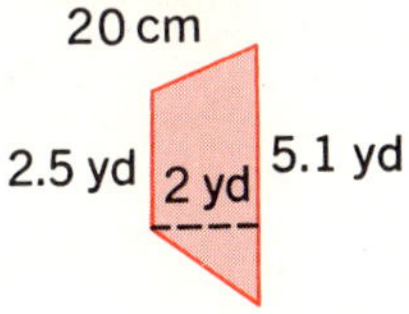

Find the circumference and area of each.

18.
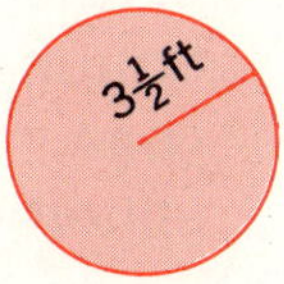

19.
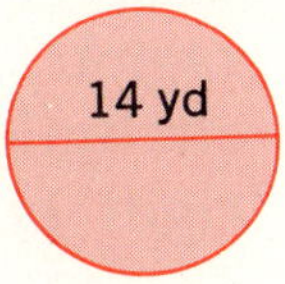

20.
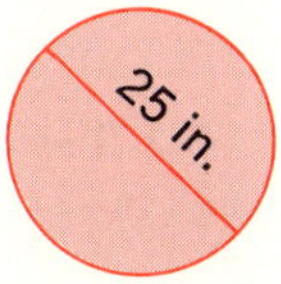

Find the lateral area and surface area of each.

21.

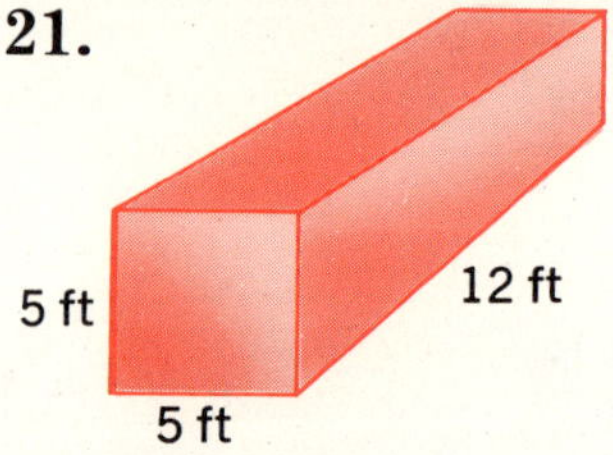

22.

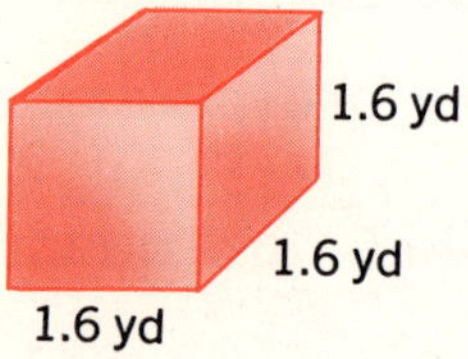

23.

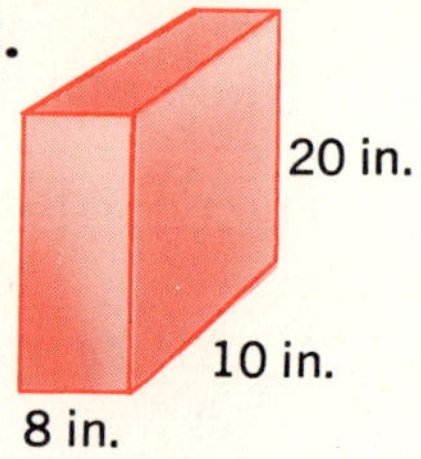

Find the lateral area and surface area of each. Use 3.14 for π.

24.

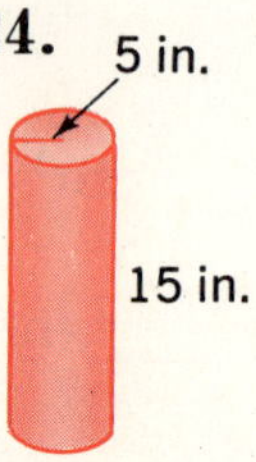

25.

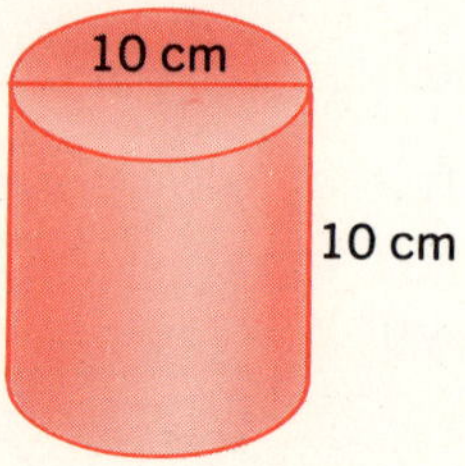

26.

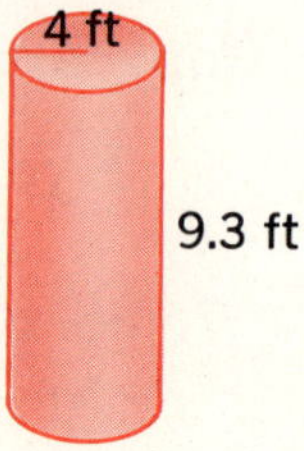

Find the volume of each.

27.

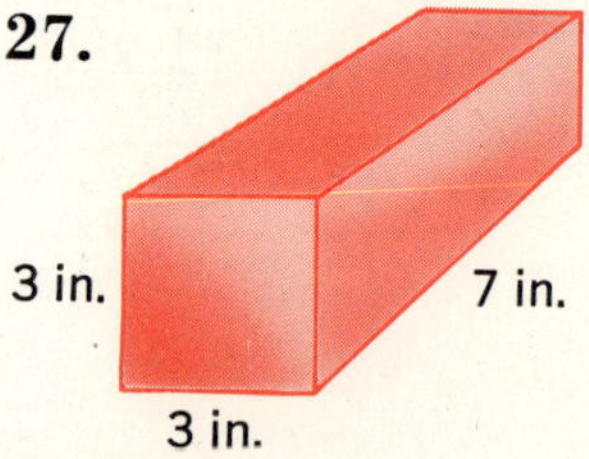

28.

29.

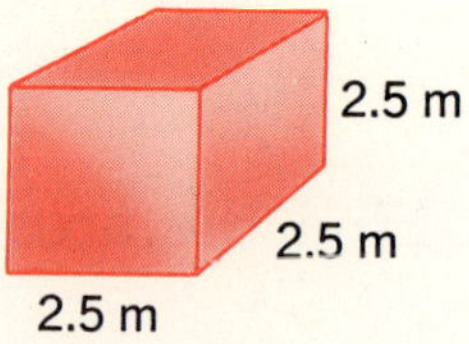

Solve this problem.

30. A cylindrical water tank is 25 feet high and 22 feet in diameter. What is its volume? Use 3.14 for π.

Terms You Should Know

quadrilateral	triangle	surface area
trapezoid	perimeter	lateral area
parallelogram	area	cylinder
rectangle	region	volume
rhombus	circumference	cone
square	right prism	sphere

CHAPTER TEST

Which are true and which are false?

1. All rectangles are trapezoids.

2. Some trapezoids are parallelograms.

3. All parallelograms are rhombi.

Copy and complete.

4. 288 in.^{2} = ____ ft^2

5. 10,000 mm^2 = ____ m^2

Find the perimeter and area of each.

6.

7.

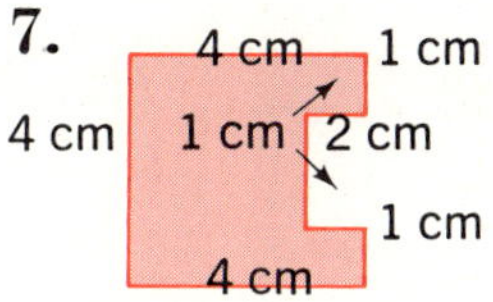

Find the area of each.

8.

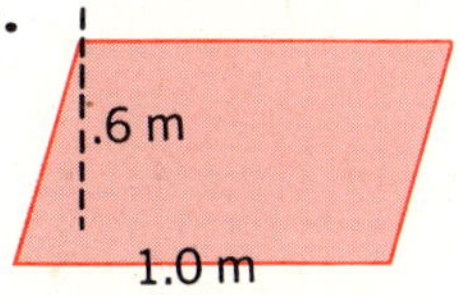

9.

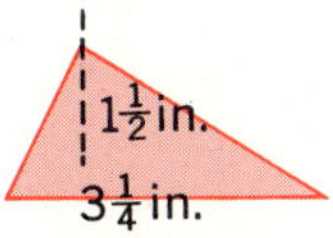

10.

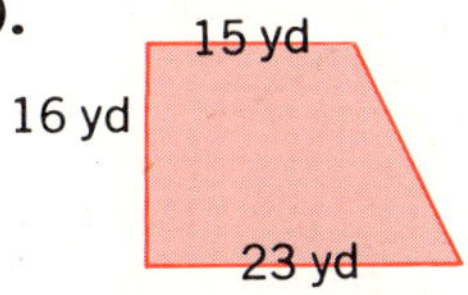

Find the circumference and area of each. Use 3.14 for π.

11.

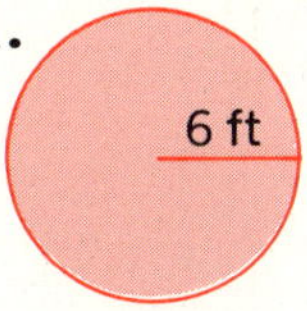

12.

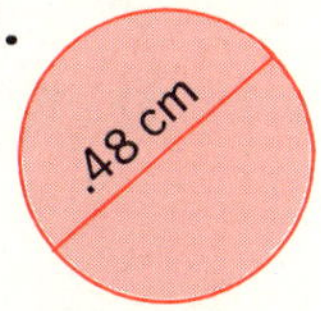

13.

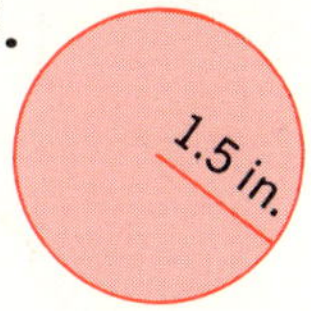

14. The bases of a rectangular prism are 20 feet by 25 feet and the height is 30 feet. Find the lateral area, surface area, and volume of the prism.

15. Find the surface area and volume of a cylinder 6 yards in diameter and 10 yards high. Use 3.14 for π.

CHAPTER NINE

Rational Numbers

RATIONAL NUMBERS

We have studied the systems of whole numbers, numbers of arithmetic, and integers.

> Any number which can be named by a fractional numeral with an integer for the numerator and a natural number for the denominator is a **rational number.**

1. Here are some rational numbers: $\frac{^-5}{7}$, $\frac{2}{3}$, $\frac{0}{3}$, 10, $^-43$. Give ten more examples of rational numbers.

2. The whole number 3 is a rational number because we can name it $\frac{3}{1}$. Show why these are rational numbers.

   a. 18 **b.** 0 **c.** 24 **d.** 103 **e.** 257

3. The integer $^-8$ is a rational number because we can name it $\frac{^-8}{1}$. Show why these are rational numbers.

   a. $^-24$ **b.** $^-34$ **c.** 56 **d.** $^-29$ **e.** 145

4. We read $\frac{^-9}{2}$ as "negative 9 over 2" or "negative nine-halves." Read these.

   a. $\frac{^-10}{7}$ **b.** $\frac{^-3}{14}$ **c.** $\frac{^-1}{5}$ **d.** $\frac{^-7}{3}$ **e.** $\frac{^-25}{8}$

5. The number .24 is a rational number because we can name it $\frac{24}{100}$. Show why these are rational numbers.

   a. .56 **b.** $^-.2$ **c.** 2.5 **d.** .008 **e.** $^-1.58$

EXERCISES

1. Give five examples of rational numbers.

Show why each is a rational number.

2. $^-82$ **3.** 19 **4.** $^-56$ **5.** .18 **6.** .003

ORDER AND THE NUMBER LINE

Let's see how we can make a number line for rational numbers.

LET'S EXPLORE

1. Draw a number line for integers. Use a unit about $\frac{1}{2}$ inch long.

2. On your number line mark and label points for these numbers.

   a. $\frac{1}{2}$ **b.** $\frac{3}{4}$ **c.** $1\frac{1}{2}$ **d.** 3.3 **e.** 4.1

3. Let's put the point for $\frac{^-1}{2}$ halfway between the points for 0 and $^-1$. Mark and label points for $\frac{^-1}{2}$ and $\frac{^-3}{4}$.

4. Mark and label points for these rational numbers.

   a. $\frac{^-1}{4}$ **b.** $\frac{^-3}{2}$ **c.** $\frac{^-9}{4}$ **d.** $^-2.5$ **e.** $\frac{2}{3}$

5. We say that one rational number is greater than another if it is named farther to the right on the number line. Which are true and which are false?

   a. $^-5 < {}^-2$ **b.** $\frac{^-3}{2} > \frac{^-1}{4}$ **c.** $\frac{^-3}{5} < \frac{2}{5}$ **d.** $\frac{1}{4} < \frac{3}{2}$

6. All positive rational numbers are greater than zero. What is true of all negative rational numbers?

EXERCISES

Draw a number line for rational numbers. Mark and label each of these on the number line.

1. $\frac{^-5}{2}$ **2.** $\frac{4}{5}$ **3.** $\frac{^-3}{5}$ **4.** $\frac{^-2}{3}$ **5.** $^-2.5$

6. $1\frac{1}{4}$ **7.** $^-3.5$ **8.** $^-4\frac{1}{2}$ **9.** $3\frac{3}{5}$ **10.** $\frac{^-1}{5}$

Which are true and which are false?

11. ${}^-\frac{2}{3} < 0$ 12. ${}^-\frac{3}{4} > \frac{1}{2}$ 13. ${}^-\frac{7}{8} > {}^-\frac{1}{2}$ 14. ${}^-10 < {}^-5$

15. $\frac{5}{6} < 1\frac{1}{2}$ 16. ${}^-\frac{1}{3} > {}^-\frac{5}{2}$ 17. ${}^-\frac{8}{12} > \frac{5}{6}$ 18. $0 < {}^-\frac{3}{7}$

19. ${}^-\frac{2}{5} > {}^-\frac{1}{8}$ 20. ${}^-8 < {}^-2$ 21. ${}^-\frac{1}{9} > {}^-\frac{3}{4}$ 22. ${}^-\frac{5}{6} < 0$

Make these true. Use > or <.

23. ${}^-\frac{1}{2} \,\underline{\quad}\, 0$ 24. ${}^-\frac{3}{4} \,\underline{\quad}\, {}^-\frac{9}{2}$ 25. $1\frac{5}{6} \,\underline{\quad}\, 1\frac{1}{2}$ 26. ${}^-\frac{3}{7} \,\underline{\quad}\, {}^-\frac{10}{3}$

27. ${}^-2.4 \,\underline{\quad}\, {}^-1.8$ 28. ${}^-\frac{7}{9} \,\underline{\quad}\, {}^-\frac{7}{3}$ 29. $\frac{5}{6} \,\underline{\quad}\, {}^-\frac{2}{3}$ 30. ${}^-1\frac{3}{4} \,\underline{\quad}\, {}^-2\frac{1}{8}$

31. ${}^-1\frac{7}{8} \,\underline{\quad}\, {}^-3\frac{1}{10}$ 32. ${}^-\frac{1}{8} \,\underline{\quad}\, {}^-\frac{2}{3}$ 33. ${}^-5.6 \,\underline{\quad}\, 0$ 34. ${}^-\frac{5}{8} \,\underline{\quad}\, {}^-1\frac{1}{3}$

35. ${}^-3.7 \,\underline{\quad}\, {}^-5.7$ 36. ${}^-\frac{6}{5} \,\underline{\quad}\, {}^-\frac{9}{2}$ 37. ${}^-\frac{2}{5} \,\underline{\quad}\, {}^-\frac{8}{3}$ 38. $\frac{7}{8} \,\underline{\quad}\, \frac{3}{4}$

Let's Practice

Solve these problems.

1. Mr. Ryan's car averages 24.6 miles per gallon. About how far can he go on 17.6 gallons?

2. Paige bought a pair of ice skates on sale. The original price was $15.95 and the discount was 20%. The sales tax was 3%. Find the total cost.

3. Nathan received a 25% commission for selling greeting cards. He sold 6 boxes at $3.00 each, 10 boxes at $2.00 each and 12 boxes at $1.50 each. How much money did he make?

4. Mr. Fisher bought 12 pounds of grass seed at 27¢ per pound, a new hose for $4.39, and a bag of fertilizer for $2.79. The sales tax was 4%. Find the total cost.

INVERSES AND ABSOLUTE VALUE

Every rational number has an inverse. The inverse of $\frac{^-5}{6}$ is $\frac{5}{6}$. The inverse of $\frac{3}{4}$ is $\frac{^-3}{4}$.

1. Name the inverse of each.

   a. $\frac{^-8}{13}$ b. $\frac{9}{10}$ c. $\frac{^-5}{12}$ d. $\frac{^-6}{11}$ e. $\frac{^-10}{23}$

2. Remember that the absolute value of a number tells how far it is from 0 on the number line.

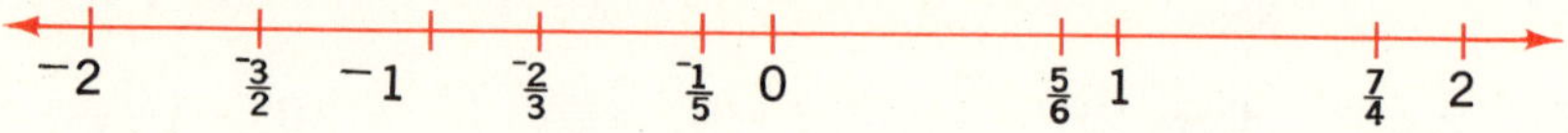

   We say $\left|\frac{^-3}{2}\right| = \frac{3}{2}$ since $\frac{^-3}{2}$ is $\frac{3}{2}$ units from 0. Write the simplest name for each.

   a. $\left|\frac{^-2}{3}\right|$ b. $\left|\frac{5}{6}\right|$ c. $\left|\frac{^-1}{5}\right|$ d. $\left|\frac{7}{4}\right|$ e. $\left|\frac{^-9}{8}\right|$

Here is a flow chart for finding the absolute value of a rational number.

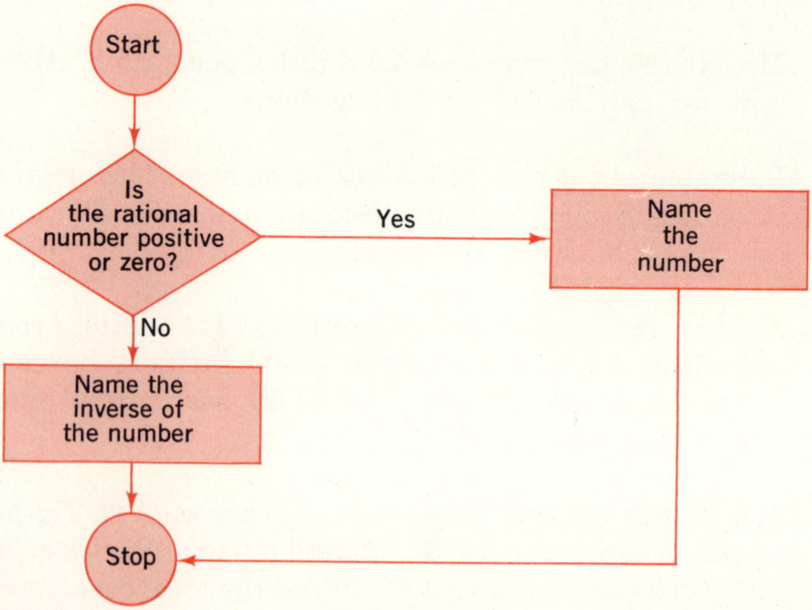

3. Find the absolute value of each number. Use the flow chart.

a. $\frac{^-3}{5}$ b. $\frac{^-8}{10}$ c. 2.7 d. $\frac{^-37}{4}$ e. $\frac{4}{9}$

4. Which are true and which are false?

a. $|^-8| < |16|$ b. $\left|\frac{^-3}{4}\right| > \left|\frac{^-1}{2}\right|$ c. $|0| < \left|\frac{^-2}{5}\right|$

d. $\left|\frac{1}{6}\right| > \left|\frac{^-9}{2}\right|$ e. $\left|\frac{^-1}{10}\right| < |0|$ f. $\left|\frac{^-4}{5}\right| > \left|\frac{^-1}{8}\right|$

EXERCISES

Name the inverse of each.

1. $\frac{^-5}{7}$ 2. $\frac{8}{9}$ 3. $\frac{14}{15}$ 4. $^-3.5$ 5. $\frac{^-8}{9}$

6. $\frac{^-6}{11}$ 7. $\frac{^-4}{20}$ 8. $\frac{^-10}{3}$ 9. 5.6 10. $\frac{9}{16}$

Write the simplest name for each.

11. $\left|\frac{1}{6}\right|$ 12. $\left|\frac{^-5}{3}\right|$ 13. $\left|\frac{^-8}{9}\right|$ 14. $|^-4.9|$ 15. $\left|\frac{^-7}{10}\right|$

16. $\left|\frac{^-10}{13}\right|$ 17. $|0|$ 18. $\left|\frac{^-7}{10}\right|$ 19. $\left|\frac{15}{32}\right|$ 20. $\left|\frac{3}{7}\right|$

21. $|3.7|$ 22. $\left|\frac{^-2}{9}\right|$ 23. $\left|\frac{7}{8}\right|$ 24. $\left|\frac{^-10}{13}\right|$ 25. $^-3.2$

Which are true and which are false?

26. $\left|\frac{^-4}{6}\right| > |^-3|$ 27. $|0| < \left|\frac{^-4}{7}\right|$ 28. $\left|\frac{^-3}{4}\right| > \left|\frac{1}{2}\right|$

29. $\left|\frac{5}{6}\right| > \left|\frac{^-2}{3}\right|$ 30. $|5.6| > \left|\frac{^-3}{8}\right|$ 31. $|^-18| > |3|$

32. $|0| > \left|\frac{^-2}{3}\right|$ 33. $\left|\frac{2}{7}\right| < |^-9|$ 34. $\left|\frac{^-3}{7}\right| > \left|\frac{^-1}{2}\right|$

35. $\left|\frac{1}{5}\right| > |^-2|$ 36. $\left|\frac{^-3}{8}\right| < \left|\frac{^-3}{4}\right|$ 37. $|5.6| > |^-8.3|$

DO YOU REMEMBER?

Compute and simplify.

1. $\frac{3}{4} + \frac{1}{2}$
2. $\frac{3}{7} + \frac{2}{5}$
3. $\frac{1}{6} + \frac{3}{7}$
4. $\frac{5}{6} + \frac{7}{12}$
5. $\frac{3}{7} + \frac{2}{3}$
6. $\frac{7}{8} + \frac{3}{4}$
7. $\frac{9}{10} + \frac{1}{3}$
8. $\frac{7}{10} + \frac{4}{5}$
9. $\frac{13}{14} + \frac{3}{4}$
10. $\frac{9}{10} - \frac{1}{10}$
11. $\frac{3}{5} - \frac{1}{10}$
12. $\frac{20}{21} - \frac{2}{7}$
13. $\frac{4}{5} - \frac{1}{2}$
14. $\frac{5}{6} - \frac{1}{9}$
15. $\frac{12}{25} - \frac{2}{5}$
16. $\frac{7}{8} - \frac{2}{3}$
17. $\frac{6}{7} - \frac{2}{3}$
18. $\frac{6}{11} - \frac{1}{3}$
19. $\frac{1}{3} \times \frac{6}{8}$
20. $\frac{5}{9} \times \frac{1}{3}$
21. $\frac{2}{5} \times \frac{10}{27}$
22. $\frac{8}{21} \times \frac{14}{25}$
23. $\frac{5}{8} \times \frac{4}{5}$
24. $\frac{3}{10} \times \frac{2}{3}$
25. $8 \times \frac{7}{8}$
26. $10 \times \frac{3}{5}$
27. $11 \times \frac{3}{8}$
28. $\frac{1}{2} \div \frac{3}{4}$
29. $\frac{4}{9} \div \frac{2}{3}$
30. $\frac{1}{2} \div \frac{5}{8}$
31. $6 \div \frac{1}{2}$
32. $\frac{4}{5} \div \frac{8}{15}$
33. $\frac{3}{4} \div 6$
34. $\frac{5}{6} \div \frac{2}{7}$
35. $\frac{1}{9} \div 10$
36. $\frac{4}{7} \div 9$
37. $32 + 14$
38. $^-25 + {}^-61$
39. $^-25 + 82$
40. $48 - 52$
41. $^-65 - 88$
42. $^-72 - {}^-87$
43. $^-14 \times 36$
44. $^-15 \times {}^-29$
45. $^-150 \times {}^-17$
46. $24 \div {}^-6$
47. $^-210 \div 70$
48. $^-450 \div {}^-15$

MULTIPLICATION

Here is the way Frank multiplied $\frac{^-2}{3}$ and $\frac{4}{5}$.

$$\frac{^-2}{3} \times \frac{4}{5} = \frac{^-2 \times 4}{3 \times 5} = \frac{^-8}{15}$$

1. Copy and complete.

   a. $\frac{^-4}{7} \times \frac{6}{11} = \frac{^-4 \times 6}{7 \times 11} = \_\_\_\_$

   b. $\frac{^-5}{6} \cdot \frac{^-8}{9} = \frac{^-5 \cdot {}^-8}{6 \cdot 9} = \frac{^-5 \cdot {}^-4 \cdot 2}{9 \cdot 3 \cdot 2} = \_\_\_\_$

To multiply rational numbers like $\frac{^-7}{8}$ and $\frac{^-6}{7}$, we multiply the numerators and the denominators.

2. Multiply and simplify.

   a. $\frac{^-4}{7} \times \frac{2}{3}$ b. $\frac{^-5}{8} \cdot \frac{^-6}{10}$ c. $\frac{^-10}{13} \times \frac{^-3}{8}$

   d. $\frac{^-4}{5} \times \frac{8}{9}$ e. $\frac{0}{17} \cdot \frac{^-5}{6}$ f. $\frac{^-3}{2} \cdot \frac{^-1}{5} \cdot \frac{^-3}{4}$

EXERCISES

Multiply and simplify.

1. $\frac{^-5}{8} \times \frac{^-6}{7}$
2. $\frac{^-6}{11} \times \frac{10}{17}$
3. $\frac{7}{16} \cdot \frac{^-2}{3}$
4. $\frac{^-8}{9} \times \frac{^-7}{16}$
5. $\frac{0}{1} \cdot \frac{^-3}{2}$
6. $\frac{^-2}{3} \cdot \frac{5}{3}$
7. $\frac{^-1}{2} \cdot \frac{^-3}{4}$
8. $\frac{^-3}{2} \cdot \frac{^-1}{5}$
9. $\frac{^-2}{5} \times \frac{^-3}{16}$
10. $\frac{^-4}{3} \cdot \frac{0}{3}$
11. $\frac{^-7}{3} \times \frac{^-5}{2}$
12. $\frac{^-4}{5} \cdot \frac{5}{6}$
13. $\frac{1}{2} \cdot \frac{^-3}{4} \cdot \frac{2}{3}$
14. $\frac{5}{7} \cdot \frac{^-3}{5} \cdot \frac{^-1}{3}$
15. $\frac{^-4}{3} \cdot \frac{0}{3} \cdot \frac{5}{4}$
16. $\frac{^-2}{1} \cdot \frac{^-3}{7} \cdot \frac{^-1}{4}$
17. $\frac{2}{3} \cdot \frac{^-3}{5} \cdot \frac{^-1}{4}$
18. $\frac{5}{6} \cdot \frac{^-2}{3} \cdot \frac{^-1}{10}$

RECIPROCALS

All positive rational numbers have reciprocals. Zero has no reciprocal. Do negative rational numbers have reciprocals?

LET'S EXPLORE

1. Recall that the reciprocal of $\frac{3}{4}$ is $\frac{4}{3}$. Name the reciprocal of each.

   a. $\frac{5}{6}$ b. $\frac{10}{9}$ c. $\frac{10}{11}$ d. $\frac{5}{17}$ e. $\frac{52}{19}$

2. Multiply and simplify.

   a. $\frac{5}{6} \cdot \frac{6}{5}$ b. $\frac{2}{3} \cdot \frac{3}{2}$ c. $\frac{1}{8} \cdot \frac{8}{1}$ d. $\frac{9}{11} \cdot \frac{11}{9}$

3. What is the product of a number and its reciprocal?

4. Multiply and simplify.

   a. $\frac{^-5}{1} \cdot \frac{^-1}{5}$ b. $\frac{^-3}{4} \cdot \frac{^-4}{3}$ c. $\frac{^-7}{5} \cdot \frac{^-5}{7}$ d. $\frac{^-3}{8} \cdot \frac{^-8}{3}$

The reciprocal of a number like $\frac{^-8}{9}$ is $\frac{^-9}{8}$.

5. Name the reciprocal of each.

   a. $\frac{^-5}{7}$ b. $\frac{^-10}{31}$ c. $\frac{^-8}{17}$ d. $\frac{^-7}{11}$ e. $\frac{^-92}{93}$

EXERCISES

Name the reciprocal of each.

1. $\frac{1}{2}$ 2. $\frac{3}{4}$ 3. $\frac{^-1}{3}$ 4. $\frac{^-3}{7}$ 5. 0

6. $\frac{^-9}{4}$ 7. 2 8. $\frac{^-3}{1}$ 9. $\frac{^-8}{5}$ 10. $\frac{^-5}{1}$

11. $\frac{^-7}{12}$ 12. $\frac{^-42}{47}$ 13. $\frac{^-8}{5}$ 14. $\frac{2}{9}$ 15. $\frac{^-8}{17}$

DIVISION

To divide one rational number by another, we can multiply by the reciprocal of the divisor.

1. Let's divide $\frac{3}{4}$ by $\frac{^-2}{3}$.

   a. What is the reciprocal of $\frac{^-2}{3}$?

   b. Multiply: $\frac{3}{4} \cdot \frac{^-3}{2}$.

   c. What is $\frac{3}{4} \div \frac{^-2}{3}$?

2. Let's divide $\frac{^-4}{5}$ by $\frac{^-7}{8}$.

   a. What is the reciprocal of $\frac{^-7}{8}$?

   b. Multiply: $\frac{^-4}{5} \cdot \frac{^-8}{7}$.

   c. What is $\frac{^-4}{5} \div \frac{^-7}{8}$?

3. Divide and simplify.

   a. $\frac{1}{6} \div \frac{^-1}{4}$ b. $\frac{2}{7} \div \frac{5}{6}$ c. $\frac{^-1}{9} \div \frac{^-3}{4}$ d. $\frac{^-12}{1} \div \frac{5}{8}$

EXERCISES

Divide and simplify.

1. $\frac{4}{5} \div \frac{^-3}{7}$
2. $\frac{^-3}{4} \div \frac{^-5}{7}$
3. $\frac{^-16}{9} \div \frac{5}{4}$
4. $\frac{4}{5} \div \frac{^-7}{12}$
5. $\frac{^-12}{10} \div \frac{^-13}{7}$
6. $\frac{^-42}{3} \div \frac{^-5}{7}$
7. $\frac{^-5}{8} \div \frac{^-1}{2}$
8. $\frac{3}{4} \div \frac{^-3}{8}$
9. $\frac{2}{3} \div \frac{^-1}{3}$
10. $\frac{8}{9} \div \frac{1}{4}$
11. $\frac{^-7}{8} \div \frac{^-1}{5}$
12. $\frac{5}{6} \div \frac{^-2}{3}$
13. $\frac{^-3}{5} \div \frac{5}{9}$
14. $\frac{11}{12} \div \frac{^-11}{13}$
15. $\frac{5}{7} \div \frac{^-4}{9}$
16. $\frac{^-9}{10} \div \frac{^-1}{6}$
17. $\frac{6}{7} \div \frac{^-1}{14}$
18. $\frac{0}{5} \div \frac{^-2}{3}$
19. $\frac{^-6}{11} \div \frac{^-11}{6}$
20. $\frac{^-5}{6} \div \frac{^-1}{5}$

ADDITION

Adding positive rational numbers is just like adding numbers of arithmetic. Here are three examples.

$\frac{1}{9} + \frac{4}{9} = \frac{5}{9}$

$$\begin{array}{rcl} \frac{4}{5} & = & \frac{24}{30} \\ + \frac{1}{6} & = & \frac{5}{30} \\ \hline & & \frac{29}{30} \end{array}$$

$$\begin{array}{r} .35 \\ .68 \\ 1.2 \\ \hline 2.23 \end{array}$$

1. Add.

   a. $\frac{3}{8} + \frac{2}{8}$ b. $1\frac{1}{16} + 2\frac{3}{8}$ c. $1.34 + 2.96 + .03$

2. Recall that the sum of two negative integers is negative. We use this idea to help add two negative rational numbers. We can add $\frac{^-5}{13}$ and $\frac{^-7}{13}$ like this:

$$\frac{^-5}{13} + \frac{^-7}{13} = \frac{^-5 + {^-7}}{13} = \frac{^-12}{13}$$

   Add.

   a. $\frac{^-4}{7} + \frac{^-2}{7}$ b. $\frac{^-6}{11} + \frac{^-4}{11}$ c. $\frac{^-4}{21} + \frac{^-6}{21}$ d. $\frac{^-8}{15} + \frac{^-3}{15}$

3. Sometimes we can simplify after we add.

$$\frac{^-10}{21} + \frac{^-4}{21} = \frac{^-10 + {^-4}}{21} = \frac{^-14}{21} = \frac{^-2 \times 7}{3 \times 7} = \frac{^-2}{3}$$

   Add and simplify.

   a. $\frac{^-5}{6} + \frac{^-5}{6}$ b. $\frac{^-1}{9} + \frac{^-5}{9}$ c. $\frac{^-6}{7} + \frac{^-1}{7}$ d. $\frac{^-4}{3} + \frac{^-2}{3}$

4. Let's add $\frac{^-1}{5}$ and $\frac{^-2}{3}$.

   a. What is the least common denominator?

   b. Give the missing numerators.

$$\frac{^-1}{5} = \frac{\triangle}{15} \qquad \frac{^-2}{3} = \frac{\square}{15}$$

   c. What is the sum $\frac{^-3}{15} + \frac{^-10}{15}$? $\frac{^-1}{5} + \frac{^-2}{3}$?

5. Let's add $\frac{-17}{10}$ and $\frac{-4}{5}$.

a. What is the least common denominator?

b. Find the missing numerator. $\frac{-4}{5} = \frac{\square}{10}$

c. Add and simplify: $\frac{-17}{10} + \frac{-8}{10}$.

6. Add and simplify.

a. $\frac{-5}{6} + \frac{-2}{3}$ b. $\frac{-1}{8} + \frac{-3}{16}$ c. $\frac{-5}{9} + \frac{-2}{3}$ d. $\frac{-7}{18} + \frac{-5}{6}$

EXERCISES

Add and simplify.

1. $\frac{2}{5} + \frac{4}{7}$
2. $\frac{5}{11} + \frac{3}{4}$
3. $\frac{9}{14} + \frac{5}{6}$
4. $1\frac{1}{3} + 2\frac{1}{2}$
5. $3\frac{1}{4} + 5\frac{3}{8}$
6. $2\frac{1}{8} + 3\frac{5}{16}$
7. $4\frac{1}{9} + 5\frac{2}{3}$
8. $\frac{-4}{7} + \frac{-2}{7}$
9. $\frac{-5}{16} + \frac{-8}{16}$
10. $\frac{-3}{20} + \frac{-17}{20}$
11. $\frac{-5}{13} + \frac{-6}{13}$
12. $\frac{-5}{8} + \frac{-1}{4}$
13. $\frac{-7}{10} + \frac{-4}{5}$
14. $\frac{-3}{8} + \frac{-9}{64}$
15. $\frac{-19}{24} + \frac{-7}{12}$
16. $\frac{-18}{30} + \frac{-3}{15}$
17. $\frac{-7}{8} + \frac{-5}{16}$
18. $\frac{-10}{21} + \frac{-5}{7}$
19. $\frac{-8}{11} + \frac{-3}{5}$
20. $\frac{-9}{10} + \frac{-2}{5}$
21. $\frac{3}{20} + \frac{5}{18}$

UNEXPLORED TERRITORY

1. Write standard numerals.

a. 1^2 b. 11^2 c. 111^2 d. $1{,}111^2$

2. Study the pattern. What comes next?

MORE ABOUT ADDITION

Here is how Craig added $\frac{^-5}{8}$ and $\frac{7}{8}$.

$$\frac{^-5}{8} + \frac{7}{8} = \frac{^-5 + 7}{8} = \frac{2}{8} = \frac{1}{4}$$

1. Copy and complete.

   a. $\frac{9}{10} + \frac{^-3}{10} = \frac{9 + {^-3}}{10} = \frac{6}{10} =$ ____

   b. $\frac{^-6}{13} + \frac{10}{13} = \frac{^-6 + 10}{13} =$ ____

2. Add and simplify.

   a. $\frac{^-16}{17} + \frac{5}{17}$ b. $\frac{6}{7} + \frac{^-4}{7}$ c. $\frac{^-15}{24} + \frac{23}{24}$ d. $\frac{^-11}{36} + \frac{27}{36}$

   e. $\frac{^-3}{8} + \frac{7}{8}$ f. $\frac{^-9}{10} + \frac{1}{10}$ g. $\frac{^-6}{11} + \frac{2}{11}$ h. $\frac{9}{28} + \frac{^-15}{28}$

3. Let's add $\frac{^-3}{4}$ and $\frac{1}{6}$.

   a. What is the least common denominator?

   b. Complete: $\frac{^-3}{4} = \frac{^-3}{4} \cdot \frac{3}{3} = \frac{\square}{12}$

   $\frac{1}{6} = \frac{1}{6} \cdot \frac{2}{2} = \frac{\triangle}{12}$

   c. What is the sum $\frac{^-9}{12} + \frac{2}{12}$? $\frac{^-3}{4} + \frac{1}{6}$?

4. Let's add $\frac{^-5}{6}$ and $\frac{3}{10}$.

   a. What is the least common denominator?

   b. Complete: $\frac{^-5}{6} = \frac{\triangle}{30}$ $\frac{3}{10} = \frac{\square}{30}$

   c. Add and simplify: $\frac{^-25}{30} + \frac{9}{30}$.

5. Add and simplify.

   a. $\frac{^-1}{3} + \frac{1}{4}$ b. $\frac{^-2}{3} + \frac{5}{6}$ c. $\frac{^-4}{5} + \frac{1}{7}$ d. $\frac{^-3}{4} + \frac{5}{6}$

   e. $\frac{5}{9} + \frac{^-3}{10}$ f. $\frac{^-7}{8} + \frac{2}{5}$ g. $\frac{^-2}{9} + \frac{1}{6}$ h. $\frac{^-3}{8} + \frac{7}{16}$

EXERCISES

Add and simplify.

1. $\frac{-2}{3} + \frac{2}{3}$
2. $\frac{4}{9} + \frac{-7}{9}$
3. $\frac{7}{10} + \frac{-3}{10}$
4. $\frac{-5}{16} + \frac{7}{16}$
5. $\frac{-3}{10} + \frac{2}{5}$
6. $\frac{-1}{3} + \frac{2}{7}$
7. $\frac{-5}{11} + \frac{3}{4}$
8. $\frac{-2}{5} + \frac{4}{7}$
9. $\frac{-1}{9} + \frac{5}{6}$
10. $\frac{1}{10} + \frac{-3}{4}$
11. $\frac{-8}{11} + \frac{4}{5}$
12. $\frac{7}{12} + \frac{-3}{4}$
13. $\frac{1}{6} + \frac{-4}{5}$
14. $\frac{5}{6} + \frac{-7}{8}$
15. $\frac{-11}{12} + \frac{3}{5}$
16. $\frac{3}{8} + \frac{-7}{20}$
17. $\frac{8}{9} + \frac{-3}{7}$
18. $\frac{3}{10} + \frac{-4}{5}$
19. $\frac{-3}{14} + \frac{2}{7}$
20. $\frac{11}{16} + \frac{-3}{4}$

Solve these problems.

21. Mr. Cole said that on Monday his stock was up $\frac{3}{8}$ but on Tuesday it was down $\frac{3}{4}$. Describe this situation using rational numbers. How could you describe the overall change in the stock?

22. In an exercise program Charlie jogged $\frac{1}{8}$ mile the first 2 days, $\frac{1}{4}$ mile the third day and $\frac{1}{2}$ mile the fourth day. How far did he jog in all?

23. For making cookies, Mrs. Dunty bought $\frac{3}{4}$ lb of almonds and $\frac{2}{3}$ lb of pecans. Find the total weight of the nuts.

24. Mr. Sherer has a piece of wood that is $4\frac{7}{8}$ inches thick. He needs a piece of wood to finish a stereo cabinet that is $4\frac{1}{16}$ inches thick. How much waste will he have?

SUBTRACTION

We can subtract a rational number by adding its inverse.

1. Let's subtract $\frac{3}{7}$ from $\frac{^-5}{7}$.

   a. What is the inverse of $\frac{3}{7}$?

   b. Add: $\frac{^-5}{7} + \frac{^-3}{7}$

   c. What is $\frac{^-5}{7} - \frac{3}{7}$?

2. Let's subtract $\frac{^-2}{5}$ from $\frac{4}{5}$.

   a. What is the inverse of $\frac{^-2}{5}$?

   b. Add: $\frac{4}{5} + \frac{2}{5}$

   c. What is $\frac{4}{5} - \frac{^-2}{5}$?

3. Let's subtract $\frac{^-4}{9}$ from $\frac{^-7}{9}$.

   a. What is the inverse of $\frac{^-4}{9}$?

   b. Add: $\frac{^-7}{9} + \frac{4}{9}$

   c. What is $\frac{^-7}{9} - \frac{^-4}{9}$?

   d. Write the simplest name.

4. Subtract by adding the inverse. Simplify.

   a. $\frac{2}{5} - \frac{^-4}{5}$ b. $\frac{5}{6} - \frac{1}{6}$ c. $\frac{^-7}{11} - \frac{^-9}{11}$ d. $\frac{^-8}{9} - \frac{4}{9}$

5. Let's subtract $\frac{^-2}{5}$ from $\frac{1}{3}$.

   a. What is the least common denominator?

   b. Solve: $\frac{^-2}{5} = \frac{\triangle}{15}$ $\frac{1}{3} = \frac{\square}{15}$

   c. Subtract: $\frac{5}{15} - \frac{^-6}{15}$

6. Subtract and simplify.

a. $\frac{^-1}{3} - \frac{1}{6}$ b. $\frac{^-2}{5} - \frac{^-1}{4}$ c. $\frac{^-7}{8} - \frac{^-3}{4}$ d. $\frac{5}{6} - \frac{2}{3}$

e. $\frac{8}{9} - \frac{^-2}{3}$ f. $\frac{^-1}{8} - \frac{5}{16}$ g. $\frac{^-4}{63} - \frac{2}{7}$ h. $\frac{7}{24} - \frac{^-3}{8}$

EXERCISES

Subtract and simplify.

1. $\frac{^-7}{12} - \frac{^-5}{12}$ 2. $\frac{5}{8} - \frac{3}{8}$ 3. $\frac{^-7}{9} - \frac{2}{9}$ 4. $\frac{9}{14} - \frac{^-7}{14}$

5. $\frac{4}{9} - \frac{^-1}{18}$ 6. $\frac{^-7}{8} - \frac{3}{4}$ 7. $\frac{^-9}{10} - \frac{^-3}{5}$ 8. $\frac{^-5}{6} - \frac{3}{4}$

9. $\frac{27}{32} - \frac{^-15}{16}$ 10. $\frac{7}{13} - \frac{1}{2}$ 11. $\frac{5}{16} - \frac{^-3}{32}$ 12. $\frac{^-11}{14} - \frac{^-10}{14}$

13. $\frac{^-5}{8} - \frac{1}{3}$ 14. $\frac{^-9}{16} - \frac{^-5}{32}$ 15. $\frac{1}{2} - \frac{^-1}{11}$ 16. $\frac{2}{3} - \frac{5}{6}$

17. $\frac{^-4}{15} - \frac{9}{10}$ 18. $\frac{^-6}{7} - \frac{13}{35}$ 19. $\frac{1}{2} - \frac{7}{8}$ 20. $\frac{^-9}{20} - \frac{^-5}{8}$

Solve these problems.

21. Mrs. Dreier bought a large fish that weighed $4\frac{1}{4}$ pounds. After she removed the head, tail and bones, it weighed only $3\frac{1}{8}$ pounds. How much waste did she pay for?

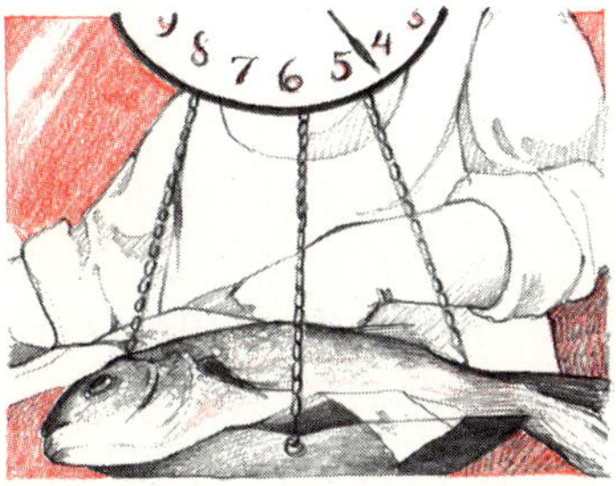

22. An alloy is $\frac{3}{5}$ copper, $\frac{1}{3}$ silver and the rest is gold. What part of the alloy is gold?

23. Robyn planned to budget her weekly allowance and babysitting money. If she planned to use $\frac{3}{7}$ for her school needs and $\frac{1}{3}$ for fun, how much could she save weekly?

SQUARE ROOTS OF RATIONAL NUMBERS

Recall that 2 is a square root of 4 since $2 \cdot 2 = 4$. Does 4 have any other square roots?

LET'S EXPLORE

1. a. Multiply: $2 \cdot 2$ $\quad ^{-}2 \cdot ^{-}2$

   b. 4 has two square roots. What are they?

2. a. Multiply: $12 \cdot 12$ $\quad ^{-}12 \cdot ^{-}12$

   b. Name two square roots of 144.

3. Find two square roots of each.

   a. 25 b. 1 c. 36 d. 100 e. 81

4. a. Multiply: $\frac{2}{3} \cdot \frac{2}{3}$ $\quad \frac{^{-}2}{3} \cdot \frac{^{-}2}{3}$

   b. Name two square roots of $\frac{4}{9}$.

5. a. Multiply: $\frac{3}{7} \cdot \frac{3}{7}$ $\quad \frac{^{-}3}{7} \cdot \frac{^{-}3}{7}$

   b. Name two square roots of $\frac{9}{49}$.

Positive rational numbers may have two square roots.

6. Find two square roots of each. Use the table on page 406.

   a. $\frac{25}{4}$ b. $\frac{16}{25}$ c. $\frac{36}{64}$ d. $\frac{1}{16}$ e. $\frac{49}{25}$

   f. $\frac{144}{196}$ g. $\frac{400}{121}$ h. $\frac{256}{81}$ i. $\frac{961}{324}$ j. $\frac{169}{576}$

7. Do negative rational numbers have square roots? Try to name $\frac{^{-}4}{9}$ as the product of two factors which are the same.

Negative rational numbers have no rational square roots.

EXERCISES

Find two square roots of each. Use the table on page 406.

1. $\frac{100}{64}$
2. $\frac{9}{16}$
3. $\frac{121}{256}$
4. $\frac{81}{100}$
5. $\frac{16}{121}$
6. $\frac{^-16}{25}$
7. $\frac{169}{225}$
8. $\frac{49}{81}$
9. $\frac{^-36}{25}$
10. $\frac{256}{49}$
11. $\frac{121}{196}$
12. $\frac{^-25}{81}$
13. $\frac{196}{289}$
14. $\frac{400}{169}$
15. $\frac{900}{144}$
16. $\frac{784}{121}$
17. $\frac{1{,}024}{441}$
18. $\frac{49}{1{,}296}$
19. $\frac{1{,}521}{529}$
20. $\frac{81}{400}$
21. $\frac{289}{441}$
22. $\frac{361}{81}$
23. $\frac{529}{49}$
24. $\frac{256}{81}$
25. $\frac{1{,}089}{1{,}681}$
26. $\frac{900}{121}$
27. $\frac{900}{1{,}600}$
28. $\frac{^-484}{1{,}024}$

Let's Practice

1. What are congruent figures?
2. Draw two congruent segments.
3. Draw two congruent angles.
4. Draw a rectangle.
5. Draw a square.
6. Draw a trapezoid.
7. Draw a parallelogram.

PROBLEM SOLVING

1. Jack's committee had to sell refreshments at the dance for $2\frac{1}{2}$ hours. Each committee member agreed to serve $\frac{1}{5}$ of the time. How long was each member on duty?

2. Tina and Alison made 5 pans of fudge for the class party. They were able to get 40 pieces of fudge from each pan. How many pieces of fudge did they have in all?

3. At the model building club, each member agreed to pay \$1.50 dues. How many \$3.25 models will they be able to buy?

4. A bathroom floor measures 5 feet by 8 feet. Wall tile is to be installed to a height of 4 feet. The door and the bathtub take up 33 square feet.

   a. Find how many square feet of tiling will be needed.

   b. Nine tiles cover one square foot. How many tiles will be needed?

   c. What will the tiles cost at 17¢ per tile?

5. Carolyn was preparing $\frac{1}{2}$ of a recipe which called for $\frac{3}{4}$ cup of orange juice. How much orange juice should she use?

DO YOU REMEMBER?

Multiply and simplify.

1. $\frac{3}{8} \times \frac{^{-}2}{7}$ 2. $\frac{^{-}5}{6} \times \frac{^{-}2}{3}$ 3. $\frac{8}{9} \times \frac{3}{4}$ 4. $\frac{^{-}7}{12} \times \frac{10}{21}$

5. $\frac{^{-}7}{10} \times \frac{^{-}4}{5}$ 6. $\frac{^{-}8}{11} \times \frac{^{-}22}{32}$ 7. $\frac{^{-}10}{7} \times \frac{6}{5}$ 8. $\frac{^{-}3}{14} \times \frac{^{-}1}{5}$

Find the reciprocal of each.

9. $\frac{^{-}6}{7}$ 10. $\frac{^{-}4}{5}$ 11. $\frac{9}{10}$ 12. $\frac{^{-}3}{8}$

13. $\frac{0}{18}$ 14. $\frac{^{-}6}{5}$ 15. $\frac{^{-}11}{13}$ 16. $\frac{18}{25}$

Divide and simplify.

17. $\frac{8}{15} \div \frac{^{-}4}{5}$ 18. $\frac{^{-}6}{13} \div \frac{^{-}3}{26}$ 19. $\frac{1}{7} \div \frac{^{-}1}{8}$ 20. $\frac{^{-}12}{13} \div \frac{^{-}3}{4}$

21. $\frac{^{-}5}{8} \div \frac{5}{2}$ 22. $\frac{^{-}34}{5} \div \frac{4}{3}$ 23. $5 \div \frac{^{-}1}{4}$ 24. $\frac{^{-}7}{8} \div {}^{-}4$

Add and simplify.

25. $\frac{5}{8} + \frac{^{-}1}{3}$ 26. $\frac{3}{10} + \frac{3}{4}$ 27. $\frac{^{-}3}{10} + \frac{^{-}1}{2}$ 28. $\frac{^{-}5}{6} + \frac{^{-}2}{5}$

29. $\frac{5}{6} + \frac{^{-}2}{9}$ 30. $\frac{4}{5} + \frac{1}{6}$ 31. $\frac{^{-}2}{9} + \frac{^{-}7}{12}$ 32. $\frac{^{-}2}{3} + \frac{1}{4} + \frac{^{-}3}{8}$

Subtract and simplify.

33. $\frac{1}{2} - \frac{^{-}2}{5}$ 34. $\frac{^{-}9}{16} - \frac{1}{8}$ 35. $\frac{^{-}3}{8} - \frac{^{-}7}{16}$ 36. $\frac{9}{14} - \frac{^{-}3}{7}$

37. $\frac{11}{16} - \frac{5}{12}$ 38. $\frac{^{-}4}{9} - \frac{^{-}1}{18}$ 39. $7 - \frac{^{-}7}{12}$ 40. $\frac{^{-}5}{7} - \frac{1}{9}$

Find two square roots of each.

41. $\frac{400}{169}$ 42. $\frac{169}{16}$ 43. $\frac{81}{25}$ 44. $\frac{^{-}64}{121}$

45. $\frac{256}{289}$ 46. $\frac{400}{441}$ 47. $\frac{400}{529}$ 48. $\frac{196}{361}$

DECIMAL NUMERALS FOR RATIONAL NUMBERS

When fractional numerals are given, we can find decimal numerals by dividing the numerator by the denominator.

$$\frac{^-7}{8} \qquad \begin{array}{r} .875 \\ 8\overline{)7.000} \\ \underline{6\,4} \\ 60 \\ \underline{56} \\ 40 \\ \underline{40} \end{array} \qquad \frac{^-7}{8} = {}^-.875$$

1. Find decimal numerals.

   a. $\frac{3}{20}$ b. $\frac{^-7}{4}$ c. $\frac{^-3}{25}$ d. $\frac{^-12}{5}$ e. $\frac{21}{8}$

2. Ted found a decimal numeral for $\frac{^-5}{6}$.

   $$\begin{array}{r} .8333 \\ 6\overline{)5.0000} \\ \underline{4\,8} \\ 20 \\ \underline{18} \\ 20 \\ \underline{18} \\ 2 \end{array}$$

   a. Will the 3's in his quotient ever end?

   b. When could Ted tell that the 3's will repeat?

3. Ted used a bar to show that the 3's repeat. $\frac{^-5}{6} = {}^-.83\overline{3}$

   a. Find a decimal for $\frac{1}{6}$.

   b. Use a bar to show that the 6's repeat.

4. Find decimal numerals.

   a. $\frac{^-2}{9}$ b. $\frac{^-7}{12}$ c. $\frac{2}{3}$ d. $\frac{^-2}{15}$ e. $\frac{5}{11}$

Decimal numerals for rational numbers either end or repeat.

Examples $\frac{1}{8} = .125$ $\frac{^-4}{15} = {}^-.26\overline{6}$ $\frac{^-1}{9} = {}^-.1\overline{1}$

EXERCISES

Find decimal numerals.

1. $\frac{5}{9}$ 2. $\frac{^-3}{5}$ 3. $\frac{^-2}{13}$ 4. $\frac{6}{14}$ 5. $\frac{^-4}{5}$

6. $\frac{^-5}{7}$ 7. $\frac{1}{12}$ 8. $\frac{3}{20}$ 9. $\frac{^-2}{3}$ 10. $\frac{1}{7}$

11. $\frac{^-7}{15}$ 12. $\frac{^-11}{6}$ 13. $\frac{3}{8}$ 14. $\frac{17}{30}$ 15. $\frac{1}{13}$

16. $\frac{7}{12}$ 17. $\frac{^-2}{15}$ 18. $\frac{5}{9}$ 19. $\frac{^-8}{3}$ 20. $\frac{7}{30}$

21. $\frac{3}{11}$ 22. $\frac{^-7}{20}$ 23. $\frac{^-13}{6}$ 24. $\frac{5}{16}$ 25. $\frac{10}{11}$

UNEXPLORED TERRITORY

1. Find decimal numerals.

   a. $\frac{1}{11}$ b. $\frac{2}{11}$ c. $\frac{3}{11}$ d. $\frac{4}{11}$ e. $\frac{5}{11}$

   f. $\frac{6}{11}$ g. $\frac{7}{11}$ h. $\frac{8}{11}$ i. $\frac{9}{11}$ j. $\frac{10}{11}$

2. What pattern do you find?

3. Find decimal numerals.

   a. $\frac{1}{7}$ b. $\frac{2}{7}$ c. $\frac{3}{7}$ d. $\frac{4}{7}$ e. $\frac{5}{7}$ f. $\frac{6}{7}$

4. What pattern do you find?

5. Find decimal numerals.

   a. $\frac{1}{13}$ b. $\frac{2}{13}$ c. $\frac{3}{13}$ d. $\frac{4}{13}$ e. $\frac{5}{13}$ f. $\frac{6}{13}$

   g. $\frac{7}{13}$ h. $\frac{8}{13}$ i. $\frac{9}{13}$ j. $\frac{10}{13}$ k. $\frac{11}{13}$ l. $\frac{12}{13}$

6. What pattern do you find?

FINDING FRACTIONAL NUMERALS

We know how to find decimal numerals for rational numbers. Suppose we start with a decimal numeral. How can we find a fractional numeral?

1. If a decimal ends, we can easily find a fractional numeral.

$$.387 = \frac{387}{1{,}000} \qquad 4.73 = \frac{473}{100}$$

Find fractional numerals.

a. .892 **b.** 6.20 **c.** .5283 **d.** 25.633

If the decimal numeral does not end, what do we do?

LET'S EXPLORE

2. Recall that when we multiply by 10, we can move the decimal point one place to the right.

$$.6\overline{6} \times 10 = 6.6\overline{6}$$

Multiply.

a. $.75\overline{5} \times 10$ **b.** $.48\overline{8} \times 10$ **c.** $.94\overline{4} \times 10$

3. If we subtract a number from 10 times the number, we get 9 times the number.

$$(10 \times 2) - 2 = 9 \times 2, \text{ or } 18$$

$$(10 \times .35\overline{5}) - .35\overline{5} = (9 \times .35\overline{5})$$

Name each as 9 times a number.

a. $(10 \times .68\overline{8}) - .68\overline{8}$ **b.** $(10 \times .73\overline{3}) - .73\overline{3}$

c. $(10 \times .46\overline{6}) - .46\overline{6}$ **d.** $(10 \times .84\overline{4}) - .84\overline{4}$

4. Let's find a fractional numeral for $.46\overline{6}$. First write a decimal numeral for $10 \times .46\overline{6}$.

5. Explain each step of this procedure.

$$\underbrace{(10 \times .46\overline{6})}_{} - .46\overline{6} = \underbrace{9 \times .46\overline{6}}_{}$$

$$4.6\overline{6} - .46\overline{6} = 4.2 \longleftarrow \begin{array}{r} 4.66\overline{6} \\ -\ .46\overline{6} \\ \hline 4.2 \end{array}$$

$$\text{Thus,} \quad 9 \times .46\overline{6} = 4.2$$

$$.46\overline{6} = \frac{4.2}{9} \leftarrow \text{Related division sentence}$$

6. Let's find the simplest fractional numeral for $\frac{4.2}{9}$.

$$\frac{4.2}{9} = \frac{4.2}{9} \cdot \frac{10}{10} = \frac{42}{90} = \frac{7}{15} \cdot \frac{6}{6} = \frac{7}{15}$$

What is the simplest fractional numeral for $.46\overline{6}$?

7. Now let's think about the decimal numeral $.23\overline{3}$. Write a decimal numeral for $10 \times .23\overline{3}$?

8. Explain each step of this procedure.

$$\underbrace{(10 \times .23\overline{3})}_{} - .23\overline{3} = \underbrace{9 \times .23\overline{3}}_{}$$

$$2.3\overline{3} - .23\overline{3} = 2.1 \longleftarrow \begin{array}{r} 2.33\overline{3} \\ -\ .23\overline{3} \\ \hline 2.1 \end{array}$$

$$\text{Thus,} \quad 9 \times .23\overline{3} = 2.1$$

$$.23\overline{3} = \frac{2.1}{9} \leftarrow \text{Related division sentence}$$

9. Find the simplest fractional numeral for $.23\overline{3}$.

10. Find simplest fractional numerals.

a. $.12\overline{2}$ **b.** $.53\overline{3}$ **c.** $.86\overline{6}$ **d.** $.61\overline{1}$

EXERCISES

Find simplest fractional numerals.

1. $.75\overline{5}$ **2.** $.97\overline{7}$ **3.** $.62\overline{2}$ **4.** $.72\overline{2}$

5. $.48\overline{8}$ **6.** $.36\overline{6}$ **7.** $.94\overline{4}$ **8.** $.65\overline{5}$

9. $.83\overline{3}$ **10.** $.59\overline{9}$ **11.** $.16\overline{6}$ **12.** $.39\overline{9}$

MORE ABOUT FRACTIONAL NUMERALS

Numbers like $.46\overline{6}$ and $.23\overline{3}$ repeat in a cycle of one. By multiplying by 10, we can subtract easily. What about a numeral like $.24\overline{24}$?

LET'S EXPLORE

1. Recall that to multiply by 100, we move the decimal point two places to the right.

$$.24\overline{24} \times 100 = 24.24\overline{24}$$

Multiply.

a. $.59\overline{59} \times 100$ **b.** $.83\overline{83} \times 100$ **c.** $.34\overline{34} \times 100$

2. If we subtract a number from 100 times the number, we get 99 times the number.

$$(100 \times 5) - 5 = 99 \times 5, \text{ or } 495$$

$$(100 \times .24\overline{24}) - .24\overline{24} = (99 \times .24\overline{24})$$

Express each as 99 times a number.

a. $(100 \times .72\overline{72}) - .72\overline{72}$ **b.** $(100 \times .56\overline{56}) - .56\overline{56}$

c. $(100 \times .91\overline{91}) - .91\overline{91}$ **d.** $(100 \times .75\overline{75}) - .75\overline{75}$

3. Let's find a fractional numeral for $.24\overline{24}$. First write a decimal numeral for $100 \times .24\overline{24}$.

4. Explain each step of this procedure.

$$\underbrace{(100 \times .24\overline{24})} - .24\overline{24} = \underbrace{99 \times .24\overline{24}}$$

$$24.24\overline{24} - .24\overline{24} = 24$$

Thus, $99 \times .24\overline{24} = 24$

$.24\overline{24} = \frac{24}{99}$ ←Related division sentence

$$\frac{24}{99} = \frac{3 \times 8}{3 \times 33} = \frac{8}{33}$$

Thus, $.24\overline{24} = \frac{8}{33}$.

5. Find simplest fractional numerals.

a. $.63\overline{63}$ b. $.28\overline{28}$ c. $.41\overline{41}$ d. $.12\overline{12}$ $\frac{4}{33}$

6. To find a fractional numeral for .634634, we first multiply by 1,000. Explain each step of this procedure.

a. $(1{,}000 \times .634\overline{634}) = 634.634\overline{634}$

b. $\underbrace{(1{,}000 \times .634\overline{634})}_{634.634\overline{634}} - .634\overline{634} = \underbrace{999 \times .634\overline{634}}_{634}$

$634.634\overline{634} - .634\overline{634} = 634$

Thus, $999 \times .634\overline{634} = 634$

$.634\overline{634} = \frac{634}{999}$ ← Related division sentence

7. Find simplest fractional numerals.

a. $.539\overline{539}$ b. $.803\overline{803}$ c. $.714\overline{714}$

EXERCISES

Find simplest fractional numerals.

1. $.8\overline{8}$ 2. $.2\overline{2}$ 3. $.57\overline{7}$ 4. $.3\overline{3}$

5. $.61\overline{1}$ 6. $.6\overline{6}$ 7. $.16\overline{16}$ 8. $.25\overline{25}$

9. $.24\overline{24}$ 10. $.63\overline{63}$ 11. $.81\overline{81}$ 12. $.75\overline{75}$

13. $.213\overline{213}$ 14. $.236\overline{236}$ 15. $.761\overline{761}$ 16. $.828\overline{828}$

Let's Practice

1. Mrs. Beeler's car holds 5 quarts of oil. How many gallons is this?

2. Mrs. Singer picked 9 pecks of blueberries to make jelly. How many bushels is this?

3. Mr. Taylor took 25 bushels of tomatoes to the market. He sells them for $1.65 per peck. How much will he make if he sells all the tomatoes?

PROBLEM SOLVING

1. Mrs. Jan's Girl Scout Troop is making blackberry jelly. The recipe calls for $7\frac{1}{2}$ cups of sugar for 3 quarts of berries. Each girl has 2 quarts of berries. How much sugar will each girl need?

2. The record for the 100 yard dash at Stricker Junior High was $11\frac{4}{10}$ seconds. Jon Chu set a new record of $10\frac{9}{10}$ seconds. What was the difference in the times?

3. For her vegetable garden, Mrs. Brant needed 24 stakes each $7\frac{1}{2}$ feet long. What will they cost at 36¢ each?

4. Tony carries groceries for $1.00 per hour. He averages about $6\frac{1}{2}$ hours per week. About how much will Tony earn per week?

5. Mark had to make a chart for science class. He used a piece of tagboard 36 inches long. How many columns, each $\frac{3}{4}$ inch wide, can he draw?

6. Ginny's aunt gave her some curtains that were $5\frac{1}{4}$ feet long. Ginny's windows were only $4\frac{3}{4}$ feet long. How much did she have to shorten the curtains?

CHAPTER REVIEW

Show that each is a rational number.

1. $^{-}12$ 2. 24 3. .14 4. 2.4

Which are true and which are false?

5. $\frac{-1}{2} > 0$ 6. $\frac{-3}{4} > {}^{-}1$ 7. $\frac{3}{2} < {}^{-}2$ 8. $0 < \frac{-1}{3}$

Name the inverse of each.

9. $\frac{-1}{2}$ 10. 6 11. 0 12. $^{-}1.3$

Write the simplest name for each.

13. $\left|\frac{5}{8}\right|$ 14. $|0|$ 15. $\left|\frac{-1}{2}\right|$ 16. $|{}^{-}2.3|$

Multiply and simplify.

17. $\frac{-1}{2} \cdot \frac{3}{4}$ 18. $\frac{-2}{3} \cdot \frac{-1}{5}$ 19. $\frac{9}{10} \cdot \frac{3}{7}$

20. $\frac{7}{10} \cdot \frac{-3}{14}$ 21. $\frac{6}{5} \cdot \frac{2}{3}$ 22. $\frac{-1}{6} \cdot \frac{-3}{4}$

23. $\frac{1}{3} \cdot \frac{-5}{8} \cdot \frac{3}{10}$ 24. $\frac{-4}{5} \cdot \frac{10}{11} \cdot \frac{-1}{2}$ 25. $\frac{-20}{21} \cdot \frac{-4}{15} \cdot \frac{-27}{28}$

Name the reciprocal of each.

26. $\frac{1}{2}$ 27. $\frac{-2}{3}$ 28. 0 29. $^{-}1\frac{1}{5}$

Divide and simplify.

30. $\frac{-5}{13} \div \frac{1}{2}$ 31. $\frac{-5}{8} \div \frac{-3}{4}$ 32. $\frac{3}{10} \div \frac{4}{5}$

33. $\frac{3}{8} \div \frac{5}{7}$ 34. $\frac{-1}{2} \div \frac{-7}{8}$ 35. $\frac{2}{9} \div \frac{-2}{3}$

36. $\frac{-5}{11} \div \frac{-10}{33}$ 37. $\frac{12}{13} \div \frac{15}{20}$ 38. $\frac{-2}{21} \div \frac{9}{10}$

Add and simplify.

39. $\frac{3}{5} + \frac{-1}{5}$ 40. $\frac{-3}{8} + \frac{-3}{8}$ 41. $\frac{-5}{9} + \frac{1}{3}$

42. $\frac{1}{2} + \frac{7}{16}$ 43. $\frac{-11}{8} + \frac{5}{8}$ 44. $\frac{-5}{6} + \frac{-1}{2}$

45. $\frac{2}{3} + \frac{-5}{9}$ 46. $\frac{3}{20} + \frac{-5}{8}$ 47. $\frac{-7}{10} + \frac{-4}{15}$

Subtract and simplify.

48. $\frac{-3}{8} - \frac{1}{8}$ 49. $\frac{-1}{12} - \frac{-3}{4}$ 50. $\frac{1}{5} - \frac{-3}{5}$

51. $\frac{5}{12} - \frac{7}{8}$ 52. $\frac{-3}{5} - \frac{-7}{10}$ 53. $\frac{-20}{21} - \frac{-1}{2}$

54. $\frac{-7}{15} - \frac{5}{8}$ 55. $\frac{6}{11} - \frac{-3}{5}$ 56. $\frac{-21}{25} - \frac{-3}{10}$

Find two square roots of each.

57. $\frac{49}{121}$ 58. $\frac{4}{64}$ 59. $\frac{25}{81}$ 60. $\frac{9}{64}$

Find decimal numerals.

61. $\frac{7}{20}$ 62. $\frac{-5}{16}$ 63. $\frac{-23}{25}$ 64. $\frac{-1}{9}$ 65. $\frac{-3}{11}$

Solve this problem.

66. Mrs. Brown bought $15\frac{1}{2}$ yards of cotton material to make curtains. The material cost $.89 per yard. She also bought two spools of thread at $.10 each and 3 packages of bias tape at $.20 per package. How much change did she receive from a $20 bill?

Terms You Should Know

rational number
absolute value
decimal numeral
inverse
reciprocal
square root

CHAPTER TEST

Show that each is a rational number.

1. $^{-}24$
2. 5.6

Which are true and which are false?

3. $\frac{^{-}3}{5} > {}^{-}1$
4. $\frac{^{-}5}{8} < \frac{^{-}1}{3}$

Name the inverse and reciprocal of each.

5. 7.3
6. $^{-}5$
7. $\frac{5}{8}$
8. $\frac{^{-}2}{3}$

Write the simplest name for each.

9. $\left|\frac{^{-}5}{9}\right|$
10. $|7|$

Multiply and simplify.

11. $\frac{^{-}5}{6} \cdot \frac{^{-}3}{4}$
12. $\frac{2}{5} \cdot \frac{^{-}1}{3}$
13. $\frac{^{-}9}{10} \cdot \frac{20}{33}$

Divide and simplify.

14. $\frac{3}{8} \div \frac{1}{2}$
15. $\frac{^{-}5}{6} \div \frac{^{-}20}{21}$
16. $\frac{^{-}1}{10} \div \frac{3}{5}$

Add and simplify.

17. $\frac{9}{10} + \frac{^{-}8}{9}$
18. $\frac{^{-}3}{4} + \frac{^{-}1}{8}$
19. $\frac{^{-}2}{5} + \frac{3}{10}$

Subtract and simplify.

20. $\frac{^{-}2}{3} - \frac{5}{8}$
21. $\frac{^{-}6}{7} - \frac{^{-}1}{3}$
22. $\frac{5}{12} - \frac{^{-}7}{8}$

Find two square roots of each.

23. $\frac{81}{25}$
24. $\frac{49}{100}$

Find decimal numerals.

25. $\frac{3}{50}$
26. $\frac{^{-}5}{12}$

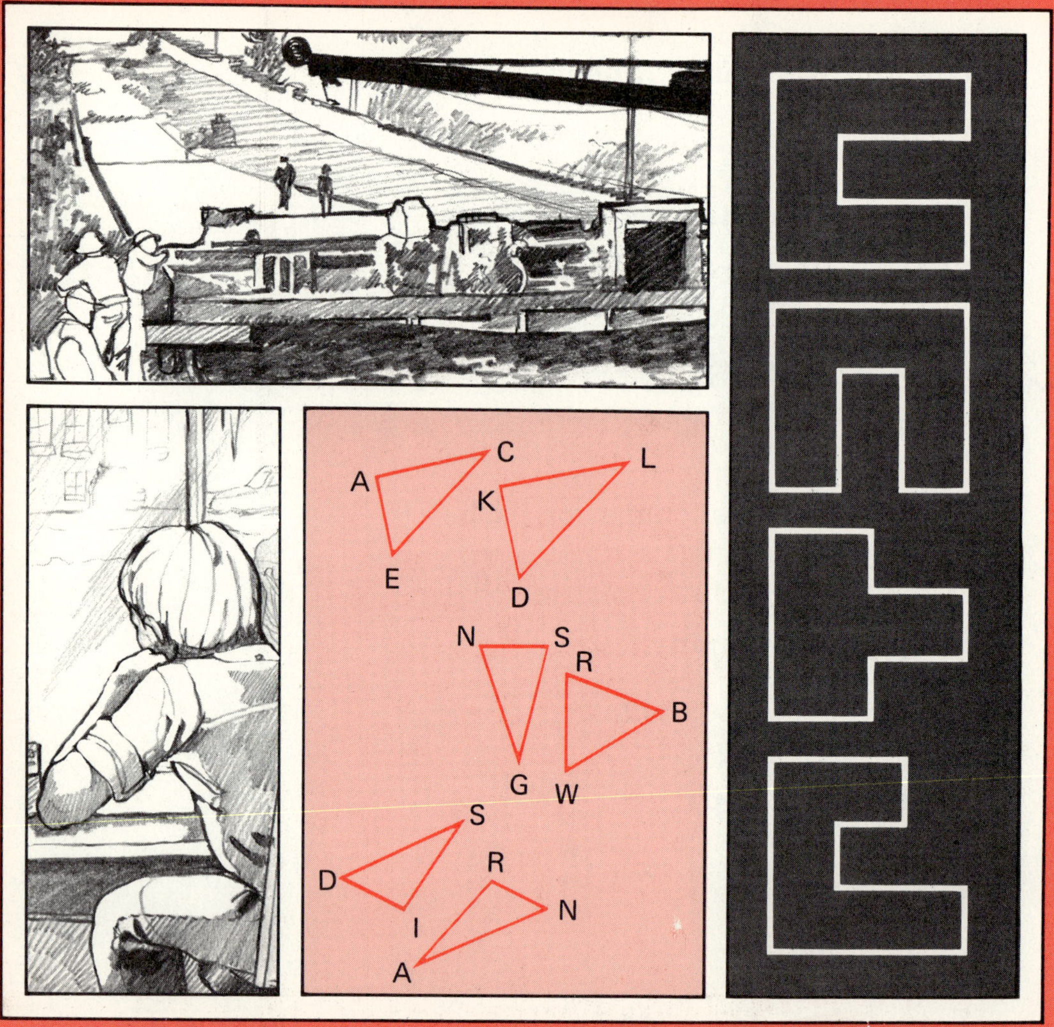

CHAPTER TEN

Similar Figures

CONGRUENT TRIANGLES

Congruent segments have the same length. Congruent angles have the same measure. Here are two congruent triangles.

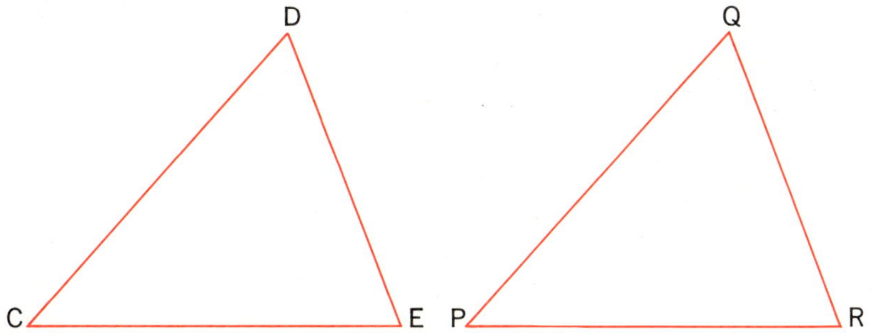

LET'S EXPLORE

1. Does $\triangle CDE$ look as if it would fit on $\triangle PQR$ exactly?
   - **a.** If so, where would point C fall?
   - **b.** Where would point D fall?
   - **c.** Where would point E fall?
   - **d.** Would the triangles fit if C were placed on Q?
   - **e.** Would the triangles fit if C were placed on R?
2. If the triangles fit exactly, $\angle C$ matches $\angle P$. Then $\angle C \cong \angle P$.
   - **a.** What angle in $\triangle PQR$ is congruent to $\angle E$?
   - **b.** What angle in $\triangle PQR$ is congruent to $\angle D$?
3. If the triangles fit, $\overline{CD}$ matches $\overline{PQ}$. Then $\overline{CD} \cong \overline{PQ}$.
   - **a.** What side in $\triangle PQR$ is congruent to $\overline{DE}$?
   - **b.** What side in $\triangle PQR$ is congruent to $\overline{CE}$?

In congruent triangles, matching sides are congruent and matching angles are congruent.

4. $\triangle ABC \cong \triangle RST$

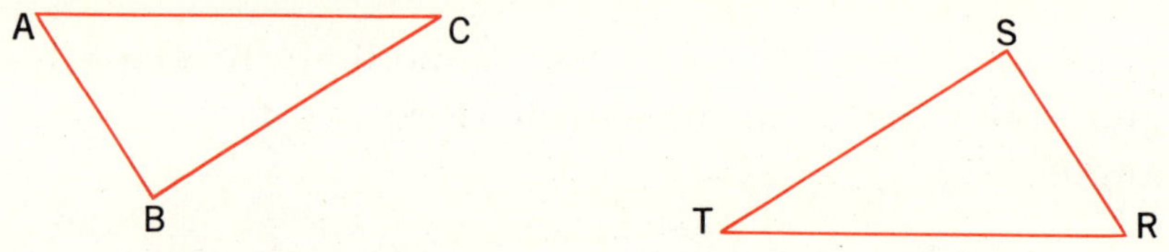

a. Which pairs of angles match?

b. Which pairs of sides match?

EXERCISES

1. Pick out the pairs of triangles which look congruent.

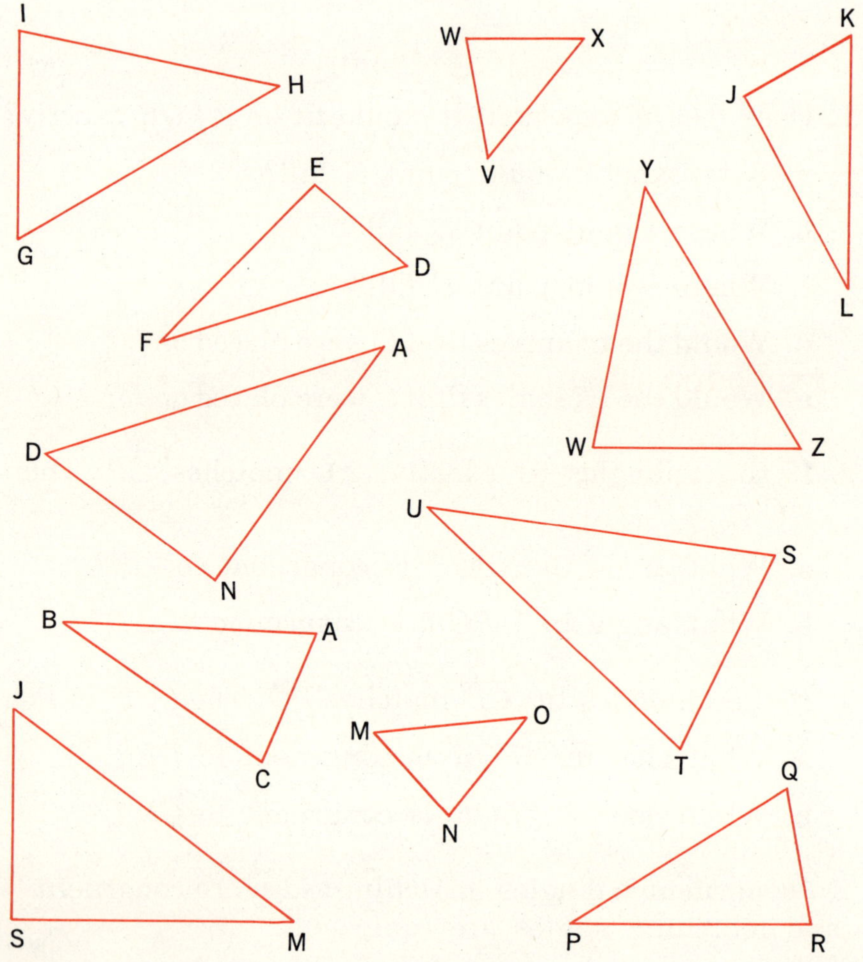

$\triangle PZJ \cong \triangle KQA$. Complete.

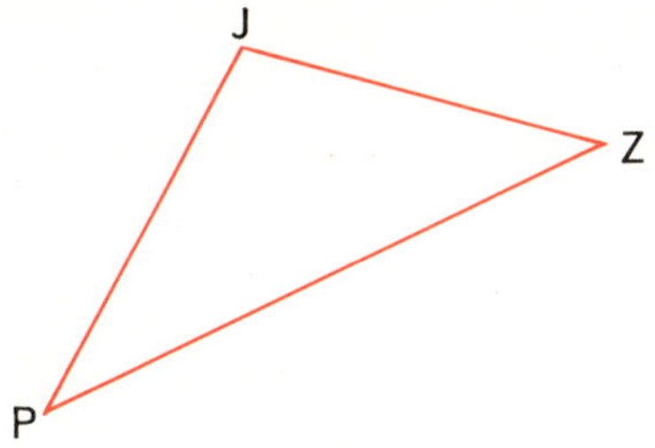

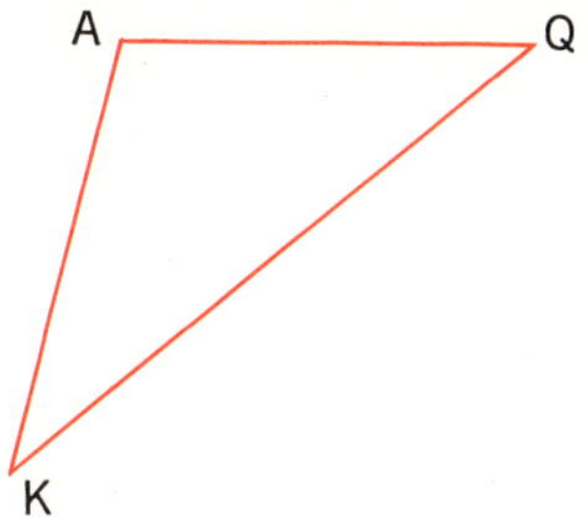

2. P matches ____

3. Q matches ____

4. J matches ____

5. A matches ____

6. $\overline{JZ} \cong$ ____

7. $\overline{KQ} \cong$ ____

8. $\overline{QK} \cong$ ____

9. $\angle P \cong$ ____

10. Draw a triangle congruent to this one.

   a. Tell which pairs of sides match.

   b. Tell which pairs of angles match.

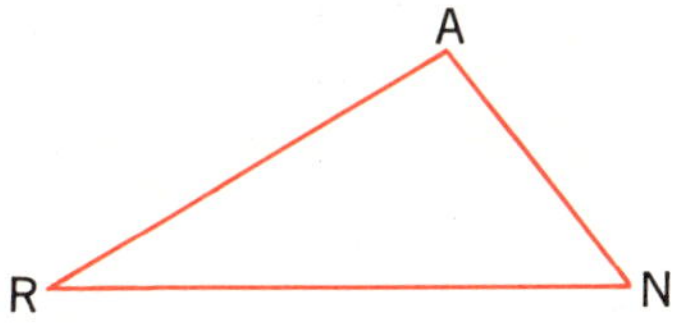

UNEXPLORED TERRITORY

1. Is every triangle congruent to itself?

2. Suppose $\triangle RST \cong \triangle GHK$. Is $\triangle GHK \cong \triangle RST$?

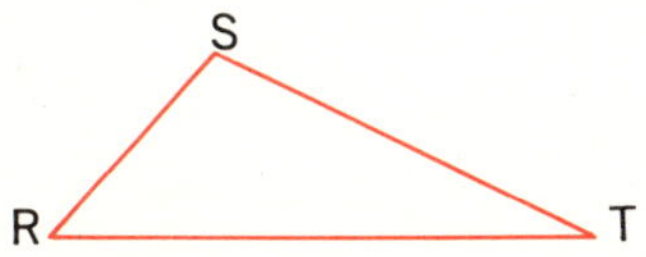

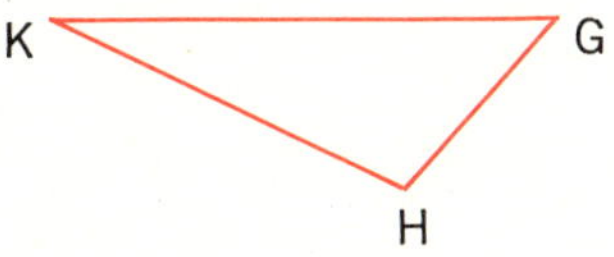

3. If $\triangle ABC \cong \triangle DEF$ and $\triangle DEF \cong \triangle GHI$, is $\triangle ABC \cong \triangle GHI$?

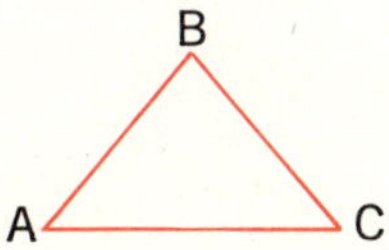

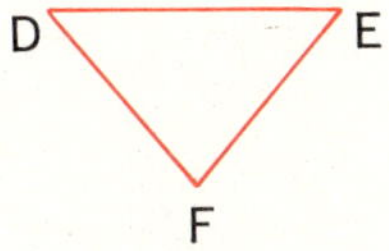

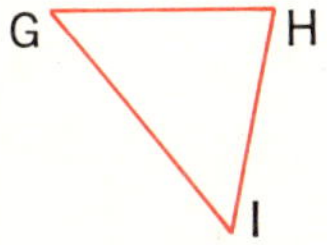

SIMILAR FIGURES

Some figures which look alike are not congruent.

1. Which pairs of figures look alike except for size?

a. b. c. d.

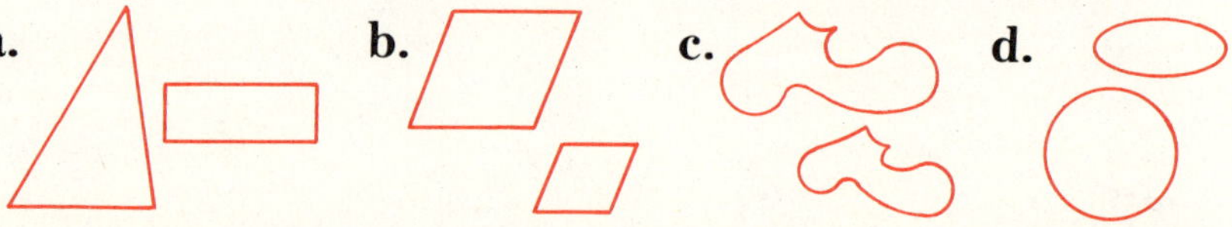

Similar figures look alike. They may or may not have the same size.

2. Which figures can be shown similar by reflection?

a. b. c. d.

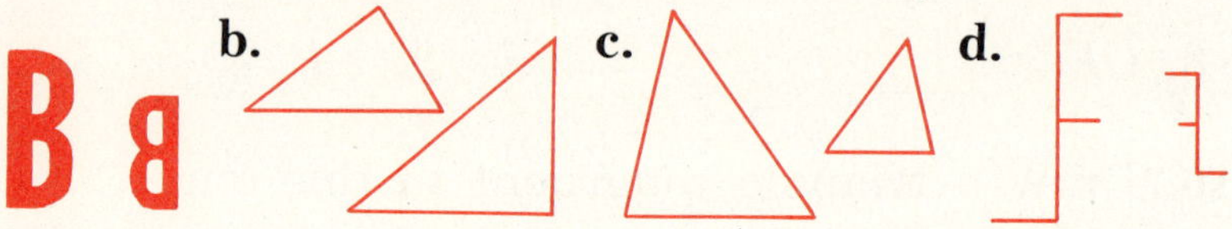

3. Which figures can be shown similar by rotation?

a. b. c. d.

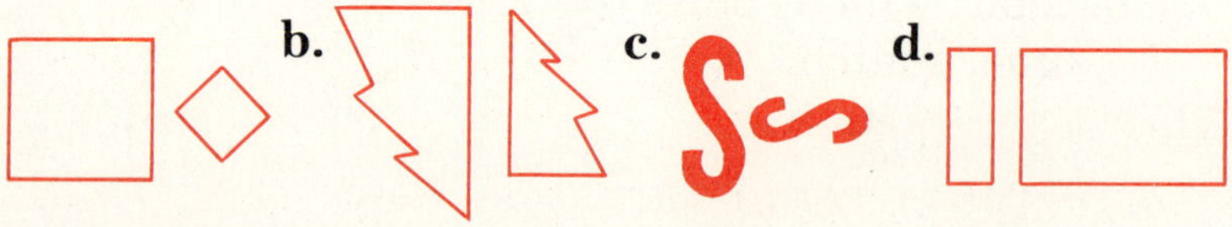

4. Which figures can be shown similar by rotation and reflection?

a. b. c. d.

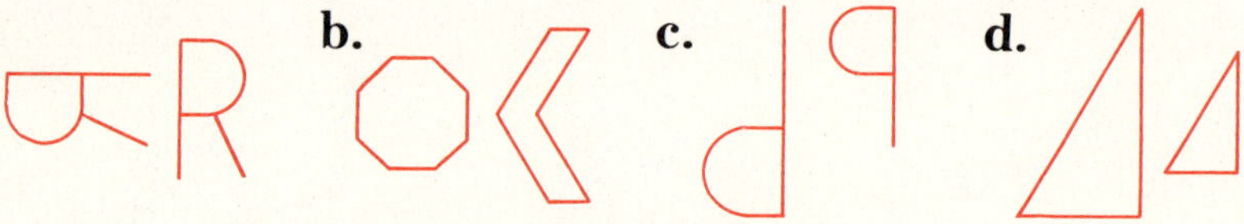

5. Which pairs of figures look similar?

a. b. c.

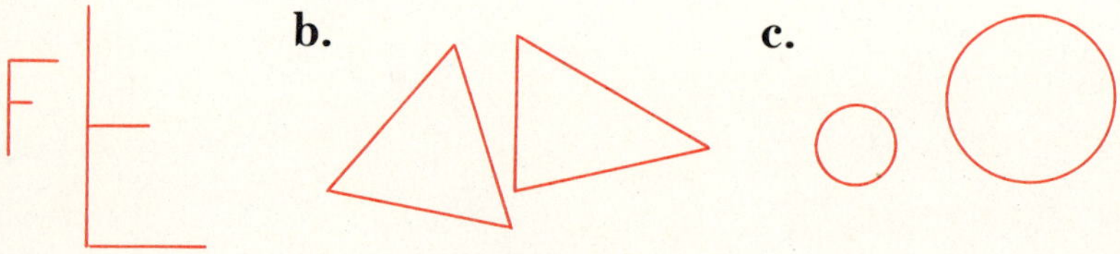

EXERCISES

Which figures in each set look similar to the black figure?

1. 

2. 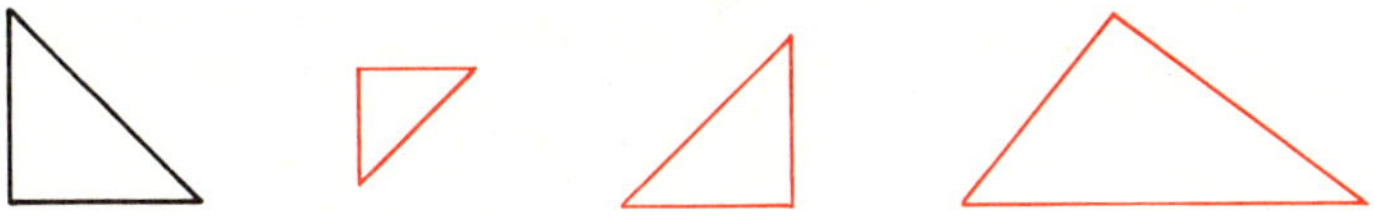

3. 

4. 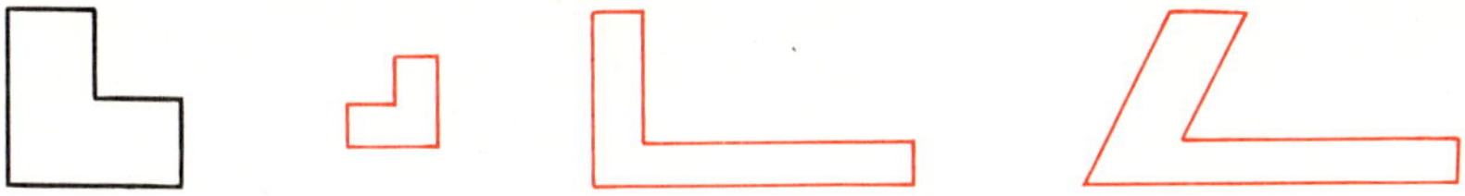

5. 

6. 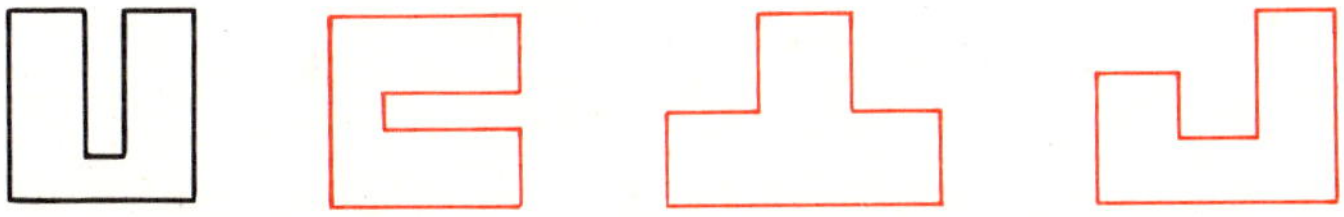

7. 

8. 

9. 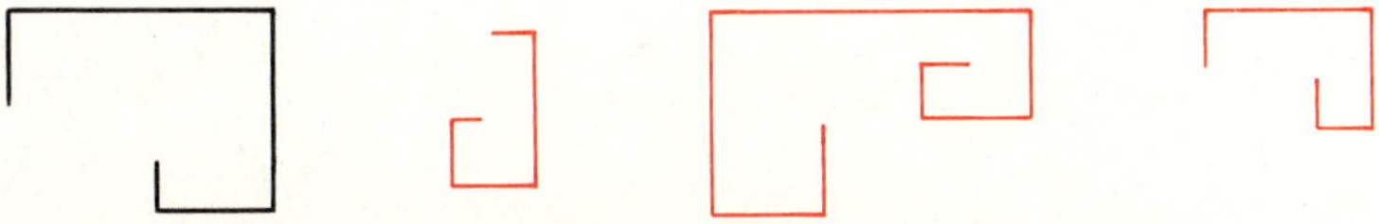

10. Are congruent figures similar?

COMPARING FIGURES

In each set which one is different? How is it different?

1.

2.
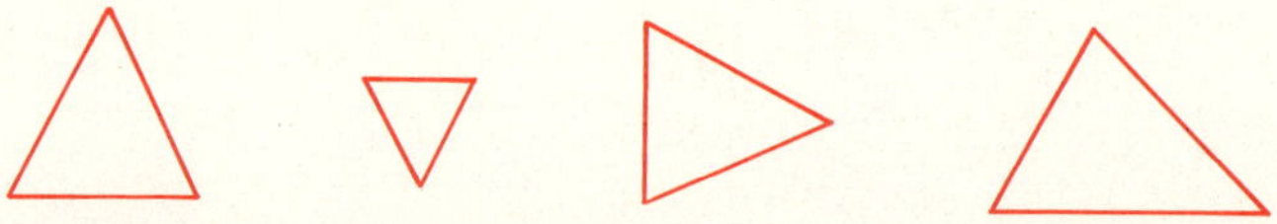

3.

4.
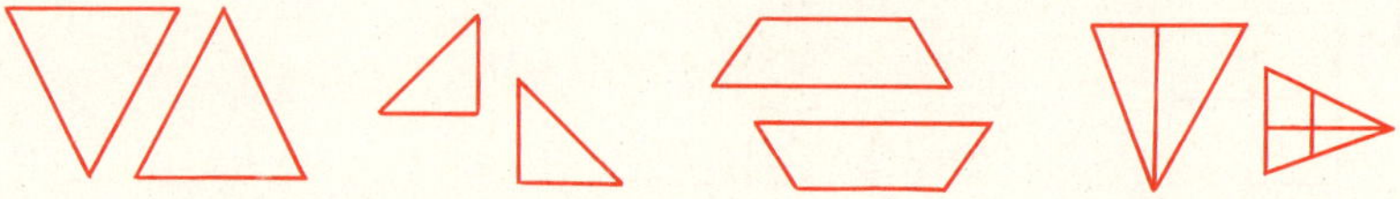

5.

6.
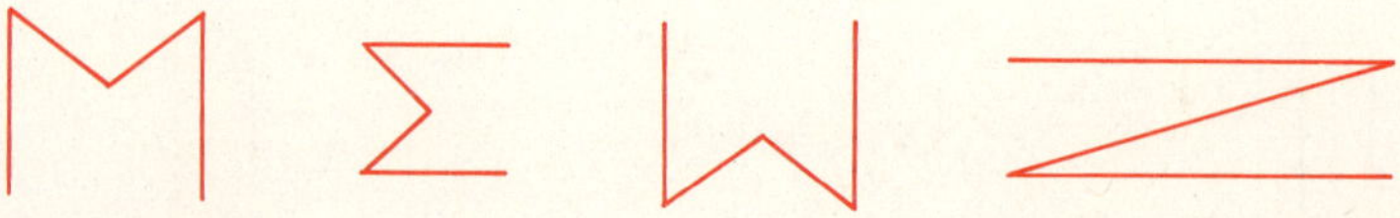

7.

8.
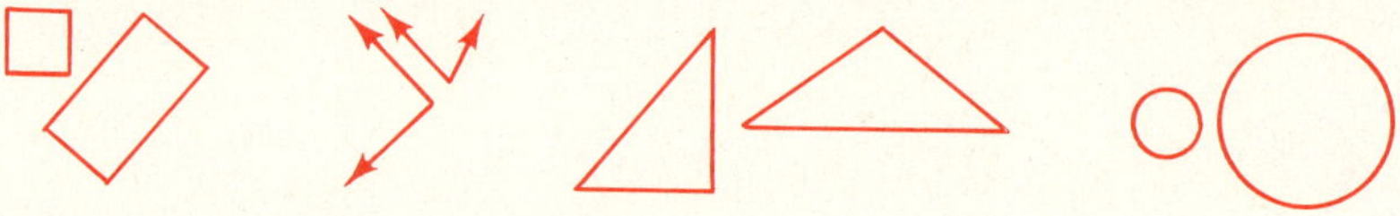

9.
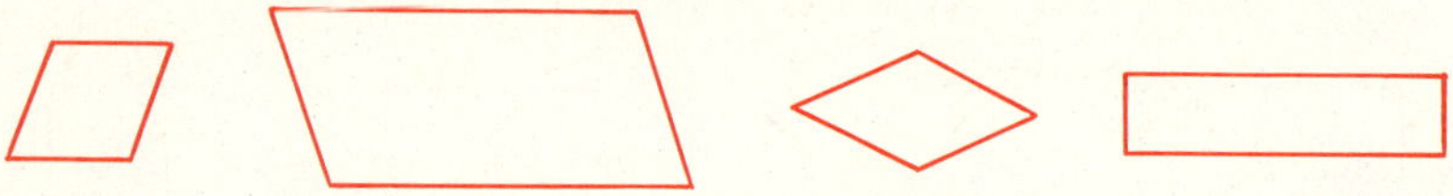

10.
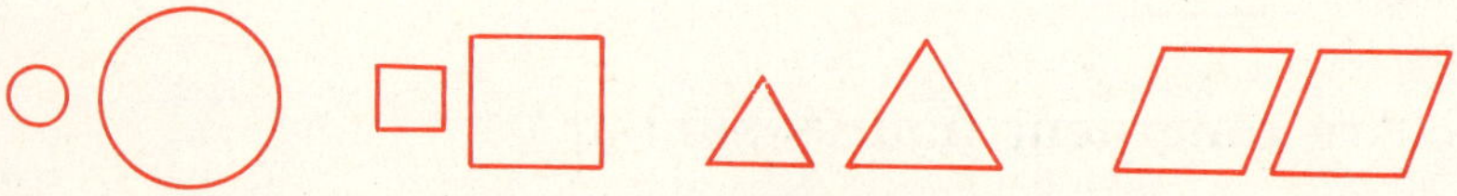

RATIOS AND PROPORTIONS

In Dave's model car collection, there are 6 antique cars and 11 modern cars. We say the *ratio* of antique cars to modern cars is 6 to 11, or $\frac{6}{11}$.

1. Write a fractional numeral for each ratio.

   a. 6 lb to 8 lb b. 11 in. to 2 in.

   c. 16 hr to 3 hr d. 4 yr to 19 yr

2. To say that two ratios are equivalent, we can write a *proportion.*

$$\frac{6}{9} = \frac{2}{3}$$

   This may be read "6 is to 9 as 2 is to 3."

   Read these.

   a. $\frac{5}{6} = \frac{20}{24}$ b. $\frac{10}{6} = \frac{5}{3}$ c. $\frac{9}{11} = \frac{27}{33}$ d. $\frac{5}{12} = \frac{15}{36}$

3. To find a missing term in a proportion, we can use *cross products.* Find the missing term.

Examples

$$\frac{2}{3} = \frac{x}{18} \qquad \frac{5}{7} = \frac{20}{y}$$

$$3 \cdot x = 2 \cdot 18 \qquad 5 \cdot y = 7 \cdot 20$$

$$3 \cdot x = 36 \qquad 5 \cdot y = 140$$

$$x = 36 \div 3 \qquad y = 140 \div 5$$

$$x = 12 \qquad y = 28$$

   a. $\frac{1}{2} = \frac{u}{8}$ b. $\frac{3}{5} = \frac{15}{y}$ c. $\frac{12}{6} = \frac{7}{x}$ d. $\frac{8}{t} = \frac{32}{2}$

EXERCISES

Solve these proportions.

1. $\frac{x}{3} = \frac{9}{27}$ 2. $\frac{1}{4} = \frac{n}{12}$ 3. $\frac{2}{3} = \frac{x}{12}$ 4. $\frac{3}{8} = \frac{9}{n}$

5. $\frac{5}{1} = \frac{20}{y}$ 6. $\frac{17}{n} = \frac{4}{3}$ 7. $\frac{4}{3} = \frac{9}{x}$ 8. $\frac{7}{x} = \frac{42}{2}$

9. $\frac{7}{x} = \frac{21}{6}$ 10. $\frac{11}{3} = \frac{5}{x}$ 11. $\frac{8}{5} = \frac{40}{y}$ 12. $\frac{12}{15} = \frac{x}{5}$

DO YOU REMEMBER?

Compute and simplify.

1. $\frac{7}{8} + \frac{1}{2}$
2. $\frac{1}{6} + \frac{5}{8}$
3. $\frac{1}{9} + \frac{5}{6}$
4. $\frac{5}{6} + \frac{3}{4}$
5. $6\frac{1}{8} + 3\frac{1}{3}$
6. $8\frac{1}{9} + 5\frac{1}{4}$
7. $3\frac{1}{2} + 4\frac{3}{4}$
8. $7\frac{1}{4} + 6\frac{2}{7}$
9. $\frac{7}{8} - \frac{3}{4}$
10. $\frac{9}{10} - \frac{2}{5}$
11. $\frac{11}{12} - \frac{2}{5}$
12. $\frac{18}{25} - \frac{2}{3}$
13. $9\frac{7}{8} - 3\frac{1}{2}$
14. $10\frac{2}{9} - 6\frac{2}{3}$
15. $9\frac{1}{2} - 4\frac{7}{8}$
16. $17\frac{1}{4} - 12\frac{5}{8}$
17. $\frac{5}{8} \times \frac{7}{9}$
18. $\frac{3}{5} \times \frac{6}{7}$
19. $\frac{9}{10} \times \frac{8}{11}$
20. $\frac{3}{7} \times \frac{5}{9}$
21. $1\frac{1}{2} \times 2\frac{5}{6}$
22. $3\frac{2}{7} \times 2\frac{4}{5}$
23. $5\frac{1}{6} \times 2\frac{4}{7}$
24. $8\frac{1}{3} \times 6\frac{3}{4}$
25. $\frac{3}{4} \div \frac{7}{8}$
26. $\frac{5}{16} \div 5$
27. $\frac{4}{5} \div \frac{8}{15}$
28. $\frac{5}{16} \div \frac{3}{8}$
29. $4\frac{2}{3} \div 7$
30. $1\frac{5}{16} \div 2\frac{1}{2}$
31. $12 \div 4\frac{3}{4}$
32. $3\frac{2}{3} \div 4\frac{1}{2}$
33. $8\frac{2}{3} \div 1\frac{1}{3}$

Solve these problems.

34. Mr. Berna's land is rectangular in shape. It is $\frac{5}{8}$ mile long and $\frac{4}{5}$ mile wide. Find its area in square miles.

35. Ginny wrapped 24 small favors. She used $\frac{2}{3}$ foot of ribbon for each. How many feet did she use in all?

PROBLEM SOLVING

1. The state had to pay 25% of the cost of improving a highway. The state paid $42,500. What was the total cost of the improvements?

2. Mr. May had a choice of working for a salary of $125 per week plus 2% commission, or no salary with a 22% commission. During one week his daily sales were $72.80, $86.42, $108.10, $94.80, and $134.20. Which method would give him the greatest pay?

3. A circular pond at the city park has a 70-foot diameter. A new fence to enclose it will cost $1.12 per foot. Find the cost of fencing the pond.

4. In the engine of Jack's car the air-fuel mixture was compressed from a volume of 42.35 in.^{3} to 6.05 in.^{3}. The ratio $\frac{42.35}{6.05}$ is called the *compression* ratio. Find a simpler name for the compression ratio in Jack's car.

5. Herb, a 4-H Club member, sold a 1,325-pound steer for $25 per 100 pounds. How much did he receive?

6. The city had a parking lot tax of 10¢ per car. Mr. Harris said that on an average day he parked about 280 cars. How much tax would he collect?

SIMILAR TRIANGLES

Let's consider two similar triangles.

LET'S EXPLORE

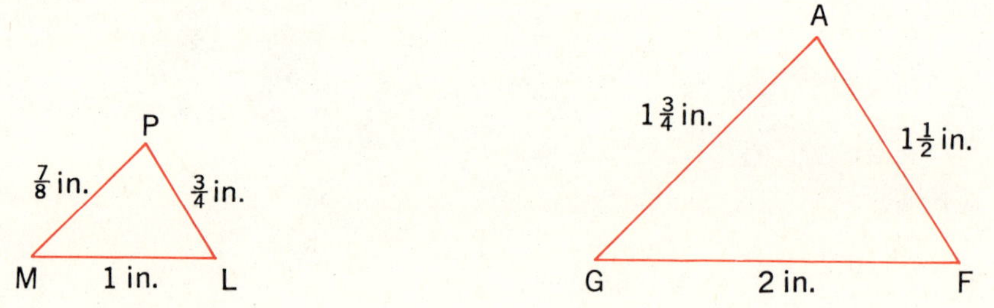

1. a. Which pairs of angles match?

   b. Measure each angle. What is true about the matching angles?

2. a. Which pairs of sides match?

   b. Measure each side.

   c. What is the ratio of $m\ \overline{MP}$ to $m\ \overline{GA}$?

   d. What is the ratio of $m\ \overline{PL}$ to $m\ \overline{AF}$?

   e. What is the ratio of $m\ \overline{ML}$ to $m\ \overline{GF}$?

   f. What is true about the measures of the matching sides?

3. Here are two more similar triangles.

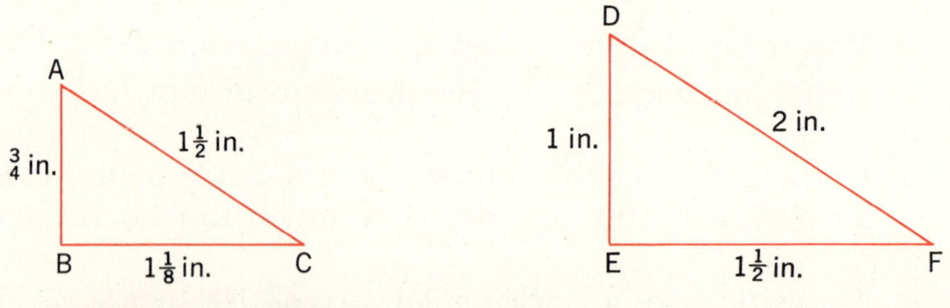

   a. Measure each angle. What is true about the matching angles?

b. Measure each side.

c. What is the ratio of $m\,\overline{AB}$ to $m\,\overline{DE}$?

d. What is the ratio of $m\,\overline{AC}$ to $m\,\overline{DF}$?

e. What is the ratio of $m\,\overline{BC}$ to $m\,\overline{EF}$?

f. What is true about the measures of the matching sides?

In similar triangles, matching angles are congruent and measures of matching sides have the same ratio.

4. To say "$\triangle ABC$ is similar to $\triangle DEF$," we write $\triangle ABC \sim \triangle DEF$. Read these.

a. $\triangle RST \sim \triangle XYZ$ b. $\triangle MNO \sim \triangle TSV$

5. $\triangle PQR \sim \triangle ABC$.

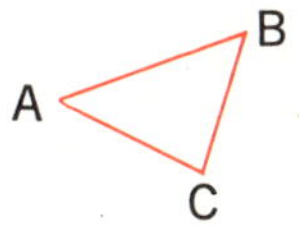

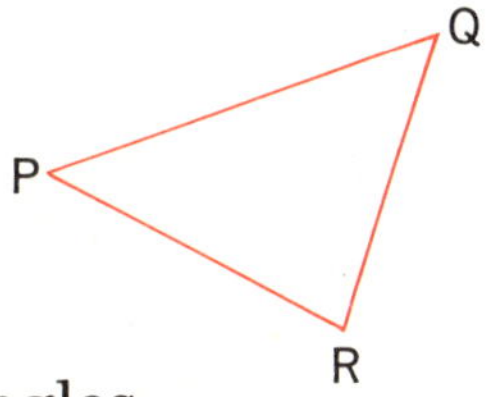

a. Name the matching sides and angles.

b. We can write this proportion about the sides.

$$\frac{m\,\overline{PQ}}{m\,\overline{AB}} = \frac{m\,\overline{PR}}{m\,\overline{AC}} = \frac{m\,\overline{QR}}{m\,\overline{BC}}$$

What can we say about the angles?

6. $\triangle RST \sim \triangle XYZ$.

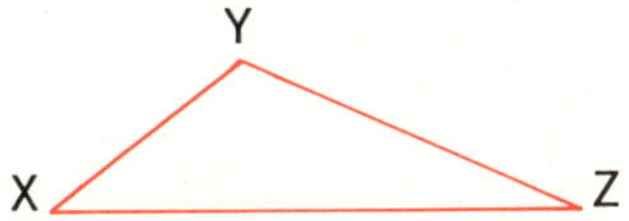

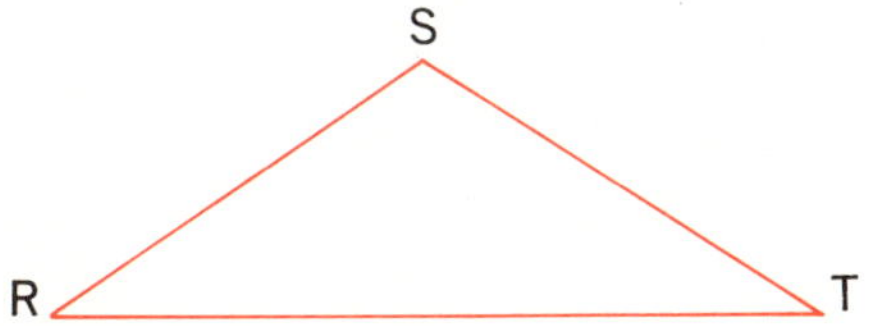

a. Name the matching sides and angles.

b. Write a proportion about the sides.

c. Which pairs of angles are congruent?

EXERCISES

For each pair of similar triangles, tell which pairs of angles are congruent and write a proportion about the sides.

1.
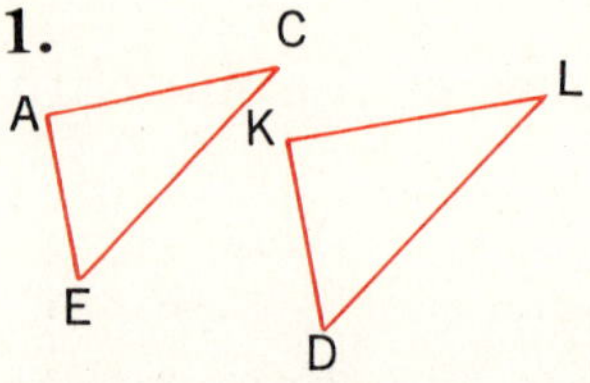

2.
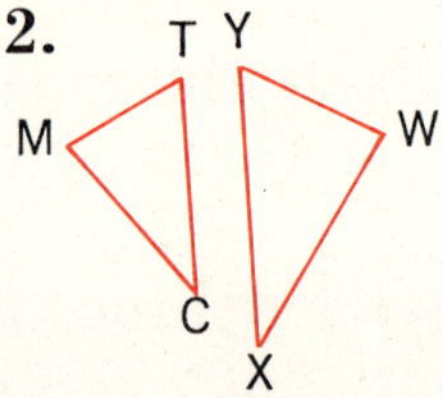

3.
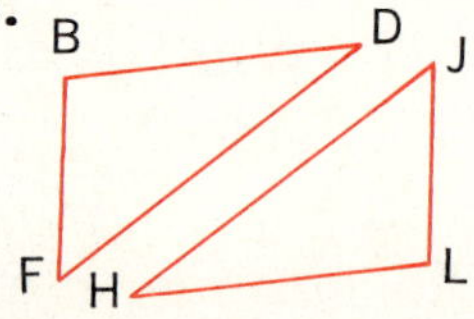

4.
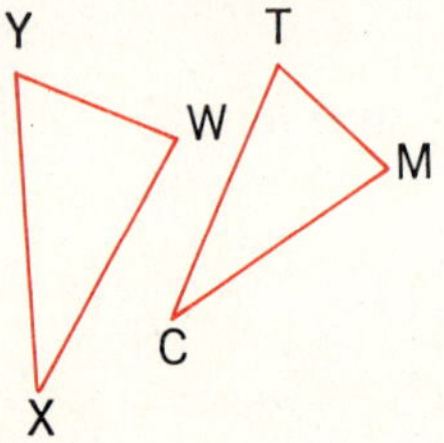

5.
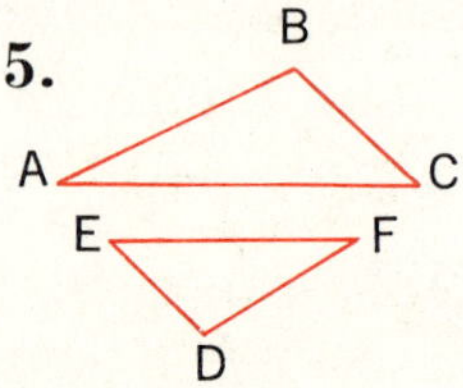

6.
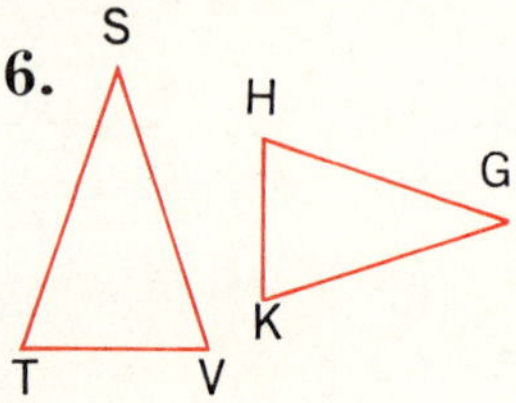

7.
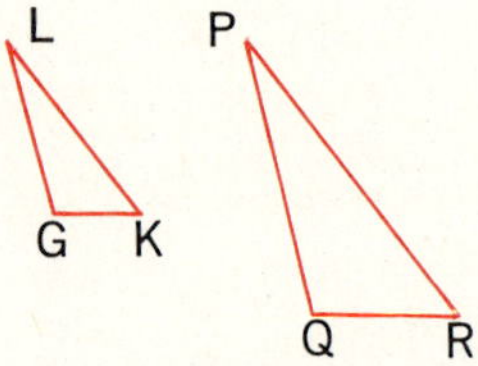

8.
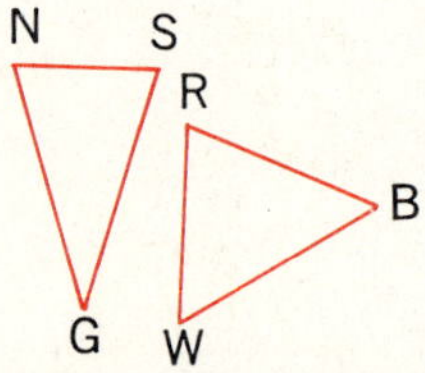

9.
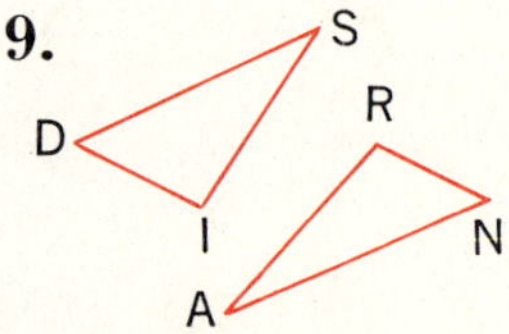

10. **a.** If two triangles are congruent, are they similar?

  b. If so, what is the ratio of the matching sides?

Let's Practice

Solve these problems.

1. In the community clean-up campaign, 87 people worked one Saturday for 6 hours. They collected about 1,500 pounds of trash. About how much trash was collected by each person.

2. Jerry needs about $75 to put racing tires on his car. He earns $17.50 each Saturday working at the drive-in. About how long will it take him to save for the tires?

FINDING MISSING MEASURES

We can find missing measures in similar triangles.

1. $\triangle DEF \sim \triangle RST$.

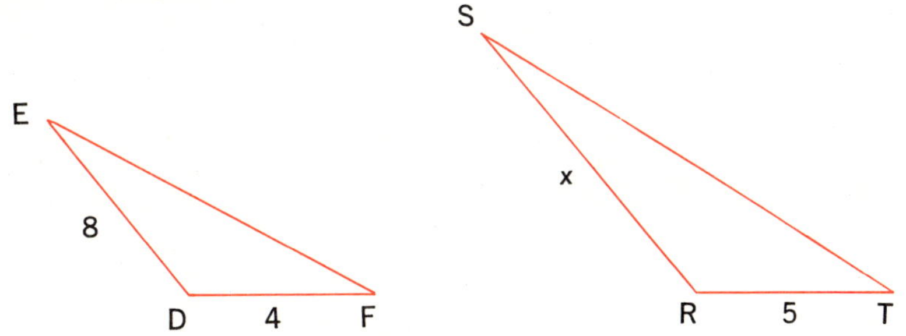

a. Which pairs of angles are congruent?

b. $m\angle D = 128°$ and $m\angle F = 29°$. Find $m\angle E$.

2. We know $m\overline{DF} = 4$, $m\overline{RT} = 5$, and $m\overline{DE} = 8$. We can find $m\overline{RS}$ this way.

$$\frac{m\overline{DF}}{m\overline{RT}} = \frac{m\overline{DE}}{m\overline{RS}}$$

$$\frac{4}{5} = \frac{8}{x}$$

$$4 \cdot x = 40$$

$$x = 40 \div 4$$

$$x = 10$$

What is $m\overline{RS}$?

3. $\triangle ABC \sim \triangle XYZ$.

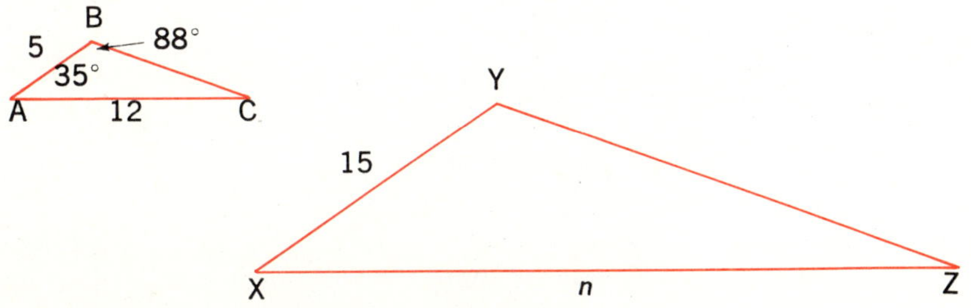

a. Find $m\angle X$, $m\angle Y$, and $m\angle Z$.

b. Write a proportion and find mXZ.

EXERCISES

1. $\triangle GHI \sim \triangle KJL$.

a. Find $m\angle L$, $m\angle H$, and $m\angle K$.

b. Find $m\,\overline{KJ}$.

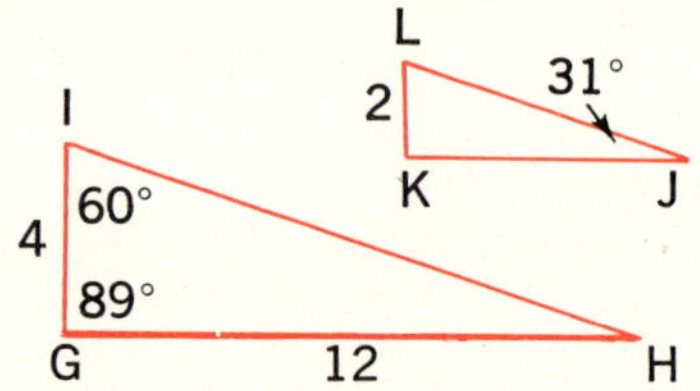

2. $\triangle DEF \sim \triangle RST$.

a. Find $m\angle S$, $m\angle T$, and $m\angle D$.

b. Find $m\overline{RT}$.

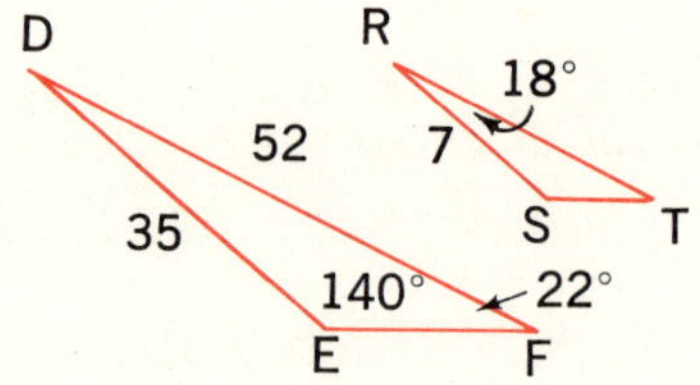

3. $\triangle KLM \sim \triangle PQR$.

a. Find $m\angle M$, $m\angle P$, $m\angle Q$, and $m\angle R$.

b. Find $m\overline{PQ}$ and $m\overline{QR}$.

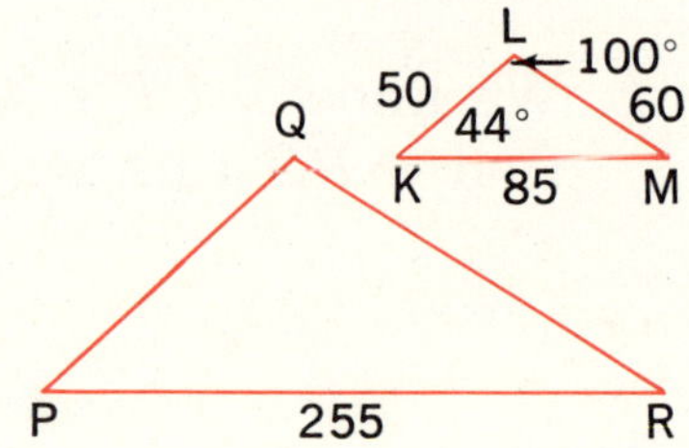

4. $\triangle GHK \sim \triangle VTU$.

a. Find $m\angle V$ and $m\angle U$.

b. Find $m\,\overline{TU}$ and $m\,\overline{VU}$.

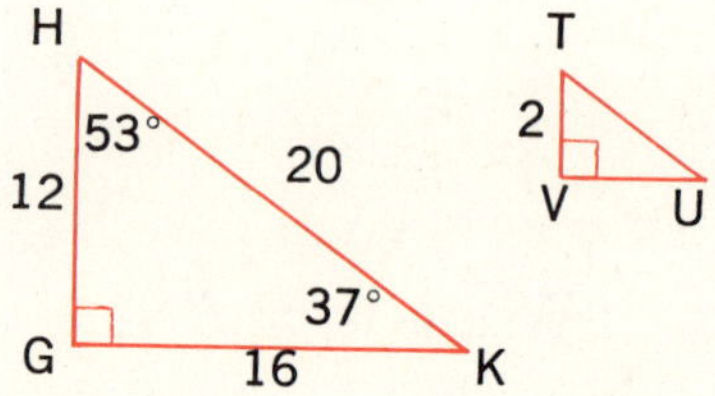

5. $\triangle XYZ \sim \triangle MNO$.

a. Find $m\angle MNO$ and $m\angle MON$.

b. Find $m\overline{MO}$ and $m\overline{NO}$.

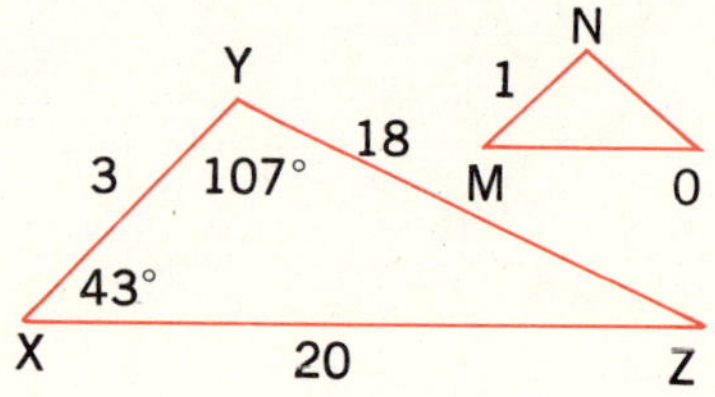

6. $\triangle GFH \sim \triangle LCT$.

Find $m\overline{LT}$ and $m\overline{FH}$.

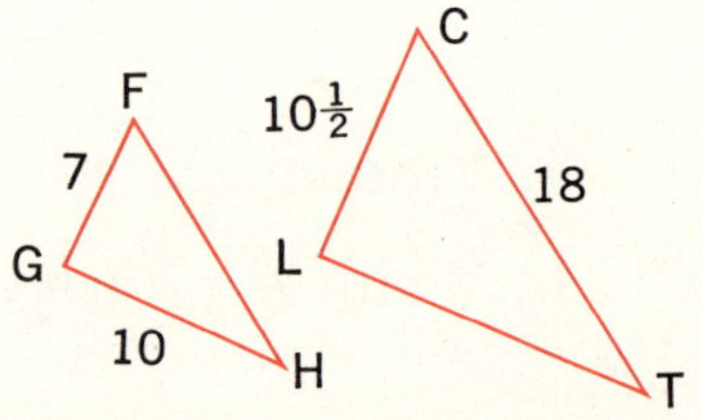

7. $\triangle PQR \sim \triangle MLN$.
Find $m\overline{LN}$ and $m\overline{MN}$.

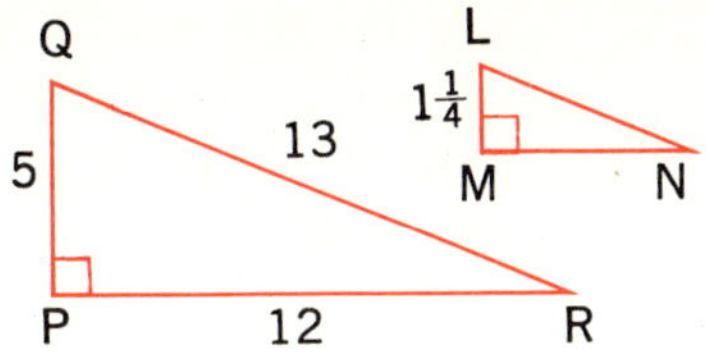

8. $\triangle GHK \sim \triangle ABC$.
Find $m\ \overline{AC}$ and $m\ \overline{BC}$

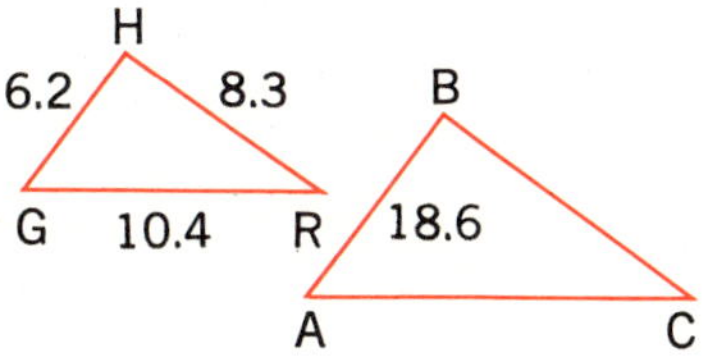

9. $\triangle RST \sim \triangle XYZ$.
Find $m\overline{XY}$ and $m\overline{ZY}$.

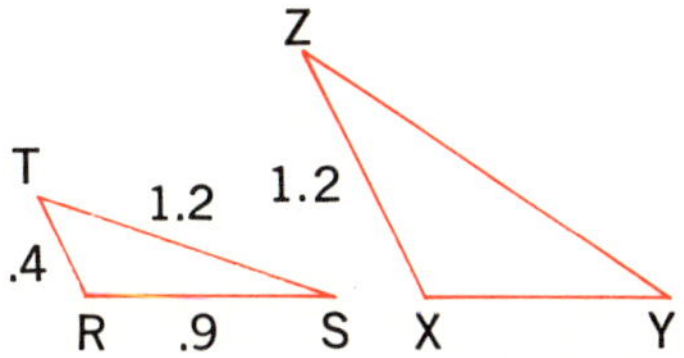

10. $\triangle KLM \sim \triangle TPQ$.
Find $m\overline{PQ}$ and $m\overline{TQ}$.

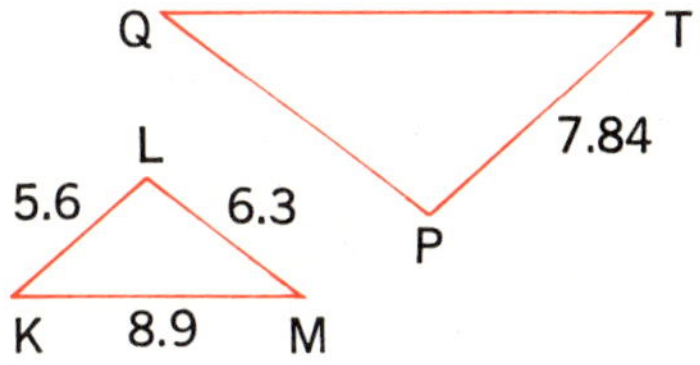

Let's Practice

Solve these problems.

1. Mr. Steigelman's investment club made the following purchases:

   28 shares of Gladstone at $\$18\frac{1}{2}$ share

   52 shares of Steel Works at $\$49\frac{3}{4}$ per share

   41 shares of Comptex at $\$21\frac{7}{8}$ per share

   Find the cost of the club's investments.

2. A finance company kept the records of charge accounts for Mr. Everly's store. The cost was $5\frac{3}{4}$¢ per $1 of charges. In July, Mr. Everly's charges totaled $18,500. How much did he have to pay to the finance company?

INDIRECT MEASUREMENT

Whenever we calculate a measurement from other measures, we are using indirect measurement. We can often use similar figures for indirect measurement.

1. Let's find the height of the flagpole using its shadow, a boy's height, and his shadow.

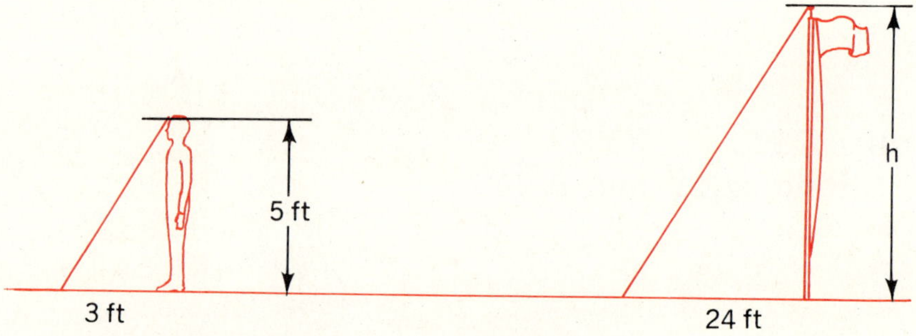

   a. How long is the boy's shadow?

   b. How long is the flagpole's shadow?

   c. What is the height of the boy?

   d. Solve this proportion.

$$\frac{3}{24} = \frac{5}{h}$$

   e. How high is the flagpole?

2. A picture of a school is 1 inch by 4 inches. The school is 140 feet long. Let's find the height of the school.

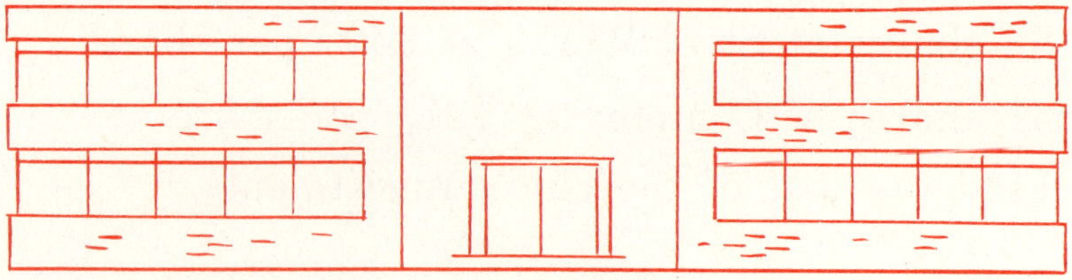

   a. Solve: $\frac{1}{n} = \frac{4}{140}$.

   b. How high is the school?

3. Aristarchus was an ancient Greek who lived in the 200's B.C. He devised the following method for finding the ratios of the distances between the earth, the moon, and the sun. He went out on a day when the half-moon and the sun were in the sky. (There are days when you can see the moon and the sun at the same time.) He measured the angle between the sun, the earth, and the moon. Then he drew a right triangle.

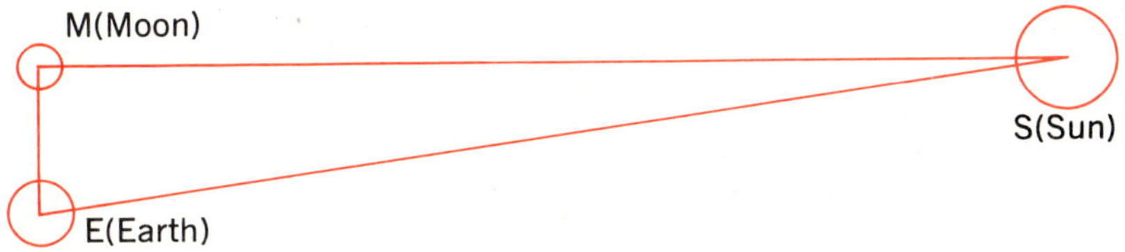

a. How did he know the angle at M was a right angle?

b. What segments would you measure to get the following ratio?

$$\frac{\textit{distance between moon and earth}}{\textit{distance between sun and earth}}$$

c. Perform Aristarchus' experiment.

EXERCISES

1. To measure the big tree in his backyard, Riley measured his shadow and the tree's shadow, as shown in the drawing. Riley is 6 feet tall. How high is the tree?

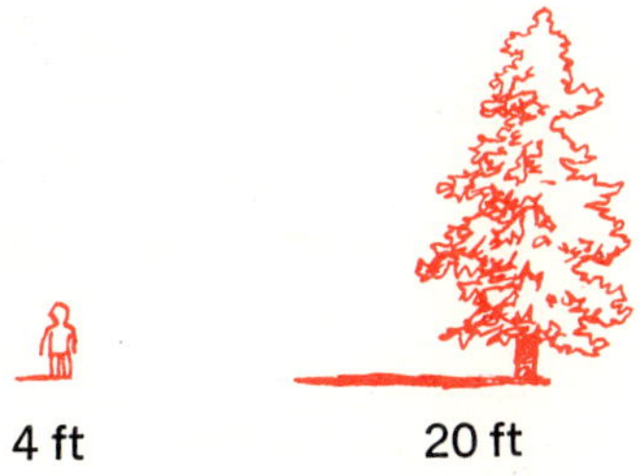

2. Terry made a scale drawing of a church. The drawing is 6 inches wide and 12 inches high. The church is 40 feet wide. How high is the church?

3. Chris used similar triangles to measure the distance across a pond in the city park. He drove stakes at A and B and chose point C so that the stakes could be seen. To make similar triangles, he measured $\overline{AC}$ and $\overline{BC}$ and chose points D and E so that $\frac{m\overline{AC}}{m\overline{CD}} = \frac{m\overline{BC}}{m\overline{CE}}$. If $m\overline{AC} = 35$ feet, $m\overline{CE} = 8$ feet, and $m\overline{DE} = 12$ feet, how far is it across the pond?

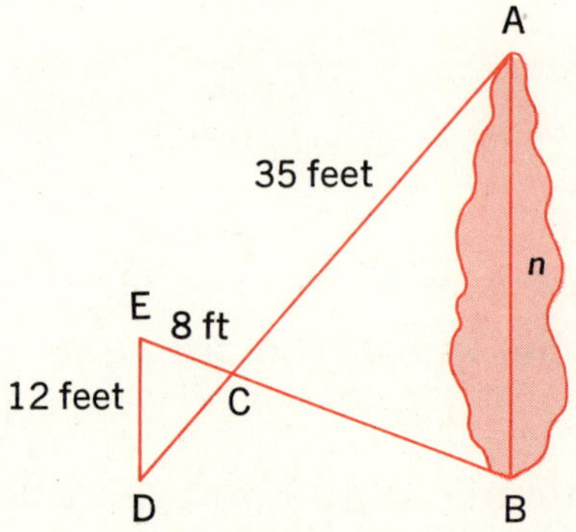

4. A yardstick held vertically casts a shadow 4.5 feet long. At the same time a television tower casts a shadow 90 feet long. How tall is the tower?

UNEXPLORED TERRITORY

You can build this instrument to find unknown distances. Through the fixed straw, sight an object whose distance is desired. Then sight the object through the movable straw. Measure the angle with the protractor. Finally, make a scale drawing of a similar triangle. Then use proportions to find the desired distance.

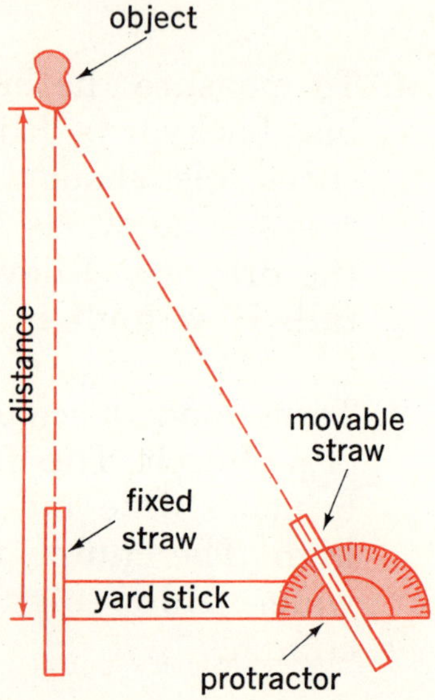

1. Build this instrument.
2. Use the instrument to measure distances.

CHAPTER REVIEW

Which pair of triangles looks congruent?

1. 

2. 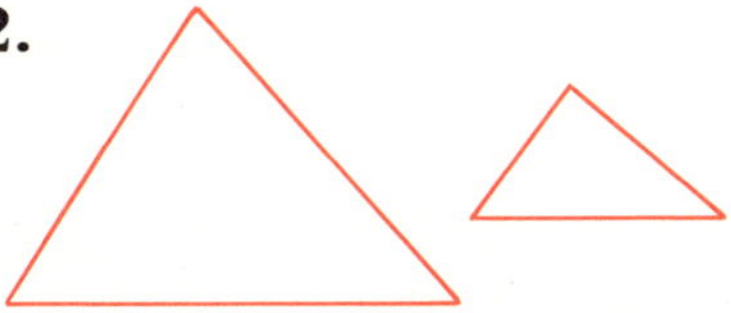

Which pair of figures can be shown similar by reflection?

3. 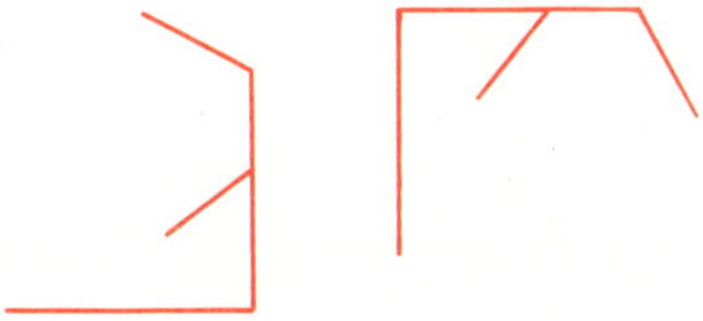

4. 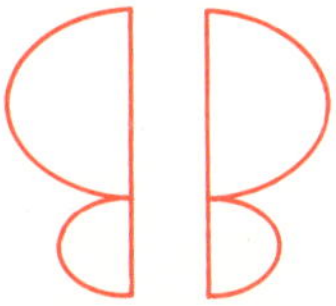

Which pair of figures can be shown similar by rotation?

5. 

6. 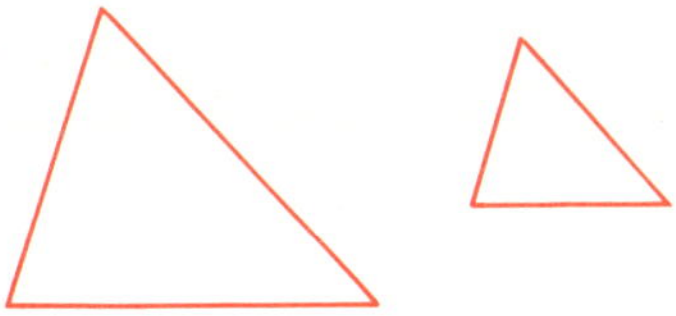

Which one is different? How is it different?

7. 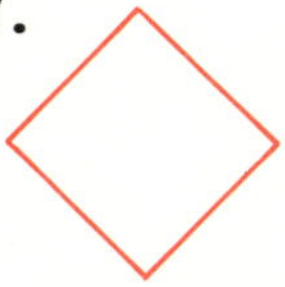

8. 

9. 

10. 

Write a fractional numeral for each ratio.

11. 3 ft to 5 ft   **12.** 7 hr:14 hr   **13.** 2 yd to 1 yd

Solve these proportions.

14. $\frac{x}{2} = \frac{5}{10}$ **15.** $\frac{2}{3} = \frac{n}{18}$ **16.** $\frac{5}{y} = \frac{20}{28}$

17. $\frac{7}{8} = \frac{x}{10}$ **18.** $\frac{5}{8} = \frac{12}{y}$ **19.** $\frac{9}{4} = \frac{y}{14}$

$\triangle DEF \sim \triangle GHI$.

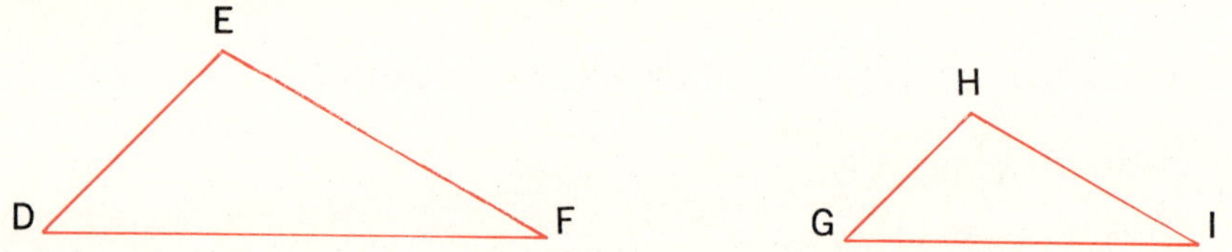

20. Name the matching sides and angles.

21. Write a true proportion about the sides.

22. Which pairs of angles are congruent?

For each pair of similar triangles, tell which pairs of angles are congruent and write a true proportion about the sides.

23. 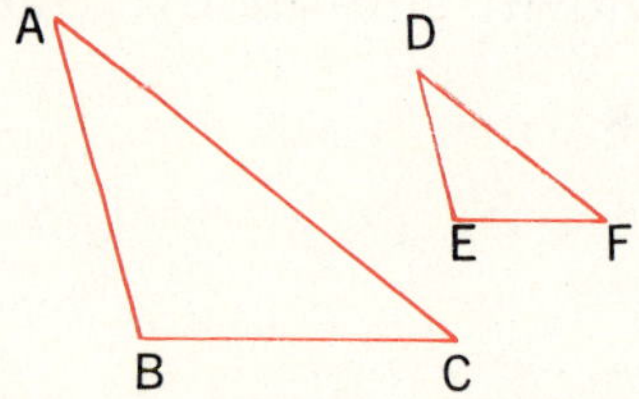

24. 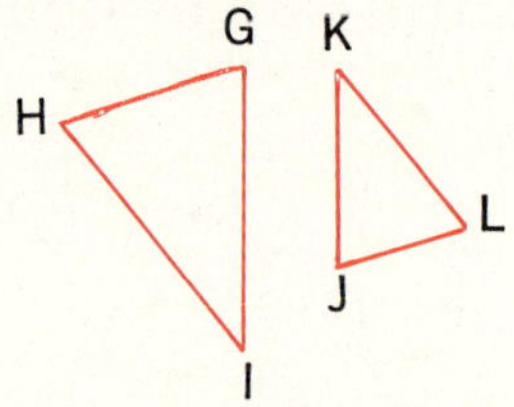

$\triangle MNO \sim \triangle PQR$.

25. Find $m\angle Q$ and $m\angle R$.

26. Find $m\ \overline{NO}$ and $m\ \overline{MO}$.

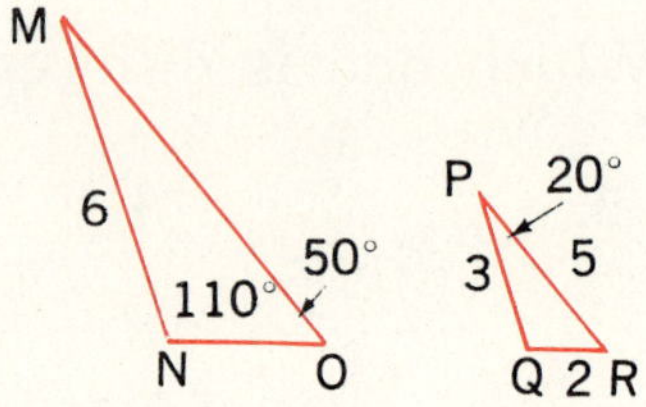

$\triangle STU \sim \triangle WXV$.

27. Find $m\angle T$ and $m\angle W$.

28. Find $m\,\overline{WX}$ and $m\,\overline{XV}$.

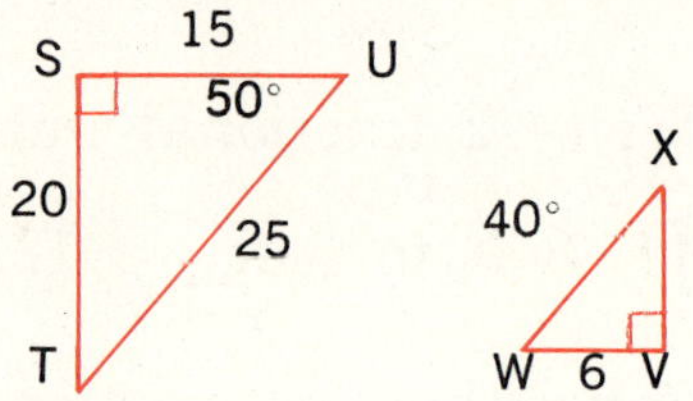

Terms You Should Know

congruent	proportion
similar	cross product
ratio	indirect measurement

CHAPTER TEST

Which pair of triangles looks congruent.

1.

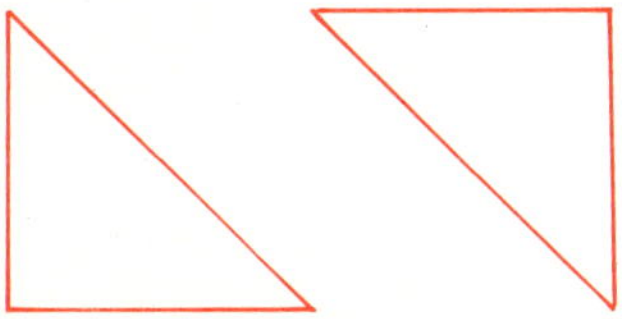

2.

Tell whether each pair can be shown similar by rotation or by reflection.

3.

4.

Write a fractional numeral for each ratio.

5. 9 yd to 13 yd

6. 30 ft:50 ft

Solve these proportions.

7. $\frac{x}{5} = \frac{15}{25}$

8. $\frac{3}{4} = \frac{y}{20}$

9. $\frac{8}{9} = \frac{n}{10}$

For each pair of similar triangles, tell which pairs of angles are congruent and write a proportion about the sides.

10.

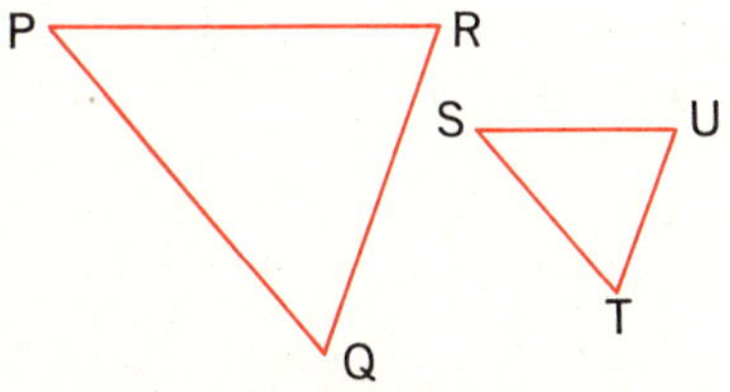

11.

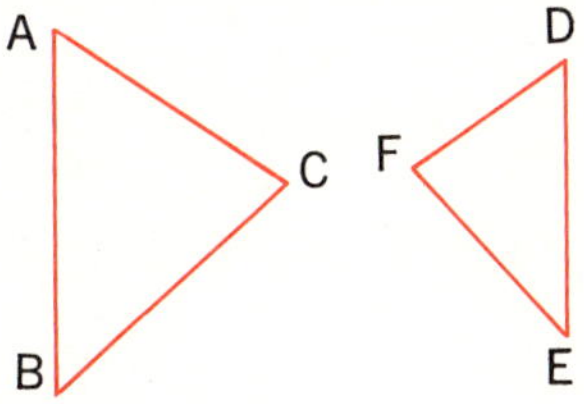

$\triangle ABC \sim \triangle DEF$.

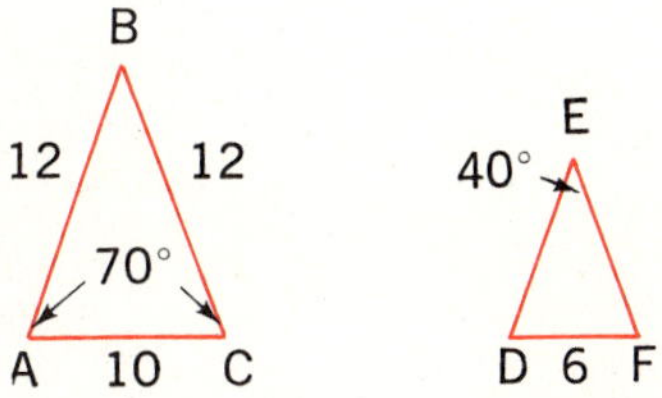

12. Find $m\angle B$, $m\angle D$, and $m\angle F$

13. Find $m\overline{DE}$ and $m\overline{EF}$.

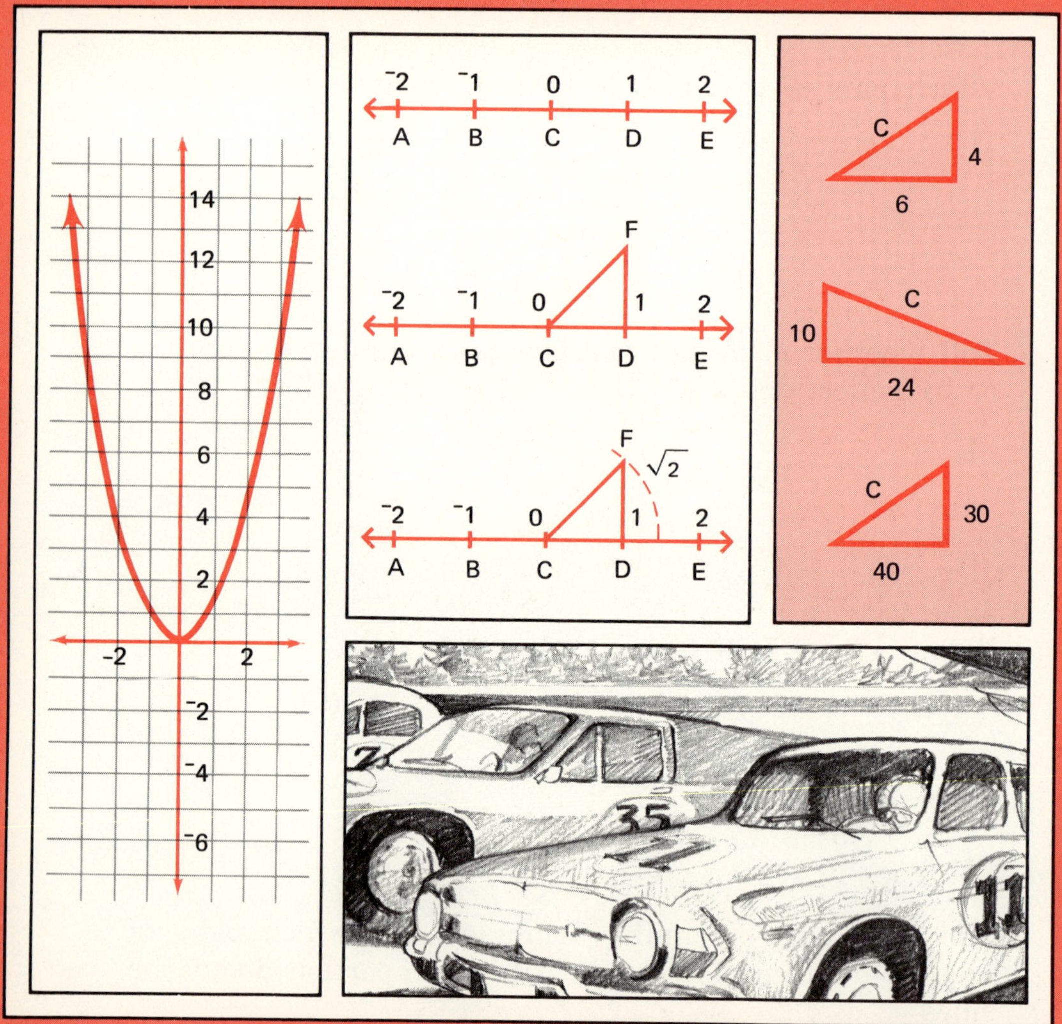

CHAPTER ELEVEN

Real Numbers

NON-REPEATING DECIMALS

We know that terminating decimals name rational numbers. Also, repeating decimals name rational numbers. Here are several examples of rational numbers.

.125	13.07	$4.56\overline{56}$	$1.237\overline{237}$
terminating decimals		*repeating decimals*	

1. Consider these non-terminating decimals. The three dots mean that the digits continue forever in the manner shown.

   .1222122212221222 . . .

   .121221222122221222221 . . .

   a. We can write the first decimal with a bar like this. $.1222\overline{1222}$ Why?

   b. We cannot write the second decimal with a bar. Why not?

2. Here is another non-terminating decimal which cannot be written with a bar.

   .651651165111651111165 . . .

   We call it a *non-repeating decimal,* since no group of digits repeats forever. Make up three more non-terminating, non-repeating decimals.

> Non-terminating, non-repeating decimals do not name rational numbers. They name **irrational numbers**.

$.122212221222\overline{1222}$	Non-terminating repeating decimal Rational number
.121221222122221 . . .	Non-terminating non-repeating decimal Irrational number

3. Which name rational and which name irrational numbers?

   a. 202002000200002 . . . b. $^{-}6.413$ c. $.576\overline{576}$

EXERCISES

1. Make up numerals for five irrational numbers.

2. A name for π is 3.1415926535 . . . At no time does a pattern repeat, no matter how far the decimal is carried out. What kind of number is π?

Which name rational and which name irrational numbers?

3. $.8\overline{8}$

4. .787887888788887 . . .

5. $70.27\overline{27}$

6. $^{-}.183$

7. $7.12345\overline{12345}$

8. π

9. $^{-}29.141441444144441 \ldots$

10. $^{-}.246246$

Here are several ways to make up numerals. In each case, start with a decimal point, and do the following:

11. Flip a coin. If heads comes up, write a 1. If tails comes up, write a 0. Do this at least 10 times.

12. Toss a die. Record the number shown on the die. Do this at least 20 times.

13. Pick a card from a deck. Record the card's value. (Use the values shown below.) Replace the card before you draw again. Do this at least 30 times.

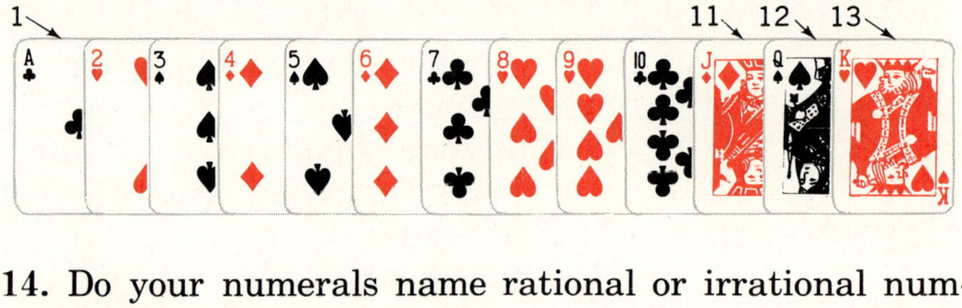

14. Do your numerals name rational or irrational numbers?

15. Suppose you continue each process in Exercises 11-13 forever. What kind of numbers would you have?

REAL NUMBERS

The set of *real numbers* contains all the rational and all the irrational numbers.

1. Name three real numbers that are rational.

2. Name three real numbers which are irrational.

3. a. Name three real numbers that are integers.

   b. Why is every integer a real number?

4. Why is every whole number a real number?

5. Which do not name whole numbers?

   a. $\sqrt{9}$ b. $\sqrt{6}$ c. $\sqrt{31}$ d. $\sqrt{16}$ e. $\sqrt{97}$

If the square root of a whole number is not a whole number, then it is an irrational number. Numbers like $\sqrt{6}$ and $\sqrt{31}$ are irrational.

6. Which of these real numbers are irrational? Use the table on page 406.

   a. $\sqrt{49}$ b. $\sqrt{52}$ c. $\sqrt{81}$ d. $\sqrt{120}$ e. $\sqrt{100}$

   f. $\sqrt{225}$ g. $\sqrt{196}$ h. $\sqrt{325}$ i. $\sqrt{448}$ j. $\sqrt{500}$

7. Name three square roots which are irrational.

EXERCISES

Which of these real numbers are irrational?

1. $\sqrt{1}$ 2. $\sqrt{18}$ 3. $\sqrt{36}$ 4. $\sqrt{87}$ 5. $\sqrt{64}$

6. $\sqrt{47}$ 7. $\sqrt{144}$ 8. $\sqrt{235}$ 9. $\sqrt{451}$ 10. $\sqrt{529}$

11. $\sqrt{602}$ 12. $\sqrt{729}$ 13. $\sqrt{796}$ 14. $\sqrt{853}$ 15. $\sqrt{900}$

16. $\sqrt{998}$ 17. $\sqrt{1,006}$ 18. $\sqrt{1,296}$ 19. $\sqrt{1,309}$ 20. $\sqrt{1,356}$

THE REAL NUMBER LINE

You may think that the rational numbers fill up all the points on a number line. This is not true. There are many gaps in a number line. These are filled by irrational numbers.

1. Copy this number line.

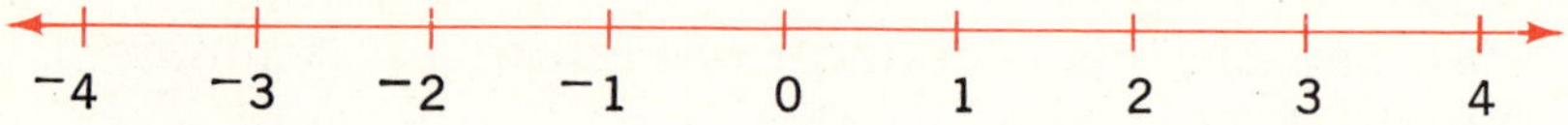

2. Locate these rational numbers on your number line.

$$\frac{1}{2}, \quad \frac{^-1}{3}, \quad \frac{3}{2}, \quad ^-.75, \quad 3.31, \quad 2\frac{1}{2}$$

3. Let's locate some irrational numbers. Look in the table on page 406.

   a. Find an approximation of $\sqrt{3}$. Locate $\sqrt{3}$ on your number line.

   b. Find an approximation of $\sqrt{5}$. Locate $\sqrt{5}$ on your number line.

   c. Find an approximation of $\sqrt{7}$. Locate $^-\sqrt{7}$ on your number line.

4. The irrational number π is about 3.14. Find an approximate location of π on your number line.

There is a point on a number line for every real number.

EXERCISES

Draw a number line. Use a unit about $\frac{1}{2}$-inch long. Find an approximate location of these real numbers.

1. $\sqrt{10}$	**2.** $\frac{2}{3}$	**3.** .125	**4.** $\sqrt{2}$
5. $\frac{7}{4}$	**6.** $\sqrt{6}$	**7.** $\sqrt{12}$	**8.** .625
9. $\sqrt{11}$	**10.** $\sqrt{13}$	**11.** $^{-}\sqrt{8}$	**12.** $^{-}.33$
13. $\frac{^{-}5}{2}$	**14.** $\sqrt{3}$	**15.** $^{-}\sqrt{14}$	**16.** $^{-}.875$
17. $^{-}\sqrt{5}$	**18.** $\sqrt{15}$	**19.** $\frac{^{-}9}{5}$	**20.** $^{-}.25$
21. $\frac{1}{8}$	**22.** $\frac{^{-}2}{2}$	**23.** $^{-}\sqrt{11}$	**24.** $\sqrt{17}$

Let's Practice

1. Use your compass and straightedge to construct copies of these segments.

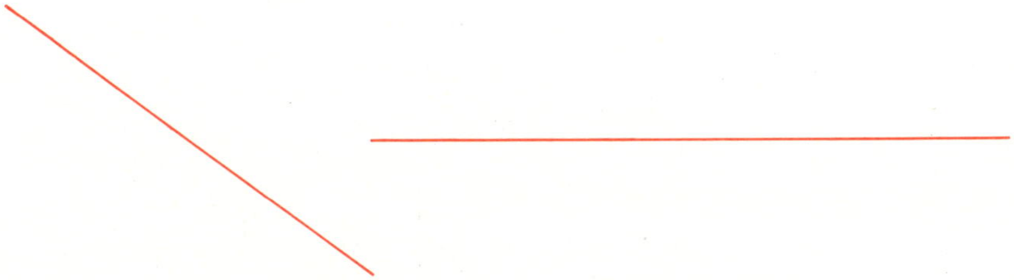

2. Copy these drawings.

a. Construct a line perpendicular to $\overleftrightarrow{ST}$ at R.

b. Construct a line perpendicular to $\overleftrightarrow{DE}$ at F.

c. Construct a line perpendicular at $\overleftrightarrow{AB}$ at C.

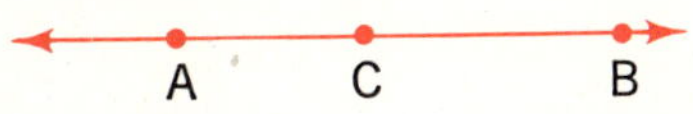

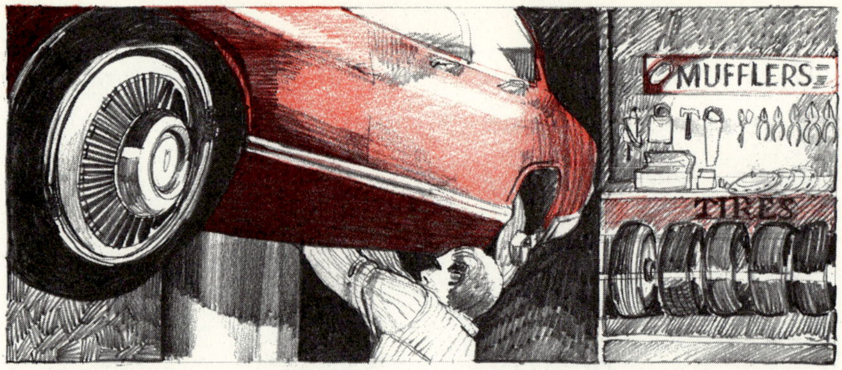

PROBLEM SOLVING

1. Mr. Gardner had 6,508 gallons of regular gasoline and 8,213 gallons of high test gasoline delivered to his service station. The price of regular gasoline was \$.274 per gallon while high test cost \$.332 per gallon. Find the total cost for this delivery.

2. Each winter Mr. Gardner has a special sale on snow tires. He sells a tire which usually costs \$21 for \$18. What per cent of discount does he give?

3. Fred installed 8 new spark plugs in a car. His labor charge was \$7.00 per hour. It took him 40 minutes to complete the job. Spark plugs cost \$1.12 each. Find the total cost of the job.

4. Mr. Gardner said that about 70% of all his business was done through credit cards. How many people used credit cards in dealing with him?

5. Cars can be rented at Mr. Gardner's garage. He charges \$5.00 per day plus 5¢ per mile. Find the approximate cost of renting a car for 5 days for a trip estimated to be 739 miles long.

6. Bill and Hank wash cars for Mr. Gardner. He charges \$2.00 per car and pays them 50¢ for each car they wash. What rate of commission do they receive?

DO YOU REMEMBER?

Add.

1. $^{-}12 + 25$	**2.** $17 + {}^{-}8$	**3.** $^{-}15 + {}^{-}30$
4. $56 + 42$	**5.** $^{-}80 + {}^{-}13$	**6.** $25 + {}^{-}52$
7. $70 + {}^{-}34$	**8.** $^{-}18 + {}^{-}12$	**9.** $100 + 420$
10. $^{-}18 + {}^{-}24 + 56$	**11.** $32 + {}^{-}18 + 70$	**12.** $^{-}12 + {}^{-}18 + {}^{-}28$

Subtract.

13. $^{-}15 - 12$	**14.** $^{-}18 - {}^{-}20$	**15.** $25 - 14$
16. $36 - 52$	**17.** $^{-}20 - 72$	**18.** $30 - {}^{-}19$
19. $62 - {}^{-}48$	**20.** $^{-}67 - {}^{-}10$	**21.** $120 - 152$
22. $^{-}324 - {}^{-}268$	**23.** $139 - 261$	**24.** $^{-}38 - {}^{-}79$

Multiply.

25. $^{-}8 \times {}^{-}12$	**26.** 5×17	**27.** $32 \times {}^{-}3$
28. $36 \times {}^{-}48$	**29.** $^{-}25 \times 70$	**30.** 142×25
31. $172 \times {}^{-}371$	**32.** $^{-}368 \times {}^{-}462$	**33.** $^{-}783 \times {}^{-}142$
34. $1{,}342 \times {}^{-}174$	**35.** $^{-}2{,}580 \times {}^{-}324$	**36.** $^{-}1{,}275 \times 355$

Divide.

37. $^{-}483 \div {}^{-}7$	**38.** $^{-}252 \div 9$	**39.** $432 \div {}^{-}6$
40. $^{-}1{,}170 \div 26$	**41.** $1{,}872 \div {}^{-}24$	**42.** $^{-}2{,}080 \div {}^{-}65$
43. $3{,}127 \div 53$	**44.** $^{-}3{,}096 \div 72$	**45.** $1{,}593 \div {}^{-}27$
46. $^{-}28{,}362 \div 58$	**47.** $32{,}305 \div {}^{-}91$	**48.** $^{-}39{,}748 \div {}^{-}76$

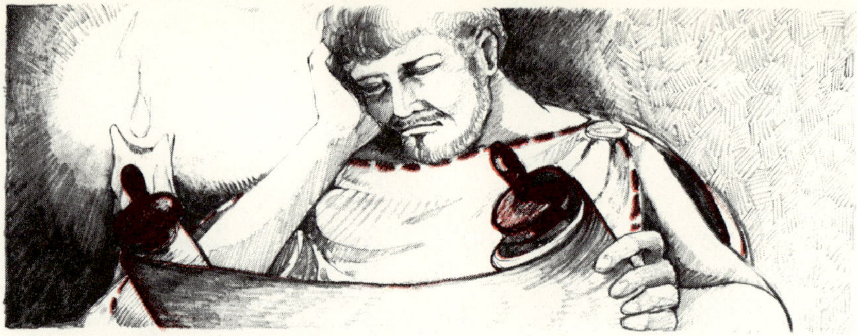

THE PYTHAGOREAN PROPERTY

In a right triangle whose legs have lengths a and b and whose hypotenuse has length c, the following is true.

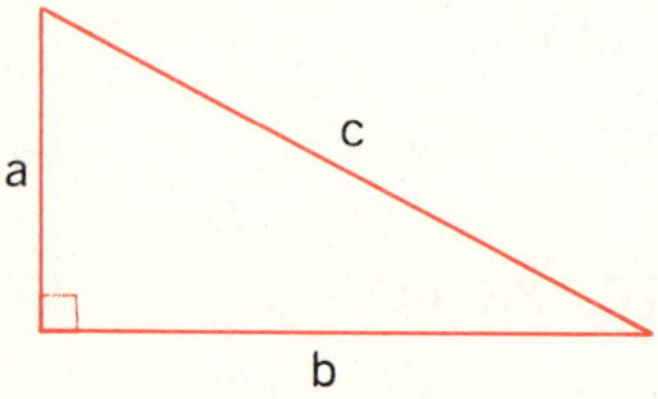

$$a^2 + b^2 = c^2$$

We call this the *Pythagorean Property.* It was discovered by Pythagoras, a famous Greek philosopher and mathematician. Pythagoras lived around the year 500 B.C.

1. Find the length of the hypotenuse in each of these right triangles. Use the table on page 406 to help find square roots.

Examples

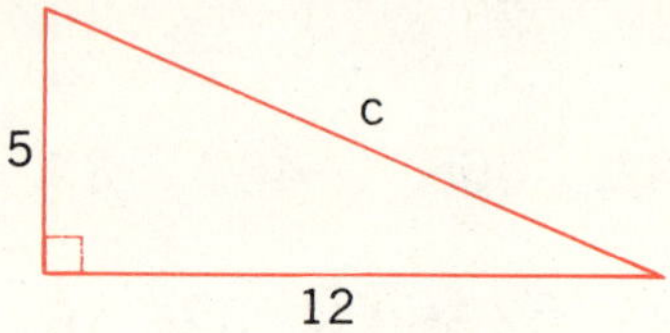

$$5^2 + 12^2 = c^2$$
$$25 + 144 = c^2$$
$$169 = c^2$$
$$\sqrt{169} = c$$
$$13 = c$$

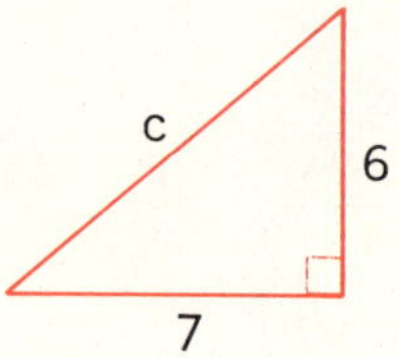

$$7^2 + 6^2 = c^2$$
$$49 + 36 = c^2$$
$$85 = c^2$$
$$\sqrt{85} = c$$
$$9.22 \approx c$$

a.

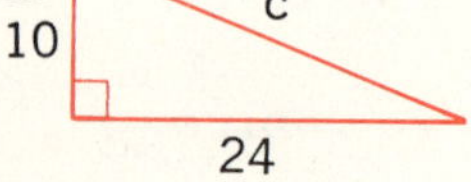

b.

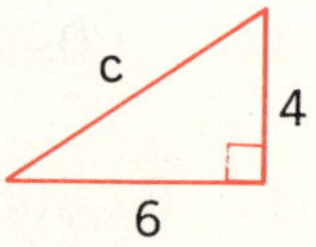

c.

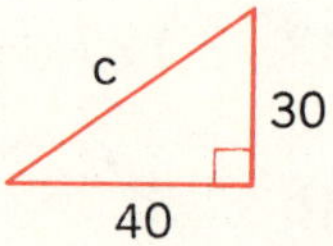

2. Find the length of the leg in each of these right triangles.

Example

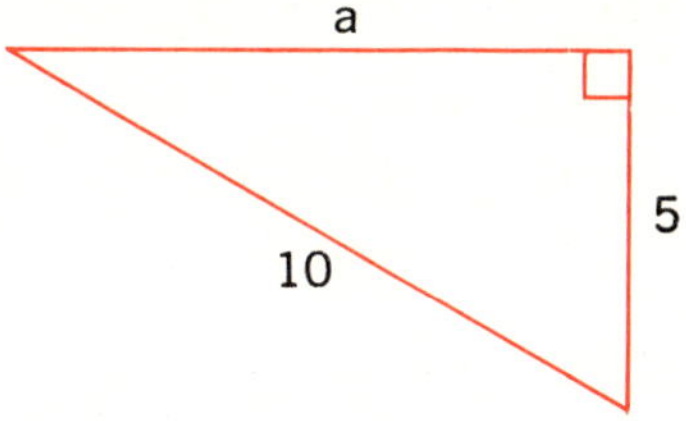

$$a^2 + 5^2 = 10^2$$
$$a^2 + 25 = 100$$
$$a^2 = 100 - 25$$
$$a^2 = 75$$
$$a = \sqrt{75}$$
$$a \approx 8.66$$

a.

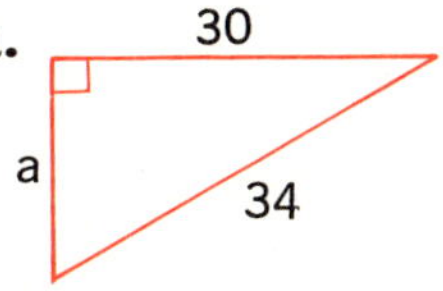

b.

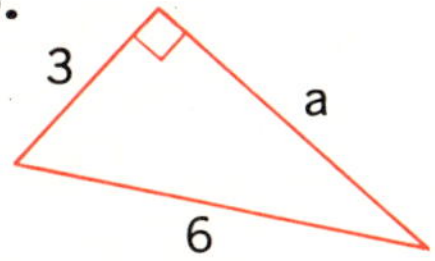

c.

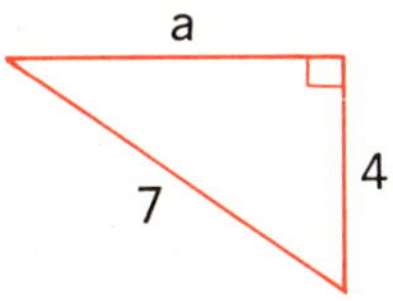

EXERCISES

Find the missing length in each of these right triangles.

1.

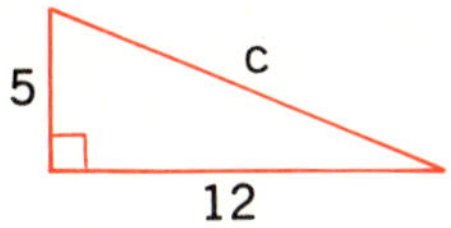

2.

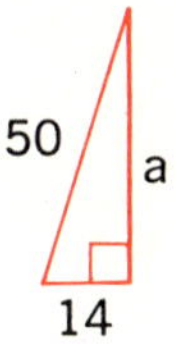

3.

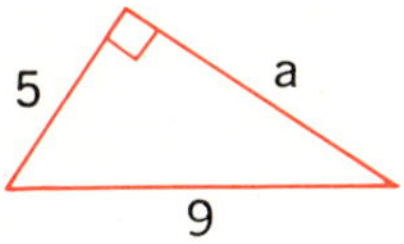

4.

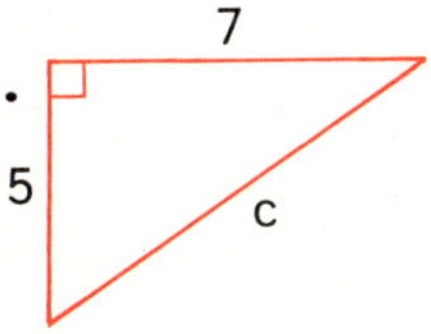

5.

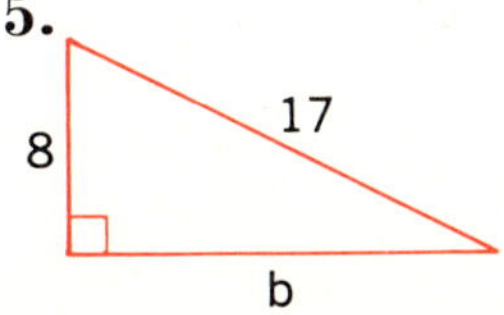

6.

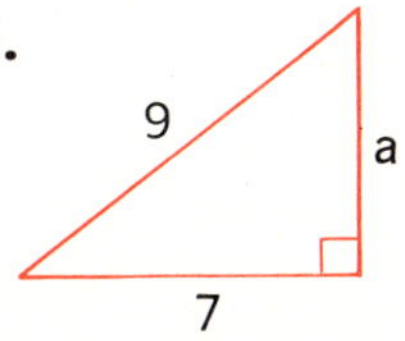

7.

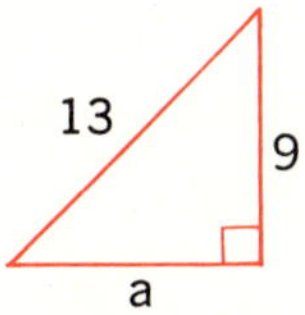

8.

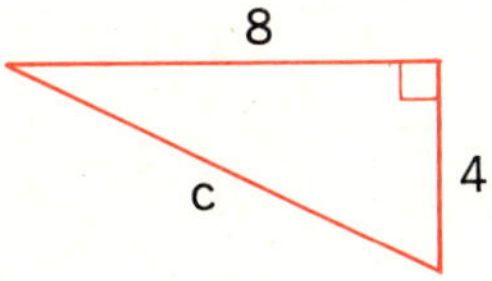

9.

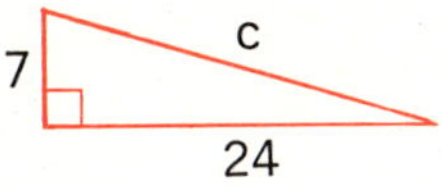

10.

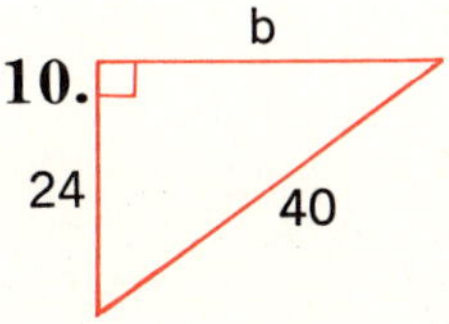

11.

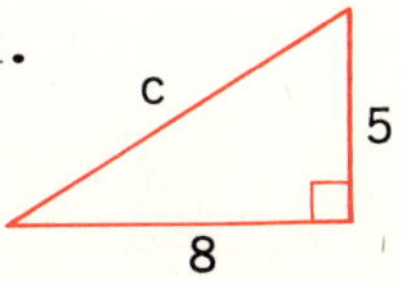

12.

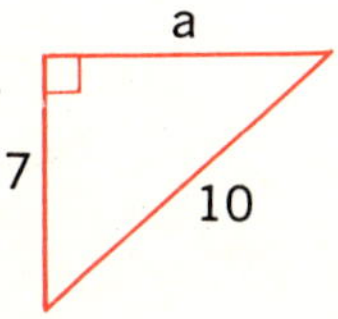

CONSTRUCTING IRRATIONAL NUMBERS

Now we will locate points for some irrational numbers by using constructions.

LET'S EXPLORE

1. Draw a number line. Use a unit 1-inch long. Label your number line as shown.

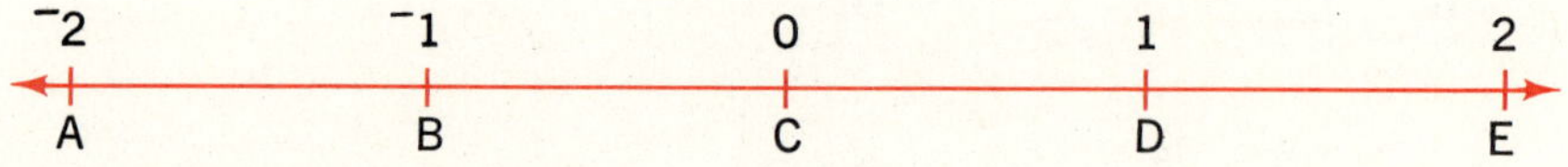

2. a. Use your compass and straightedge. Draw a line perpendicular to $\overline{AE}$ at D.

   b. Use your compass to locate point F so that $\overline{DF}$ is 1 unit long.

   c. Draw $\overline{FC}$.

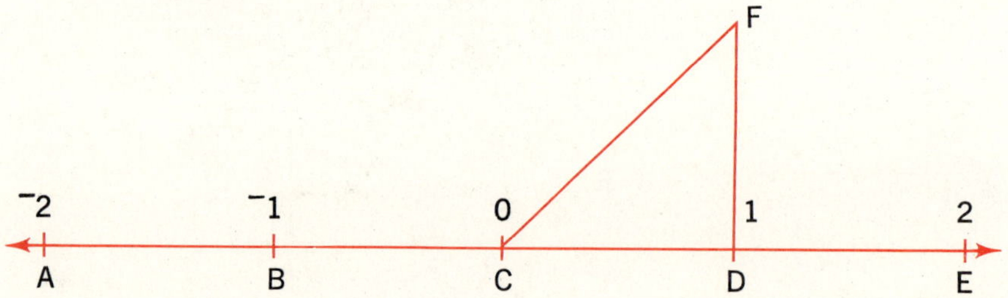

3. Use the Pythagorean Property. Find $m\overline{FC}$.

4. We know that $\overline{FC}$ is $\sqrt{2}$ units long. Use your compass and copy $\overline{FC}$ on the number line. We have now located $\sqrt{2}$ on the number line. Label the point G.

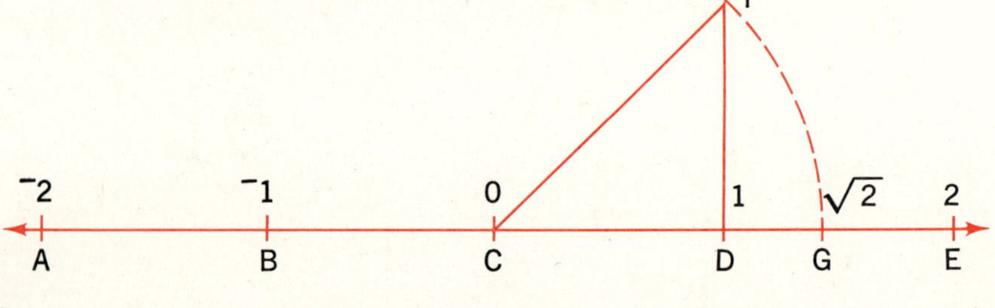

5. Repeat this procedure by constructing a perpendicular at point E. You will then have located $\sqrt{5}$.

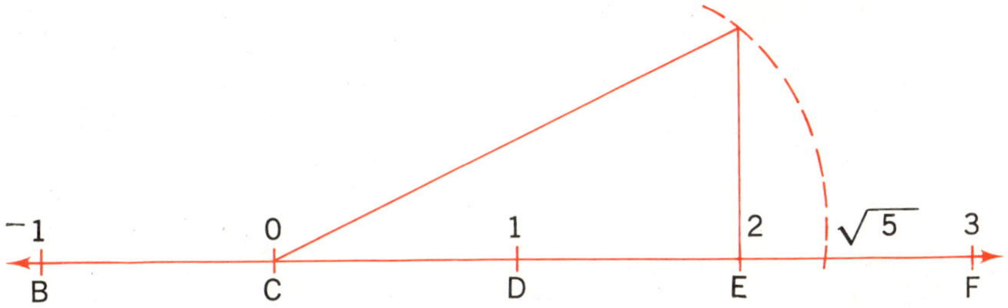

6. We can locate $^{-}\sqrt{2}$ by marking off $\sqrt{2}$ units in the negative direction.

   a. Locate $\sqrt{2}$.

   b. Locate $\sqrt{5}$.

EXERCISES

1. Locate $\sqrt{10}$ and $^{-}\sqrt{10}$ on your number line.

2. Locate $\sqrt{17}$ and $^{-}\sqrt{17}$.

3. Locate $\sqrt{13}$ and $^{-}\sqrt{13}$. (*Hint* Start by constructing a perpendicular at the point for 3. Then mark off a 2-unit segment on the perpendicular.)

4. Locate $\sqrt{3}$ and $^{-}\sqrt{3}$. (*Hint* Start by constructing a perpendicular at the point for $\sqrt{2}$.)

UNEXPLORED TERRITORY

Now let's locate π on your number line.

1. Draw a circle with a diameter which is the same as the unit of your number line. Cut out the circle.

2. What is the circumference of the circle?

3. Use the circle to locate π on your number line.

APPROXIMATING SQUARE ROOTS

1. The square root of a whole number is sometimes a whole number. Find the simplest name for each.

   a. $\sqrt{25}$ b. $\sqrt{144}$ c. $\sqrt{16}$ d. $\sqrt{225}$

   e. $\sqrt{36}$ f. $\sqrt{256}$ g. $\sqrt{400}$ h. $\sqrt{100}$

2. If the square root of a whole number is not a whole number, then it is an irrational number. Find a decimal approximation for each of these square roots. Use the table on page 406.

   a. $\sqrt{22}$ b. $\sqrt{51}$ c. $\sqrt{19}$ d. $\sqrt{43}$

   e. $\sqrt{63}$ f. $\sqrt{85}$ g. $\sqrt{90}$ h. $\sqrt{73}$

3. We know that $\sqrt{22} \approx 4.69$. This means:

   $$4.69 \times 4.69 \approx 22$$

   Find the exact product 4.69×4.69.

We can approximate square roots without using a table.

LET'S EXPLORE

4. Let's approximate $\sqrt{45}$.

   $$6 \times 6 = 36 \qquad 7 \times 7 = 49$$

   Between what two whole numbers is $\sqrt{45}$?

5. Let's take a guess. $\sqrt{45} \approx 6.5$. Now do this division: $6.5\overline{)45.00}$. Round your answer to the nearest tenth.

6. We know that $6.5 \times 6.9 \approx 45$. We want our two factors to be the same. Find the number halfway between 6.5 and 6.9. That is, find the average of 6.5 and 6.9.

7. Do this division: $6.7\overline{)45.00}$. Round your answer to the nearest tenth.

8. Notice that the divisor and quotient are the same in the ones and tenths place. What is an approximation of $\sqrt{45}$ to the nearest tenth?

9. Multiply 6.7 by 6.7 to show that $\sqrt{45} \approx 6.7$.

We can approximate a square root by guessing, dividing, and averaging. We repeat this process as many times as we wish.

Example Approximate $\sqrt{18}$ to the nearest tenth.

$4 \times 4 = 16 \qquad 5 \times 5 = 25$

$\sqrt{18}$ is between 4 and 5. Guess 4.5.

Divide:
$$
\begin{array}{r}
4.0 \\
4.5\overline{)18.0\,0} \\
\underline{18\,0} \\
0 \\
\underline{0}
\end{array}
$$

$\sqrt{18}$ is between 4.5 and 4.0.

Average:
$$
\begin{array}{r}
4.5 \\
+\ \underline{4.0} \\
8.5
\end{array}
\qquad
\begin{array}{r}
4.25 \\
2\overline{)8.50} \\
\underline{8} \\
5 \\
\underline{4} \\
10 \\
\underline{10}
\end{array}
$$

Divide:
$$
\begin{array}{r}
4.23 \\
4.25\overline{)18.00\,00} \\
\underline{17\,00} \\
1\,00\,0 \\
\underline{85\,0} \\
15\,00 \\
\underline{12\,75} \\
2\,25
\end{array}
$$

$\sqrt{18} \approx 4.2$

10. Approximate these square roots to the nearest tenth.

a. $\sqrt{23}$ b. $\sqrt{43}$ c. $\sqrt{58}$ d. $\sqrt{47}$

EXERCISES

Approximate these square roots to the nearest tenth. Use the guess, divide, and average method.

1. $\sqrt{18}$	2. $\sqrt{32}$	3. $\sqrt{78}$	4. $\sqrt{97}$
5. $\sqrt{50}$	6. $\sqrt{56}$	7. $\sqrt{12}$	8. $\sqrt{43}$
9. $\sqrt{19}$	10. $\sqrt{61}$	11. $\sqrt{87}$	12. $\sqrt{77}$
13. $\sqrt{15}$	14. $\sqrt{68}$	15. $\sqrt{48}$	16. $\sqrt{91}$
17. $\sqrt{41}$	18. $\sqrt{74}$	19. $\sqrt{29}$	20. $\sqrt{59}$
21. $\sqrt{13}$	22. $\sqrt{89}$	23. $\sqrt{52}$	24. $\sqrt{69}$

Let's Practice

Solve these problems.

1. Mr. Meadowcroft took this inventory at his service station.

Gasoline	\$ 947	Batteries	\$ 268
Oil	\$ 237	Tools	\$1,576
Tires	\$1,198	Miscellaneous	\$ 629

   Find the total value of his inventory.

2. Mr. Meadowcroft said that he sells about 2,800 gallons of gasoline per day. Estimate the number of gallons he sells in a month.

3. A lubrication and oil change usually cost \$6.00 at Mr. Meadowcroft's. Several times a year, he offers a special price of \$4.00. What per cent of discount is he giving?

4. Mr. Meadowcroft installed a rebuilt windshield wiper motor which cost \$18.50. The sales tax was 4%. The labor charge, which was not taxed, was \$3.80. Find the cost of the job.

DO YOU REMEMBER?

Recall that we can graph integer pairs using perpendicular number lines.

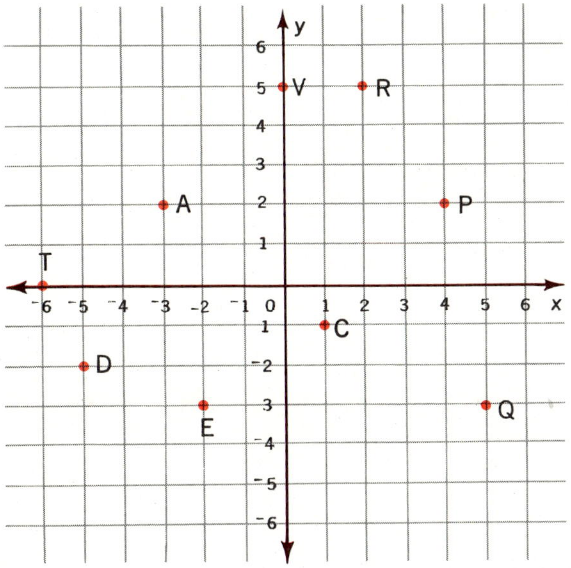

1. The coordinates of A are $(^-3, 2)$. Tell the coordinates of each point.

   a. C **b.** D **c.** P **d.** Q

   e. R **f.** T **g.** V **h.** E

2. Draw two perpendicular number lines. Graph these ordered pairs.

   a. $(5, 3)$ **b.** $(^-2, 1)$ **c.** $(^-3, ^-4)$ **d.** $(4, ^-2)$

   e. $(6, 0)$ **f.** $(^-4, ^-5)$ **g.** $(1, 3)$ **h.** $(2, 5)$

3. Graph six solutions of each sentence.

   a. $x + y = 3$ **b.** $x + 4 = y$ **c.** $x - 2 = y$

   d. $y = 3 \cdot x$ **e.** $y + {^-1} = x$ **f.** $y = {^-1} \cdot x$

4. What do you notice about each of your graphs in Exercise 3?

GRAPHING EQUATIONS ON A PLANE

When we graph sentences using ordered pairs of integers, we get a set of points. What happens when we use ordered pairs of real numbers?

LET'S EXPLORE

x	y
0	3
1	2
$^-1$	4
$1\frac{1}{2}$	$1\frac{1}{2}$
$2\frac{1}{4}$	
$3\frac{1}{2}$	
	$\frac{1}{2}$

1. Let's graph the equation $x + y = 3$.
   a. Copy and complete the chart at the right.
   b. Name six more ordered pairs of real numbers which are solutions of $x + y = 3$.
   c. Draw two perpendicular number lines. Graph your solutions of $x + y = 3$.
   d. What do you notice about the solutions you graphed?

2. For every real number, there is a point on a number line. When we graph equations on a plane using real numbers, there will be no gaps. The graph of an equation like $x + y = 3$ is a straight line.
   a. The line shown below is the graph of $x + y = 3$. Check to see that the ordered pairs in the chart lie on the line.
   b. The graph of $x + y = 3$ contains the point $(5, ^-2)$. Check to see that $(5, ^-2)$ is a solution of $x + y = 3$.

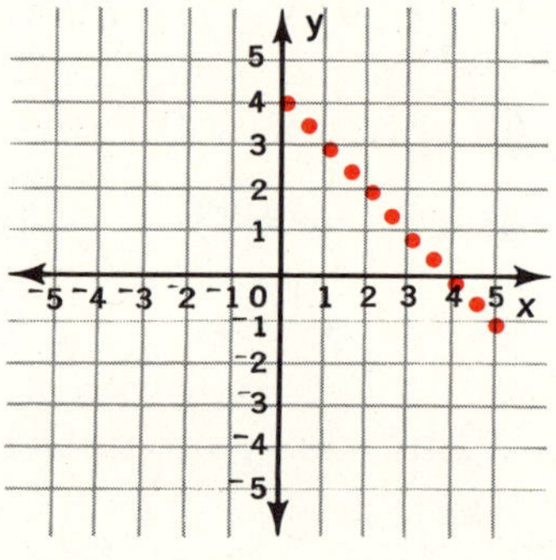

3. Jack said that since two points determine a line, we can graph an equation like $x + y = {}^-2$ by graphing only two of its solutions.

   a. Graph any two solutions of $x + y = {}^-2$.

   b. Draw a line containing the two points.

4. Cheri said that it is a good idea to graph a third point to use as a check.

   a. Graph a third solution of $x + y = {}^-2$.

   b. Does the third point lie on the line you graphed?

To graph an equation like $x + y = 4$ using real numbers, we can graph two solutions and draw a line through the two points. Then we graph a third solution as a check.

5. Graph these equations.

   a. $x + y = {}^-3$ b. $x - y = 5$ c. $2 \cdot x = y$

EXERCISES

Graph these equations.

1. $x + y = 4$	2. $x + y = {}^-2$	3. $y - x = 2$
4. $3 \cdot x = y$	5. $x - y = 5$	6. $x - y = {}^-2$
7. $y - x = 4$	8. $y = x$	9. ${}^-3 \cdot x = y$
10. $3 \cdot y = x$	11. $x + y = {}^-6$	12. $y - x = {}^-3$
13. $x + y = 0$	14. ${}^-5 \cdot x = y$	15. $y = {}^-2 \cdot x$
16. $y - x = 6$	17. $y = 4 \cdot x$	18. $x + 4 = y$
19. ${}^-5 + x = y$	20. $x - y = {}^-6$	21. $4 \cdot x = y$
22. $x - 4 = y$	23. $x + y = 8$	24. ${}^-5 + x = y$

CURVED GRAPHS

When we graphed $y = x^2$ using integer pairs, we got a curved pattern. What happens when we use ordered pairs of real numbers?

LET'S EXPLORE

1. **a.** Complete this chart for $y = x^2$.

   b. Graph the ordered pairs (x, y) from the chart.

   c. Draw in the curve. You now have a graph of $y = x^2$. It should look like the graph below.

x	y
0	0
1	1
$^{-}1$	1
2	
$^{-}2$	
3	
$^{-}3$	

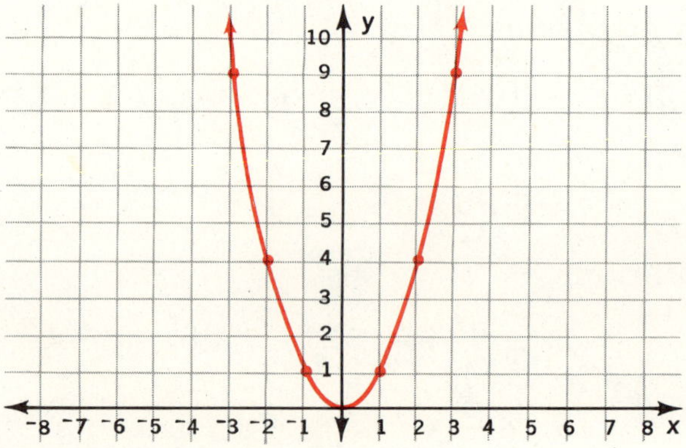

2. Let's graph $y = {}^{-}1 \cdot x^2$.

   a. Copy and complete this chart.

   b. Graph the ordered pairs (x, y) from the chart.

   c. Draw in the curve.

   d. Compare the graphs of $y = x^2$ and $y = {}^{-}1 \cdot x^2$. What do you observe?

x	y
0	0
1	$^{-}1$
$^{-}1$	$^{-}1$
2	$^{-}4$
$^{-}2$	
3	
$^{-}3$	

3. Let's graph $y = 2 \cdot x^2$.

a. Make a chart of solutions like the one on page 304.

b. Graph the ordered pairs from the chart.

c. Sketch in the curve.

4. Suppose we graphed $y = {}^{-}2 \cdot x^2$.

a. What do you think its graph would look like?

b. Test your conclusion by graphing $y = {}^{-}2 \cdot x^2$.

5. Graph $x = y^2$. It should look like the graph below.

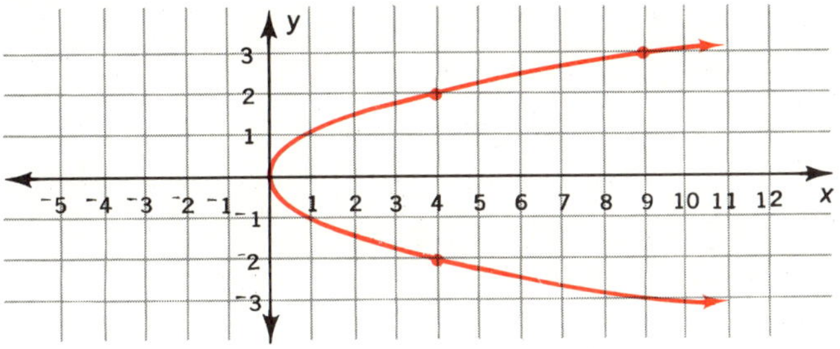

6. Graph $y = \frac{1}{2} \cdot x^2$.

EXERCISES

Graph these equations.

1. $y = 3 \cdot x^2$
2. $y = {}^{-}3 \cdot x^2$
3. $y = \frac{^{-}1}{2} \cdot x^2$
4. $y = 4 \cdot x^2$
5. $y = {}^{-}4 \cdot x^2$
6. $x = {}^{-}1 \cdot y^2$
7. $x = 2 \cdot y^2$
8. $x = {}^{-}2 \cdot y^2$
9. $x = 3 \cdot y^2$
10. $y = x^2 + 1$
11. $y = x^2 + 2$
12. $y = x^2 - 1$

UNEXPLORED TERRITORY

Graph the equation $y = \sqrt{x}$. Use the table on page 406 to help find square roots.

GRAPHICAL APPROXIMATION

We have graphed the equation $y = x^2$. We can use this graph to approximate square roots.

LET'S EXPLORE

1. Look at the graph of $y = x^2$.

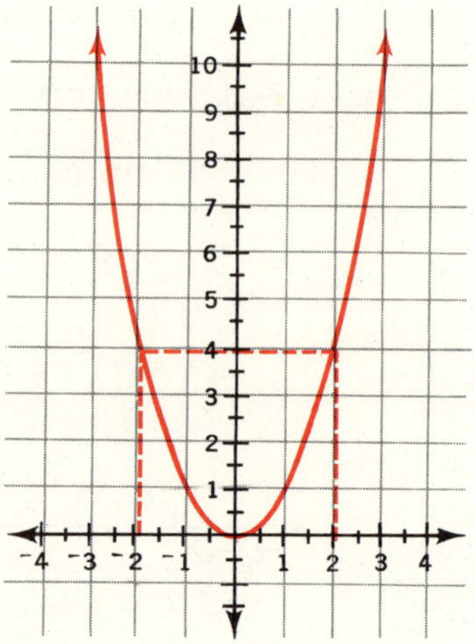

   a. On the vertical number line find the point for 4. From this point, draw a segment to the right and left meeting the graph.

   b. From these points, draw segments downward to meet the horizontal number line. What numbers on the horizontal number line do you find?

   c. What are two square roots of 4?

2. On the vertical number line find 9.

   a. Draw a segment which crosses the graph.

   b. From there, draw segments downward meeting the horizontal number line. What numbers do you find?

   c. What are two square roots of 9?

3. On the vertical number line find 3.

   a. Draw a segment which crosses the graph.

   b. From there, draw segments downward meeting the horizontal number line. What are the approximate numbers for these points?

   c. What are two approximate square roots of 3?

4. Use your graph of $y = x^2$. Approximate two square roots for each of these numbers.

a. 2 b. 6 c. 1 d. $3\frac{1}{2}$

EXERCISES

Use your graph of $y = x^2$. Approximate two square roots for each of these numbers.

1. 5
2. 7
3. 8
4. 10
5. 5
6. .5
7. $6\frac{1}{2}$
8. 4.4
9. 3.6
10. 1.4

UNEXPLORED TERRITORY

1. Let's graph $y = x^3$.

a. Copy and complete this chart.

x	y
0	0
1	
−1	
2	
−2	
3	
−3	

b. Graph the ordered pairs.

c. Sketch in the curve.

2. Let's graph $y = |x|$.

a. Make a chart of ordered pairs.

b. Graph the ordered pairs.

c. Draw the graph.

DO YOU REMEMBER?

1. What operation is the opposite of addition?

2. What operation is the opposite of division?

Write two related sentences for each.

Example

$$3 + 2 = 5 \begin{cases} 5 - 3 = 2 \\ 5 - 2 = 3 \end{cases}$$

1. $6 + 4 = 10$
2. $8 + 5 = 13$
3. $9 + 1 = 10$
4. $13 = 10 + 3$
5. $15 = 7 + 8$
6. $25 = 10 + 15$
7. $4 + {}^-2 = 2$
8. ${}^-5 + 8 = 3$
9. ${}^-3 + {}^-2 = {}^-5$
10. $4 + x = 7$
11. $y + 2 = 4$
12. $7 = x + 5$

Write two related sentences for each.

Example

$$4 \cdot 2 = 8 \begin{cases} \frac{8}{4} = 2 \\ \frac{8}{2} = 4 \end{cases}$$

13. $10 \cdot 2 = 20$
14. $3 \cdot 8 = 24$
15. $5 \cdot 9 = 45$
16. $30 = 6 \cdot 5$
17. $35 = 7 \cdot 5$
18. $40 = 5 \cdot 8$
19. $5 \cdot {}^-2 = {}^-10$
20. ${}^-3 \cdot {}^-5 = 15$
21. ${}^-7 \cdot 4 = {}^-28$
22. $5 \cdot x = 45$
23. ${}^-4 \cdot y = 16$
24. $7 \cdot x = 45$

Subtract by adding inverses.

Example $5 - 2 = 5 + {}^-2 = 3$

25. $7 - 4$
26. $4 - {}^-3$
27. ${}^-4 - 7$
28. ${}^-5 - {}^-2$
29. ${}^-12 - 4$
30. $12 - {}^-2$
31. ${}^-6 - {}^-6$
32. $100 - {}^-50$
33. ${}^-5 - {}^-6$

CHAPTER REVIEW

Tell whether each numeral names a rational or an irrational number.

1. $.12\overline{66}$
2. $.121231234...$
3. $.010010001...$
4. $.5\overline{5}$
5. $2.121\overline{21}$
6. π
7. $12.364\overline{4}$
8. $20.25\overline{5}$
9. $38.12345...$
10. $.12712771277712\ldots$
11. $\sqrt{47}$
12. $\sqrt{169}$
13. $\sqrt{3}$
14. $\sqrt{400}$
15. $\sqrt{100}$
16. $\sqrt{1{,}000}$
17. $^{-}.182$
18. $6.12345\overline{12345}$

Which of these real numbers are irrational?

19. $\sqrt{11}$
20. $\sqrt{99}$
21. $\sqrt{235}$
22. $\sqrt{6{,}084}$
23. $\sqrt{60}$
24. $\sqrt{100}$
25. $\sqrt{1{,}007}$
26. $\sqrt{2{,}500}$
27. $\sqrt{16}$
28. $\sqrt{413}$
29. $\sqrt{529}$
30. $\sqrt{797}$
31. $\sqrt{1}$
32. $\sqrt{36}$
33. $\sqrt{1{,}356}$
34. $\sqrt{64}$
35. $\sqrt{900}$
36. $\sqrt{451}$
37. $\sqrt{1{,}309}$
38. $\sqrt{19}$

Draw a number line. Use a unit about $\frac{1}{2}$-inch long. Find the approximate location of these real numbers.

39. $\sqrt{8}$
40. $.275$
41. $^{-}\sqrt{16}$
42. $\sqrt{5}$
43. $^{-}1.58$
44. $\frac{9}{4}$
45. $^{-}\sqrt{2}$
46. $\sqrt{0}$

Find the missing length in each of these right triangles.

47.

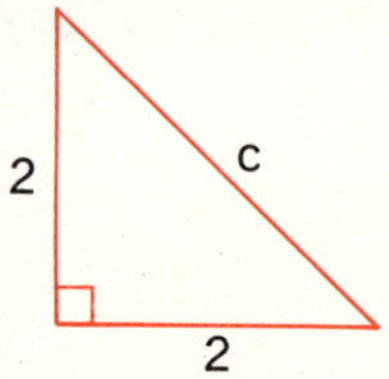

48.

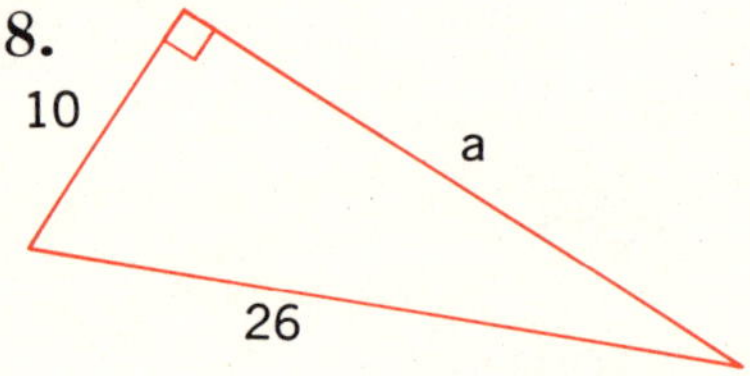

49.

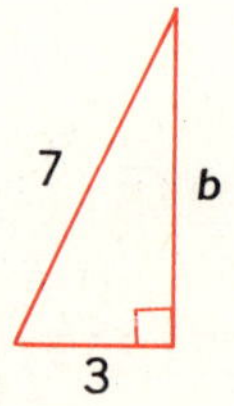

50.

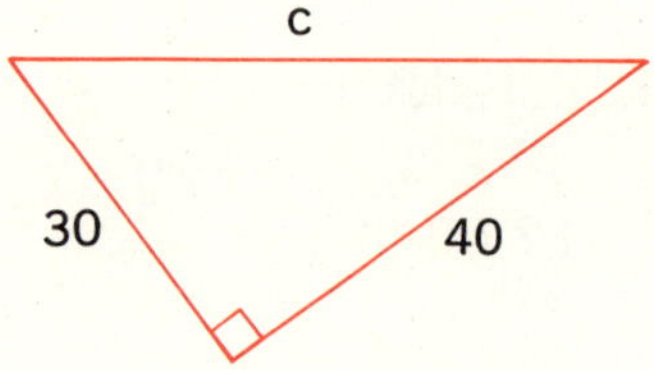

51.

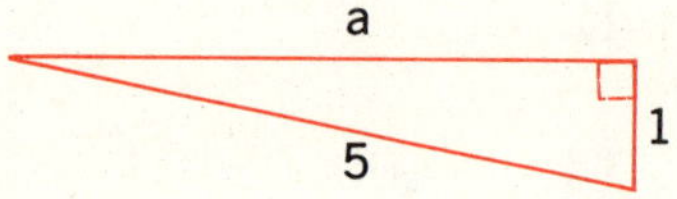

52.

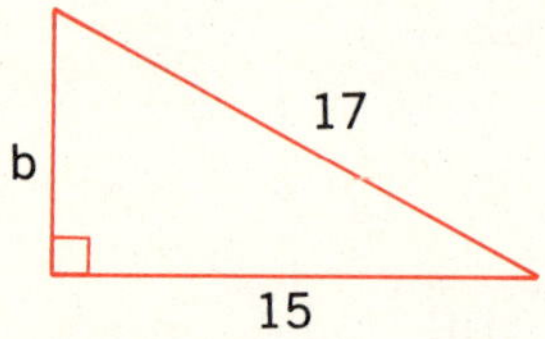

Approximate these square roots to the nearest tenth. Use the guess, divide, and average method.

53. $\sqrt{26}$ **54.** $\sqrt{52}$ **55.** $\sqrt{175}$ **56.** $\sqrt{425}$

57. $\sqrt{12}$ **58.** $\sqrt{15}$ **59.** $\sqrt{1.7}$ **60.** $\sqrt{230}$

Graph these equations.

61. $x + y = 8$ **62.** $x - y = 3$ **63.** $2 \cdot x = y$

64. $\frac{^-1}{3} \cdot x = y$ **65.** $x = {}^-1 \cdot y$ **66.** $x + y = {}^-1$

Terms You Should Know

terminating decimal
repeating decimal
rational number
non-terminating decimal
irrational number
real number
Pythagorean Property
hypotenuse
square root
ordered pair

CHAPTER TEST

Tell whether each numeral names a rational or an irrational number.

1. $.35\overline{5}$
2. .565675678...
3. $.12\overline{12}$
4. 13.020020002...

Which of these real numbers are irrational?

5. $\sqrt{49}$
6. $\sqrt{853}$
7. $\sqrt{796}$
8. $\sqrt{100}$
9. $\sqrt{38}$
10. $\sqrt{8,100}$

Draw a number line. Use a unit about $\frac{1}{2}$-inch long. Find the approximate location of these real numbers.

11. $\sqrt{12}$
12. $^{-}\sqrt{7}$
13. 1.58
14. $\frac{5}{3}$
15. $^{-}2.7$
16. $^{-}\sqrt{4}$

Find the missing length in each of these right triangles.

17.

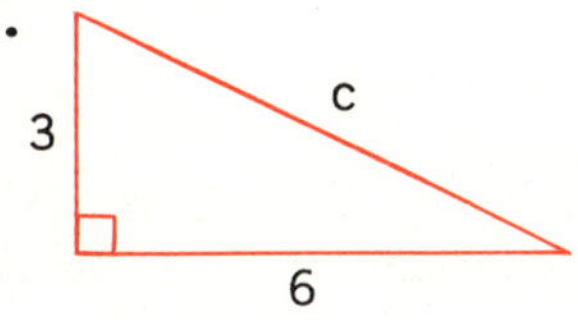

18.

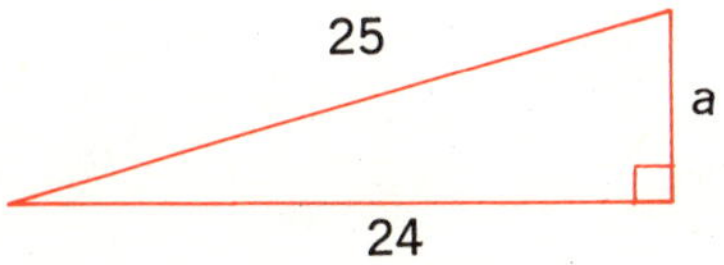

19.

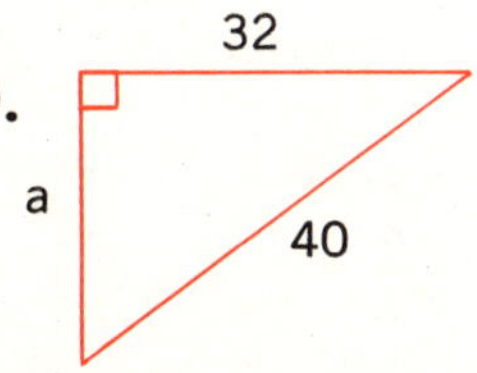

20.

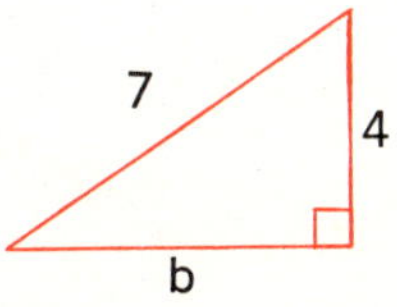

Approximate these square roots to the nearest tenth. Use the guess, divide, and average method.

21. $\sqrt{24}$
22. $\sqrt{13}$
23. $\sqrt{30}$

Graph these equations.

24. $x + y = {}^{-}3$
25. $y = x^2$

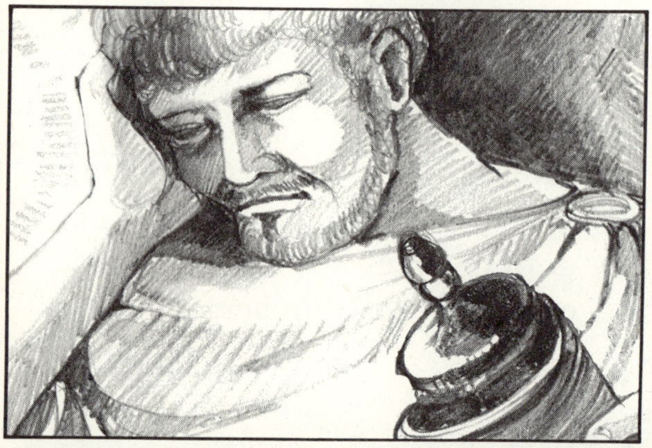

$$x - {}^{-}2 \cdot x = {}^{-}6$$
$$x + 2 \cdot x = {}^{-}6$$
$$1 \cdot x + 2 \cdot x = {}^{-}6$$
$$3 \cdot x = {}^{-}6$$
$$x = {}^{-}2$$

CHECK:

$x - {}^{-}2 \cdot x$	$= {}^{-}6$
${}^{-}2 - {}^{-}2 \cdot {}^{-}2$	${}^{-}6$
${}^{-}2 - 4$	
${}^{-}6$	

$3 + 7 = 10$ $(3 + 7) + 2 = 10 + 2$

ADD 2. $10 + 2 = 10 + 2$

$12 = 12$

CHAPTER TWELVE

Solving Equations

SOLUTIONS OF EQUATIONS

A number sentence with = is called an *equation.* Some equations are true and some are false.

1. Which are true and which are false?

   a. $3 + 2 = 5$ **b.** $5 = 7 - 4$ **c.** $4 + 5 = 7$

2. Some equations with variables are neither true nor false. Find numbers that make these true.

   a. $3 + 4 = x$ **b.** $4 + y = 5$ **c.** $t + 2 = 4$

   d. $7 + r = 9$ **e.** $n + 6 = 11$ **f.** $3 + 9 = b$

A number that makes an equation true is called a **solution** of the equation.

3. Find solutions of these equations.

   a. $10 - 2 = x$ **b.** $8 - 10 = y$ **c.** $3 + x = 5$

To **solve** an equation, we find all of its solutions.

4. Solve.

   a. $5 - 10 = x$ **b.** $4 + y = 8$ **c.** $x + x = 6$

   d. $x = x \cdot x$ **e.** $5 \cdot y = 0$ **f.** $^-4 \cdot y = 0$

EXERCISES

Solve. Use your imagination.

1. $5 + 4 = x$ **2.** $4 - 6 = y$ **3.** $3 \cdot {}^-2 = t$

4. $^-5 \cdot {}^-3 = y$ **5.** $3 + x = 6$ **6.** $5 + y = 7$

7. $^-2 + s = 5$ **8.** $3 + t = {}^-1$ **9.** $5 + w = 3$

10. $x + x = 10$ **11.** $y + y = {}^-8$ **12.** $t + t = 0$

13. $y \cdot y = y$ **14.** $x + x = x$ **15.** $w \cdot w = 1$

$$a + 3 = 9$$

$$^-4 = f + {^-7}$$

SOLVING EQUATIONS $x + a = b$

Remember that addition and subtraction are opposite operations. For the equation $x + 3 = 5$, there are two related sentences.

$$5 - x = 3 \qquad\qquad 5 - 3 = x$$

1. Write two related sentences for $x + 4 = 7$.

2. Which is easiest to solve?

   a. $x + 4 = 7$ **b.** $7 - 4 = x$ **c.** $7 - x = 4$

3. **a.** For $^-5 + x = 7$, write two related sentences.

   b. Choose the one with x alone on one side. Solve it.

   c. Is the solution also a solution of $^-5 + x = 7$?

4. **a.** For $y + 2 = {^-5}$, write a related sentence with y alone on one side.

   b. Solve your related sentence.

   c. Is the solution also a solution of $y + 2 = {^-5}$?

We can solve an equation like $x + 3 = {^-7}$ this way:
(a) Write a related sentence with x alone on one side.
(b) Solve the related sentence.
(c) Check the solution in the original equation.

5. Solve.

a. $x + {}^{-}2 = 5$ b. $3 + y = {}^{-}2$ c. $t + 3 = {}^{-}7$

d. $10 = 5 + w$ e. ${}^{-}1 = x + {}^{-}7$ f. $y + {}^{-}3 = {}^{-}3$

EXERCISES

Solve.

1. $x + {}^{-}2 = 6$
2. $x + {}^{-}3 = 5$
3. $y + {}^{-}3 = 0$
4. $y + {}^{-}4 = 0$
5. $x + {}^{-}2 = {}^{-}3$
6. $x + {}^{-}3 = {}^{-}2$
7. $t + 2 = {}^{-}5$
8. $t + 3 = {}^{-}4$
9. $w + {}^{-}4 = {}^{-}6$
10. $w + {}^{-}3 = {}^{-}5$
11. ${}^{-}3 + x = {}^{-}1$
12. ${}^{-}4 + x = {}^{-}2$
13. $y + {}^{-}5 = {}^{-}5$
14. $y + {}^{-}6 = {}^{-}6$
15. $2 = x + 1$
16. $4 = x + 2$
17. $5 = x + {}^{-}1$
18. $7 = x + {}^{-}2$
19. $7 = 3 + y$
20. $5 = 3 + y$
21. ${}^{-}2 = 1 + y$
22. ${}^{-}3 = 2 + y$
23. ${}^{-}5 = t + 3$
24. ${}^{-}7 = t + 4$
25. ${}^{-}2 = x + {}^{-}3$
26. ${}^{-}2 = w + {}^{-}2$
27. ${}^{-}10 = x + {}^{-}5$

Solve these problems.

28. Bill sold popcorn at the baseball game. Each box sold for $.15. He sold 312 boxes. How much did he take in?

29. On January 5 the high temperature in Chicago was 18°. The low temperature was ${}^{-}12°$. What was the difference in temperature?

30. Mr. Carpenter needs boards measuring $2\frac{1}{2}$ feet, $3\frac{3}{4}$ feet, and 12 feet. What is the total length he needs?

$$5 \cdot h = 15$$
$$2 \cdot q = 16$$

SOLVING EQUATIONS $a \cdot x = b$

Remember that multiplication and division are opposite operations. For the equation $3 \cdot x = 6$, there are two related sentences.

$$\frac{6}{x} = 3 \qquad \frac{6}{3} = x$$

1. Write two related sentences for $4 \cdot x = 12$.

2. Which is easiest to solve?

   a. $4 \cdot x = 12$ b. $\frac{12}{x} = 4$ c. $\frac{12}{4} = x$

3. a. For $5 \cdot x = 20$, write two related sentences.

   b. Choose the one with x alone on one side. Solve it.

   c. Is the solution also a solution of $5 \cdot x = 20$?

4. a. For $3 \cdot y = 21$, write a related sentence with y alone on one side.

   b. Solve your related sentence.

   c. Is the solution also a solution of $3 \cdot y = 21$?

We can solve an equation like $4 \cdot x = 16$ this way:
(a) Write a related sentence with x alone on one side.
(b) Solve the related sentence.
(c) Check the solution in the original equation.

5. Solve.

a. $5 \cdot x = 15$ b. $3 \cdot x = 21$ c. $2 \cdot x = {}^-10$

d. $12 = 6 \cdot y$ e. ${}^-16 = {}^-2 \cdot t$ f. $4 \cdot y = {}^-4$

EXERCISES

Solve.

1. $3 \cdot x = 9$
2. $4 \cdot x = 16$
3. $7 \cdot y = 14$
4. $8 \cdot y = 24$
5. $2 \cdot x = {}^-8$
6. $3 \cdot x = {}^-12$
7. ${}^-2 \cdot x = {}^-8$
8. ${}^-4 \cdot x = {}^-12$
9. ${}^-5 \cdot y = 15$
10. ${}^-3 \cdot y = 12$
11. $14 \cdot t = 14$
12. $12 \cdot t = {}^-12$
13. $6 \cdot y = {}^-6$
14. ${}^-5 \cdot y = 5$
15. $32 = 4 \cdot x$
16. $35 = 5 \cdot x$
17. $25 = 5 \cdot w$
18. $30 = 6 \cdot w$
19. ${}^-12 = {}^-3 \cdot x$
20. ${}^-15 = {}^-3 \cdot x$
21. $14 = 7 \cdot u$
22. $20 = 5 \cdot u$
23. ${}^-30 = 6 \cdot y$
24. ${}^-35 = 5 \cdot y$
25. $10 \cdot y = 70$
26. ${}^-1 \cdot y = 5$
27. ${}^-1 \cdot x = {}^-7$

UNEXPLORED TERRITORY

Try to solve these equations. You will have to use your imagination and trial and error.

1. $x \cdot x = x$
2. $x \cdot x = 0$
3. $x + x = 2 \cdot x$
4. $x \cdot x \cdot x = 8$
5. $|x + 2| = 3$
6. $|x - 6| = 4$
7. $x \cdot x = {}^-1$
8. $x \cdot x \cdot x = {}^-1$
9. $x \cdot x = x + x$
10. $x \div x = 1$

THE ADDITION PRINCIPLE

We can use addition to help us solve some equations.

LET'S EXPLORE

1. a. Is this sentence true? $3 + 1 = 4$

   b. Add 2 to $3 + 1$. Add 2 to 4.
   Is $(3 + 1) + 2 = 4 + 2$ true?

   c. Add 6 to $3 + 1$. Add 6 to 4.
   Is $(3 + 1) + 6 = 4 + 6$ true?

   d. Add 0 to $3 + 1$. Add 0 to 4.
   Is $(3 + 1) + 0 = 4 + 0$ true?

   e. Add $^-6$ to $3 + 1$. Add $^-6$ to 4.
   Is $(3 + 1) + {}^-6 = 4 + {}^-6$ true?

   f. Add $^-1$ to $3 + 1$. Add $^-1$ to 4.
   Is $(3 + 1) + {}^-1 = 4 + {}^-1$ true?

2. a. Is this sentence true? $9 = 10 + {}^-1$

   b. Let's add 2. Simplify $9 + 2$. Simplify $(10 + {}^-1) + 2$.
   Is $9 + 2 = (10 + {}^-1) + 2$ true?

   c. Let's add $^-10$. Is $9 + {}^-10 = (10 + {}^-1) + {}^-10$ true?

   d. Let's add 1. Is $9 + 1 = (10 + {}^-1) + 1$ true?

3. What pattern do you see? Try to describe it.

> If an equation $a = b$ is true, we get another true equation when we add any number c.
>
> $$a + c = b + c$$
>
> We call this the **Addition Principle.**

Example $3 + 7 = 10$ Add 2.

$$(3 + 7) + 2 = 10 + 2$$
$$10 + 2 = 10 + 2$$
$$12 = 12$$

EXERCISES

For each equation add the given number to get a new equation. Then simplify both sides of the new equation.

1. $4 + 2 = 6$
Add 3.

2. $5 = 7 + {}^-2$
Add $^-1$.

3. $8 = 9 - 1$
Add 2.

4. $5 \cdot 2 = 10$
Add $^-1$.

5. $6 = {}^-3 \cdot {}^-2$
Add 3.

6. $13 = 6 + 7$
Add $^-12$.

7. $x + 4 = 9$
Add $^-4$.

8. $y + 3 = 10$
Add $^-3$.

9. $t + 4 = 8$
Add $^-4$.

10. $x + {}^-5 = 7$
Add 5.

11. $y + {}^-7 = 2$
Add 7.

12. $w + 12 = {}^-3$
Add $^-12$.

13. $3 + x = 7$
Add $^-3$.

14. ${}^-5 + y = 2$
Add 5.

15. $10 + u = {}^-3$
Add $^-10$.

16. $8 = x + 5$
Add $^-5$.

17. $13 = y + {}^-4$
Add 4.

18. ${}^-12 = t + {}^-4$
Add 4.

19. $7 = 5 + x$
Add $^-5$.

20. ${}^-7 = {}^-3 + y$
Add 3.

21. ${}^-9 = x + 3$
Add $^-3$.

22. $x + 5 = 9$
Add $^-5$.

23. $y + {}^-6 = 8$
Add 6.

24. $14 + t = {}^-5$
Add $^-14$.

25. $12 = x + 7$
Add $^-7$.

26. ${}^-10 = y + {}^-7$
Add 7.

27. ${}^-16 = t + 4$
Add $^-4$.

Solve these problems.

28. In a card game, George scored 20, $^-15$, 10, and $^-5$. What was his total score?

29. Bill is 2 inches taller than Robert. Robert is $4\frac{1}{2}$ inches taller than Nancy. How much taller than Nancy is Bill?

USING THE ADDITION PRINCIPLE

Now let's see how to use the addition principle to help solve equations.

LET'S EXPLORE

1. Think about $x + 2 = 11$.
   a. What is the inverse of 2?
   b. Use the addition principle. Add $^{-}2$.
   c. What is the solution of your new equation?
   d. Is it also a solution of $x + 2 = 11$?

2. Think about $^{-}12 = y + {}^{-}3$.
   a. What is the inverse of $^{-}3$?
   b. Use the addition principle. Add 3.
   c. What is the solution of your new equation?
   d. Is it also a solution of the original equation?

3. Think about $4 + x = {}^{-}9$.
   a. What should be added to get x alone?
   b. Add that number.
   c. What is the solution of your new equation?
   d. Is it also a solution of $4 + x = {}^{-}9$?

4. Let's try $4 + y = {}^{-}5$.
   a. Use the addition principle to get y alone.
   b. What is the solution of $4 + y = {}^{-}5$?

We can solve an equation like $x + 7 = 12$ this way:
(a) Use the addition principle to get the variable alone on one side of the equation.
(b) Check the solution in the original equation.

5. Solve.

Example

$$x + {}^-5 = 2$$
$$(x + {}^-5) + 5 = 2 + 5$$
$$x = 7$$

Check

$x + {}^-5 = 2$	
$7 + {}^-5$	2
2	

Thus, 7 is a solution of $x + {}^-5 = 2$.

a. $x + 2 = 14$ **b.** ${}^-2 + z = 3$

c. $17 = {}^-3 + v$ **d.** $42 = y + 4$

EXERCISES

Solve. Use the addition principle.

1. $x + 3 = 12$

2. $y + {}^-5 = 8$

3. $t + {}^-2 = {}^-9$

4. $x + {}^-8 = {}^-3$

5. $y + {}^-7 = 0$

6. $u + {}^-17 = {}^-3$

7. $3 + x = 13$

8. ${}^-2 + y = 12$

9. $6 + w = {}^-5$

10. ${}^-5 + x = {}^-31$

11. ${}^-23 + y = 4$

12. $47 + u = {}^-12$

13. $14 = x + 2$

14. $12 = y + {}^-7$

15. $17 = v + {}^-17$

16. ${}^-19 = x + {}^-3$

17. ${}^-2 = y + {}^-27$

18. $4 = u + {}^-25$

19. $11 = {}^-2 + x$

20. ${}^-6 = {}^-23 + y$

21. ${}^-1 = w + {}^-24$

22. $33 = w + {}^-11$

23. ${}^-9 = r + {}^-18$

24. $15 = {}^-25 + w$

25. $8 + x = 18$

26. ${}^-15 = 25 + y$

27. $42 = t + {}^-12$

28. ${}^-17 = 13 + u$

THE MULTIPLICATION PRINCIPLE

We can also use multiplication to help us solve some equations.

LET'S EXPLORE

1. a. Is this sentence true? $3 \cdot 5 = 15$

   b. Multiply $3 \cdot 5$ by 2. Multiply 15 by 2. Is $(3 \cdot 5) \cdot 2 = 15 \cdot 2$ true?

   c. Multiply $3 \cdot 5$ by 4. Multiply 15 by 4. Is $4 \cdot (3 \cdot 5) = 4 \cdot 15$ true?

   d. Multiply $3 \cdot 5$ by 0. Multiply 15 by 0. Is $0 \cdot (3 \cdot 5) = 0 \cdot 15$ true?

2. a. Is this sentence true? $2 \cdot 3 = 6$

   b. Multiply $2 \cdot 3$ by $\frac{1}{2}$. Multiply 6 by $\frac{1}{2}$. Is $\frac{1}{2} \cdot (2 \cdot 3) = \frac{1}{2} \cdot 6$ true?

   c. Multiply $2 \cdot 3$ by $^{-}2$. Multiply 6 by $^{-}2$. Is $^{-}2 \cdot (2 \cdot 3) = {}^{-}2 \cdot 6$ true?

   d. Multiply $2 \cdot 3$ by $\frac{1}{3}$. Multiply 6 by $\frac{1}{3}$. Is $(2 \cdot 3) \cdot \frac{1}{3} = 6 \cdot \frac{1}{3}$ true?

3. What pattern do you see? Try to describe it.

If an equation $a = b$ is true, we get another true equation when we multiply by any number c.

$$a \cdot c = b \cdot c$$

We call this the **Multiplication Principle.**

Example $4 \cdot {}^{-}2 = {}^{-}8$
Multiply by 2.

$$(4 \cdot {}^{-}2) \cdot 2 = {}^{-}8 \cdot 2$$
$$^{-}8 \cdot 2 = {}^{-}8 \cdot 2$$
$$^{-}16 = {}^{-}16$$

EXERCISES

For each equation multiply by the given number to get a new equation. Then simplify both sides of the new equation.

1. $4 \cdot 3 = 12$
   Multiply by 4.

2. $2 \cdot 5 = 10$
   Multiply by 3.

3. $8 = 2 \cdot 4$
   Multiply by 2.

4. $^-2 \cdot 6 = {}^-12$
   Multiply by 3.

5. $^-9 = {}^-3 \cdot 3$
   Multiply by 2.

6. $^-4 = 2 \cdot {}^-2$
   Multiply by 3.

7. $3 \cdot 4 = 12$
   Multiply by $\frac{1}{3}$.

8. $3 \cdot 4 = 12$
   Multiply by $\frac{1}{4}$.

9. $2 \cdot {}^-5 = {}^-10$
   Multiply by $\frac{1}{2}$.

10. $2 \cdot {}^-5 = {}^-10$
    Multiply by $\frac{^-1}{5}$.

11. $10 = {}^-2 \cdot {}^-5$
    Multiply by $\frac{^-1}{2}$.

12. $10 = {}^-2 \cdot {}^-5$
    Multiply by $\frac{^-1}{5}$.

13. $3 \cdot x = 21$
    Multiply by $\frac{1}{3}$.

14. $5 \cdot x = 50$
    Multiply by $\frac{1}{5}$.

15. $7 \cdot y = 42$
    Multiply by $\frac{1}{7}$.

16. $^-3 \cdot x = 33$
    Multiply by $\frac{^-1}{3}$.

17. $^-5 \cdot y = {}^-35$
    Multiply by $\frac{^-1}{5}$.

18. $^-12 \cdot w = 48$
    Multiply by $\frac{^-1}{12}$.

19. $75 = \frac{1}{5} \cdot x$
    Multiply by 5.

20. $36 = \frac{^-1}{9} \cdot y$
    Multiply by $^-9$.

21. $^-42 = \frac{1}{7} \cdot u$
    Multiply by 7.

22. $^-3 \cdot x = {}^-18$
    Multiply by $\frac{^-1}{3}$.

USING THE MULTIPLICATION PRINCIPLE

Now let's see how to use the multiplication principle to help solve equations.

LET'S EXPLORE

1. Think about $3 \cdot x = 15$.

   a. What is the reciprocal of 3?

   b. Use the multiplication principle. Multiply by $\frac{1}{3}$.

   c. What is the solution of your new equation?

   d. Is it also a solution of $3 \cdot x = 15$?

2. Think about $\frac{1}{5} \cdot y = {}^{-}30$.

   a. What is the reciprocal of $\frac{1}{5}$?

   b. Use the multiplication principle. Multiply by 5.

   c. What is the solution of your new equation?

   d. Is it also a solution of the original equation?

3. Let's try ${}^{-}4 \cdot x = 20$.

   a. What is the reciprocal of $^{-}4$?

   b. Use the multiplication principle. Multiply by $\frac{1}{4}$.

   c. What is the solution of your new equation?

   d. Is it also a solution of the original equation?

We can solve an equation like $4 \cdot x = 16$ this way:
(a) Use the multiplication principle to get the variable alone on one side of the equation.
(b) Check the solution in the original equation.

4. Solve.

Example

$$^{-}7 \cdot x = 42$$
$$\frac{^{-}1}{7} \cdot (^{-}7 \cdot x) = \frac{^{-}1}{7} \cdot 42$$
$$1 \cdot x = {^{-}6}$$
$$x = {^{-}6}$$

Check

$^{-}7 \cdot x = 42$	
$^{-}7 \cdot {^{-}6}$	42
42	

a. $3 \cdot y = 12$ b. $21 = {^{-}7} \cdot x$ c. $^{-}3 \cdot z = {^{-}3}$

d. $\frac{1}{6} \cdot w = 7$ e. $^{-}9 = \frac{1}{2} \cdot x$ f. $^{-}12 = \frac{1}{5} \cdot v$

EXERCISES

Solve. Use the multiplication principle.

1. $3 \cdot x = 12$
2. $^{-}5 \cdot y = 40$
3. $^{-}2 \cdot t = {^{-}28}$
4. $^{-}8 \cdot x = 64$
5. $^{-}6 \cdot y = 60$
6. $9 \cdot w = {^{-}36}$
7. $^{-}3 \cdot x = {^{-}9}$
8. $^{-}15 \cdot y = {^{-}45}$
9. $12 \cdot u = {^{-}48}$
10. $\frac{1}{4} \cdot x = 3$
11. $\frac{1}{5} \cdot y = 4$
12. $\frac{1}{3} \cdot t = 7$
13. $15 = {^{-}3} \cdot x$
14. $^{-}15 = 3 \cdot y$
15. $^{-}20 = {^{-}4} \cdot u$
16. $7 = \frac{1}{4} \cdot x$
17. $2 = \frac{1}{10} \cdot y$
18. $^{-}3 = \frac{1}{6} \cdot u$
19. $5 = \frac{^{-}1}{3} \cdot x$
20. $^{-}4 = \frac{1}{2} \cdot y$
21. $^{-}12 = \frac{^{-}1}{3} \cdot t$
22. $^{-}4 \cdot x = 16$
23. $25 = {^{-}5} \cdot y$
24. $^{-}36 = 6 \cdot t$
25. $^{-}7 \cdot x = {^{-}49}$
26. $^{-}64 = 8 \cdot y$
27. $100 = {^{-}10} \cdot n$

UNEXPLORED TERRITORY

Solve these sentences. (They are called inequalities.) Remember, you must find *all* the integers which are solutions.

1. $|x| < 5$
2. $|x - 3| < 6$

USING THE PRINCIPLES TOGETHER

Now that you know how to use the addition and multiplication principles, let's see how to use them together.

LET'S EXPLORE

1. Let's think about the equation $3 \cdot x + 5 = 14$.
   - **a.** Use the addition principle. Add $^{-}5$.
   - **b.** Now use the multiplication principle. Multiply by $\frac{1}{3}$.
   - **c.** What is the solution of your last equation? Check the solution in the original equation.
2. Solve $^{-}4 + 4 \cdot y = 8$.
   - **a.** Use the addition principle. Add 4.
   - **b.** Now use the multiplication principle. Multiply by $\frac{1}{4}$.
   - **c.** What is the solution of your last equation? Check the solution in the original equation.
3. Solve $\frac{^{-}1}{2} \cdot x + 3 = 1$.
   - **a.** Use the addition principle. What will you add?
   - **b.** Use the multiplication principle. By what will you multiply?
   - **c.** What is the solution of your last equation? Check the solution in the original equation.

We can solve an equation like $3 \cdot x + 2 = 11$ this way:
(a) First use the addition principle. Add $^{-}2$.
(b) Then use the multiplication principle. Multiply by $\frac{1}{3}$.
(c) Check the solution in the original equation.

EXERCISES

Solve. Use the addition and multiplication principles.

1. $6 \cdot x + 5 = 17$
2. $4 \cdot y + 7 = {}^{-}5$
3. $2 \cdot t + 6 = 10$
4. $3 \cdot y + {}^{-}5 = 7$
5. $7 \cdot x + {}^{-}5 = 9$
6. $7 \cdot w + 3 = 38$
7. $\frac{1}{2} \cdot x + 2 = 7$
8. $\frac{1}{3} \cdot y + 1 = 4$
9. $\frac{1}{5} \cdot u + {}^{-}1 = 1$
10. $\frac{1}{4} \cdot x + {}^{-}3 = {}^{-}2$
11. $\frac{1}{7} \cdot y + {}^{-}1 = 0$
12. $\frac{1}{9} \cdot t + {}^{-}1 = {}^{-}1$
13. $13 = 3 \cdot x + {}^{-}2$
14. ${}^{-}12 = 4 + 2 \cdot y$
15. ${}^{-}4 = {}^{-}1 + 3 \cdot y$
16. $4 \cdot x - 3 = 5$
17. ${}^{-}5 \cdot y + 5 = 10$
18. ${}^{-}7 \cdot t + 1 = 15$

Let's Practice

For all numbers a, b, and c, this sentence is true.

$$a \cdot (b + c) = a \cdot b + a \cdot c$$

We call this the **distributive property**. Use the distributive property to simplify.

Example $3 \cdot x + 2 \cdot x = (3 + 2) \cdot x = 5 \cdot x$

1. $4 \cdot x + 5 \cdot x$
2. $5 \cdot x + 2 \cdot x$
3. $6 \cdot y + {}^{-}2 \cdot y$
4. $9 \cdot y + {}^{-}5 \cdot y$
5. $4 \cdot x + 5 \cdot x + {}^{-}3 \cdot x$
6. $3 \cdot y + 7 \cdot y + {}^{-}4 \cdot y$
7. ${}^{-}4 \cdot x + {}^{-}3 \cdot x$
8. ${}^{-}7 \cdot x + {}^{-}5 \cdot x$
9. $5 \cdot y + {}^{-}3 \cdot y + 4 \cdot y$
10. ${}^{-}8 \cdot y + {}^{-}4 \cdot y + 3 \cdot y$

PROBLEM SOLVING

1. On her 10-day trip to London, Melissa paid \$330.86 for round-trip air fare, \$250.75 for hotel and food, and \$150.95 for tours, souvenirs, and miscellaneous items. What was the total cost of her trip?

2. On Monday, Wood Travel Agency booked these travel expenses: \$5,942.50, \$285, \$1,030.75, \$562, and \$896.45. They receive a 15% commission on all bookings. What did the agency earn on Monday?

3. Robert rented a car during his 3-day visit to Italy. He drove 100.6 miles the first day, 85.4 miles the second, and 216.7 miles the third day.

   a. What was his total mileage?

   b. It cost \$5 a day and 8¢ a mile to rent the car. How much did it cost Robert?

4. Elizabeth bought a set of bone china in Germany for \$215. Upon her return, she paid a 24% custom duty. What did the china cost her in all?

5. Mrs. Greenberg works for a travel agency. She receives a 10% discount on all her personal trips. She paid \$1,386 for a trip to Paris. What would such a trip usually cost?

DO YOU REMEMBER?

Add.

1. $38.6 + 47.9$
2. $75.3 + 8.4$
3. $47.23 + {}^{-}27.7$
4. ${}^{-}8.62 + 23.156$
5. ${}^{-}15.9 + {}^{-}.5$
6. ${}^{-}.17 + 28.3 + 1.572$

7. $\begin{array}{r} 18.43 \\ 6.642 \\ \underline{257.8} \end{array}$
8. $\begin{array}{r} 81.62 \\ 7.645 \\ \underline{370.001} \end{array}$
9. $\begin{array}{r} 135.05 \\ 7.643 \\ \underline{73.08} \end{array}$
10. $\begin{array}{r} 237.08 \\ 91.492 \\ \underline{73.3} \end{array}$

Subtract.

11. $.9 - .3$
12. $28.6 - 3.8$
13. $29.4 - {}^{-}18.65$
14. $142.07 - {}^{-}2.561$
15. $42 - 2.06$
16. $8.904 - .0542$

17. $\begin{array}{r} 5.004 \\ \underline{-\ 1.35} \end{array}$
18. $\begin{array}{r} 18.2 \\ \underline{-\ 13.476} \end{array}$
19. $\begin{array}{r} 84.00 \\ \underline{-\ 38.543} \end{array}$
20. $\begin{array}{r} 3.8 \\ \underline{-\ .002} \end{array}$

Multiply.

21. $1.7 \times .34$
22. 58.5×7.4
23. 81.47×3.07
24. $1.046 \times .045$
25. $.035 \times 8.7$
26. 9.23×165.7
27. 82.7×1.4
28. $.063 \times 36.831$

Divide.

29. $18\overline{)7.74}$
30. $.05\overline{).04}$
31. $1.32\overline{)26.40}$
32. $7.2\overline{)1.872}$
33. $.16\overline{).1136}$
34. $1.2\overline{).048}$
35. $.003\overline{).000663}$
36. $2.8\overline{)1.456}$
37. $.032\overline{).000128}$

SOLVING EQUATIONS $a \cdot x + b \cdot x = c$

In an equation like $3 \cdot x + 2 \cdot x = 10$, the variable x occurs twice. Let's see how to solve such equations.

LET'S EXPLORE

1. Think about $3 \cdot x + 2 \cdot x = 10$.

   a. Use the distributive property on the left side. Copy and complete. $3 \cdot x + 2 \cdot x = (\_\_\_\_ + \_\_\_\_) \cdot x$

   b. Simplify your answer to part a.

   c. You now have a simpler equation $5 \cdot x = 10$. Solve this equation using the multiplication principle.

   d. Check in the original equation.

2. Let's try $4 \cdot x + {}^{-}2 \cdot x = 12$.

   a. Use the distributive property on the left side.

   b. Simplify.

   c. Write a simpler equation. Then solve it.

   d. Check in the original equation.

3. Let's try ${}^{-}9 \cdot x + {}^{-}2 \cdot x = 22$.

   a. Use the distributive property on the left side.

   b. Simplify.

   c. Write a simpler equation. Then solve it.

   d. Check in the original equation.

We can solve an equation like $3 \cdot x + 2 \cdot x = 5$ this way:
(a) Use the distributive property to get $5 \cdot x = 5$.
(b) Use the multiplication principle. Multiply by $\frac{1}{5}$.

When we use the distributive property to simplify, we say that we are *factoring*.

4. Solve.

Example

$$^{-}9 \cdot x + 3 \cdot x = 18$$
$$(^{-}9 + 3) \cdot x = 18 \text{ (factoring)}$$
$$^{-}6 \cdot x = 18$$
$$x = {}^{-}3$$

Check

$$\begin{array}{r|l} ^{-}9 \cdot x \;\; + 3 \cdot x & = 18 \\ \hline ^{-}9 \cdot {}^{-}3 + 3 \cdot {}^{-}3 & 18 \\ 27 + {}^{-}9 & \\ 18 & \end{array}$$

a. $2 \cdot x + 2 \cdot x = 16$

b. $6 \cdot y + {}^{-}3 \cdot y = 9$

c. $^{-}4 \cdot w + {}^{-}12 \cdot w = 32$

d. $^{-}8 \cdot v + {}^{-}12 \cdot v = {}^{-}100$

e. $\frac{1}{2} \cdot x + \frac{1}{2} \cdot x = 9$

f. $14 = {}^{-}3 \cdot y + {}^{-}4 \cdot y$

EXERCISES

Solve.

1. $4 \cdot x + 2 \cdot x = 12$

2. $9 \cdot x + 2 \cdot x = 22$

3. $^{-}1 \cdot x + {}^{-}2 \cdot x = 33$

4. $3 \cdot y + 4 \cdot y = {}^{-}7$

5. $7 \cdot x + {}^{-}3 \cdot x = {}^{-}24$

6. $5 \cdot x + 2 \cdot x = 21$

7. $4 \cdot y + {}^{-}2 \cdot y = 6$

8. $9 \cdot x + 3 \cdot x = {}^{-}48$

9. $8 \cdot y + {}^{-}11 \cdot y = 3$

10. $10 \cdot t + {}^{-}12 \cdot t = 16$

11. $7 \cdot x + {}^{-}2 \cdot x = 10$

12. $^{-}6 \cdot y + 2 \cdot y = 8$

13. $6 \cdot x + {}^{-}2 \cdot y = 8$

14. $9 \cdot y + {}^{-}11 \cdot y = {}^{-}6$

15. $4 = 8 \cdot x + {}^{-}10 \cdot x$

16. $^{-}18 = {}^{-}4 \cdot x + {}^{-}2 \cdot x$

17. $7 \cdot y + {}^{-}1 \cdot y = 42$

18. $14 = {}^{-}6 \cdot x + {}^{-}1 \cdot x$

19. $\frac{1}{2} \cdot y + \frac{1}{2} \cdot y = {}^{-}3$

20. $2 \cdot y + 2 \cdot y + 6 \cdot y = 10$

21. $^{-}12 = {}^{-}5 \cdot x + {}^{-}7 \cdot x$

22. $20 = \frac{1}{2} \cdot x + \frac{1}{3} \cdot x$

EQUATIONS $a \cdot x + x = b$

To solve an equation like $3 \cdot x + x = 8$, we first factor the left side.

LET'S EXPLORE

1. Let's think about $3 \cdot x + x = 8$.

   a. Is $1 \cdot x$ always equal to x? Why?

   b. Rewrite the equation with $1 \cdot x$ in place of x.

   c. Factor the left side and solve.

2. Let's solve $^-4 \cdot y + y = 6$.

   a. Rewrite the equation with $1 \cdot y$ in place of y.

   b. Factor the left side and solve.

3. Let's solve $25 = {}^-6 \cdot x + x$.

   a. Rewrite the equation with $1 \cdot x$ in place of x.

   b. Factor the right side and solve.

> To solve an equation like $3 \cdot x + x = 8$, we can write $1 \cdot x$ in place of x. Then we factor and solve.

4. Solve.

Example

$$\begin{aligned} {}^-5 \cdot x + x &= {}^-8 \\ {}^-5 \cdot x + 1 \cdot x &= {}^-8 \\ ({}^-5 + 1) \cdot x &= {}^-8 \\ {}^-4 \cdot x &= {}^-8 \\ x &= 2 \end{aligned}$$

Check $\quad {}^-5 \cdot x + x = {}^-8$

$^-5 \cdot 2 + 2$	$^-8$
$^-10 + 2$	
$^-8$	

a. $^-6 \cdot q + q = 20$ b. $^-5 \cdot s + s = 0$

c. $9 = 2 \cdot t + t$ d. $30 = {}^-7 \cdot x + x$

EXERCISES

Solve.

1. $5 \cdot x + x = 6$
2. $^{-}5 \cdot x + x = 4$
3. $3 \cdot y + y = {}^{-}12$
4. $^{-}3 \cdot x + x = 12$
5. $4 \cdot t + t = {}^{-}15$
6. $6 \cdot y + y = 21$
7. $x + 3 \cdot x = 16$
8. $y + {}^{-}2 \cdot y = 3$
9. $w + {}^{-}6 \cdot w = 10$
10. $x + 12 \cdot x = 0$
11. $y + {}^{-}7 \cdot y = 36$
12. $u + 6 \cdot u = {}^{-}42$
13. $60 = 4 \cdot x + x$
14. $^{-}25 = {}^{-}6 \cdot x + x$
15. $30 = {}^{-}7 \cdot x + x$
16. $20 = x + 4 \cdot x$
17. $^{-}24 = y + 5 \cdot y$
18. $17 = x + 16 \cdot x$
19. $x + 13 \cdot x = {}^{-}28$
20. $y + y = 10$
21. $^{-}20 = x + x$
22. $3 \cdot x + x + x = 15$
23. $13 = y + {}^{-}3 \cdot y + y$
24. $x + 3 \cdot x + x = {}^{-}25$
25. $9 = t + t + t$
26. $y + 2 \cdot y + y + {}^{-}2 \cdot y = 50$

UNEXPLORED TERRITORY

Try to solve these equations.

1. $x^2 - 2 = {}^{-}1$
2. $x^2 + 4 = 0$
3. $x^3 + x = 30$
4. $x^3 - x^2 = 100$
5. $x^2 - 2 \cdot x + 1 = 0$
6. $x^2 + 2 \cdot x + 1 = 0$
7. $\frac{1}{2} \cdot x^3 + 3 = 7$
8. $\frac{1}{8} \cdot x^2 - x = 12$

EQUATIONS CONTAINING ^{-}x

When an equation contains an x alone, we can write $1 \cdot x$ in place of x. What can we do if there is ^{-}x alone?

LET'S EXPLORE

1. **a.** Multiply 3 by $^{-}1$.

   b. What is the inverse of 3?

2. **a.** Multiply $^{-}5$ by $^{-}1$.

   b. What is the inverse of $^{-}5$?

3. **a.** Multiply 0 by $^{-}1$.

   b. What is the inverse of 0?

4. Multiply each number by $^{-}1$.

   a. 7 **b.** 13 **c.** 23 **d.** $^{-}2$ **e.** $^{-}16$

5. Name the inverse of each number in Item 4.

When we multiply any number by $^{-}1$, we get the inverse of that number.

$$^{-}1 \cdot x = {}^{-}x$$

6. Let's solve $3 \cdot x + {}^{-}x = 4$.

   a. Rewrite the equation, but write $^{-}1 \cdot x$ in place of ^{-}x.

   b. Now factor and solve.

7. Let's solve $^{-}x = 14$.

   a. Rewrite the equation, but write $^{-}1 \cdot x$ in place of ^{-}x.

   b. Now use the multiplication principle and solve.

When an equation contains ${}^-x$, we can write ${}^-1 \cdot x$ in place of ${}^-x$ and then solve.

8. Solve.

Example

$$
\begin{aligned}
{}^-3 \cdot x + {}^-x &= 8 \\
{}^-3 \cdot x + {}^-1 \cdot x &= 8 \\
({}^-3 + {}^-1) \cdot x &= 8 \\
{}^-4 \cdot x &= 8 \\
x &= {}^-2
\end{aligned}
$$

Check

$$
\begin{array}{r|l}
\multicolumn{2}{c}{{}^-3 \cdot x + {}^-x = 8} \\
\hline
{}^-3 \cdot {}^-2 + 2 & 8 \\
6 + 2 & \\
8 &
\end{array}
$$

a. ${}^-w = 9$

b. ${}^-r = {}^-11$

c. ${}^-x + 2 \cdot x = 8$

d. $5 \cdot y + {}^-y = {}^-12$

e. ${}^-9 \cdot y + {}^-y = 20$

f. ${}^-7 \cdot y + {}^-y = {}^-16$

EXERCISES

Solve.

1. ${}^-x = 4$

2. ${}^-y = 0$

3. ${}^-x = {}^-13$

4. $13 = {}^-y$

5. ${}^-27 = {}^-x$

6. $0 = {}^-t$

7. $4 \cdot x + {}^-x = 12$

8. $5 \cdot x + {}^-x = 8$

9. $3 \cdot x + {}^-x = 14$

10. ${}^-5 \cdot y + {}^-y = 12$

11. ${}^-7 \cdot y + {}^-y = 16$

12. ${}^-8 \cdot y + {}^-y = 18$

13. ${}^-4 \cdot x + {}^-x = {}^-25$

14. ${}^-11 \cdot x + {}^-x = {}^-24$

15. ${}^-7 \cdot x + {}^-x = {}^-24$

16. $18 = {}^-x + {}^-8 \cdot x$

17. $36 = {}^-y + {}^-5 \cdot y$

18. $17 = {}^-x + {}^-16 \cdot x$

EQUATIONS WITH SUBTRACTIONS

How can we handle equations with subtractions in them?

LET'S EXPLORE

1. Think about $x - 13 = {}^{-}7$.

   a. Remember that we can subtract by adding an inverse. Copy and complete. $x - 13 = x + \_\_\_\_$.

   b. Write your answer to part a in place of $x - 13$ in the original equation.

   c. Now you have an equation with an addition. Solve it.

2. Let's try $5 \cdot x - {}^{-}3 = 13$.

   a. What is the inverse of ${}^{-}3$?

   b. Write an equation with an addition instead of the subtraction.

   c. Solve the equation.

3. Think about $5 \cdot x - 3 \cdot x = 4$.

   a. Instead of subtracting $3 \cdot x$ we want to add its inverse. What is $3 \cdot x + {}^{-}3 \cdot x$? Factor and simplify.

   b. Is ${}^{-}3 \cdot x$ the inverse of $3 \cdot x$?

   c. In the original equation write $5 \cdot x + {}^{-}3 \cdot x$ instead of $5 \cdot x - 3 \cdot x$.

   d. Now you have an equation with an addition. Solve it.

4. Let's try $7 \cdot x - x = 18$.

   a. Write an equation with an addition instead of the subtraction.

   b. In place of ${}^{-}x$, write ${}^{-}1 \cdot x$.

   c. Solve the equation.

To solve an equation with subtractions in it, we can rewrite all subtractions as additions. (Subtracting is the same as adding an inverse.)

5. Solve.

Example

$$x - {}^{-}2 \cdot x = {}^{-}6$$
$$x + 2 \cdot x = {}^{-}6 \text{ (adding the inverse)}$$
$$1 \cdot x + 2 \cdot x = {}^{-}6$$
$$3 \cdot x = {}^{-}6$$
$$x = {}^{-}2$$

Check

$x - {}^{-}2 \cdot x$	$= {}^{-}6$
${}^{-}2 - {}^{-}2 \cdot {}^{-}2$	${}^{-}6$
${}^{-}2 - 4$	
${}^{-}6$	

a. $x - 8 = 7$

b. $2 \cdot x - 4 = 12$

c. ${}^{-}3 \cdot x - {}^{-}2 \cdot x = 13$

d. ${}^{-}4 \cdot x - {}^{-}7 \cdot x = {}^{-}9$

EXERCISES

Solve.

1. $3 \cdot y - 5 = 7$
2. $7 \cdot x - 5 = 9$
3. $4 \cdot x - 3 = 5$
4. $5 \cdot y - 5 = {}^{-}10$
5. $7 \cdot x - 1 = {}^{-}15$
6. $\frac{1}{5} \cdot y - 1 = 1$
7. $7 \cdot x - 3 \cdot x = {}^{-}24$
8. $8 \cdot y - 11 \cdot y = 3$
9. $10 \cdot x - 12 \cdot x = 16$
10. $15 = 7 \cdot y - 2 \cdot y$
11. $16 = 2 \cdot y - 6 \cdot y$
12. $8 = 6 \cdot x - 2 \cdot x$
13. $6 = 9 \cdot y - 11 \cdot y$
14. ${}^{-}30 = {}^{-}4 \cdot x - 2 \cdot x$
15. $9 \cdot x - {}^{-}3 \cdot x = 48$
16. ${}^{-}6 \cdot y - {}^{-}2 \cdot y = 16$
17. ${}^{-}10 \cdot x - {}^{-}8 \cdot x = 4$
18. $14 = 3 \cdot y - y$

EQUATIONS WITH VARIABLES ON BOTH SIDES

When an equation has variables on both sides, we can find another equation with all variables on one side. It will have the same solutions as the original one.

LET'S EXPLORE

1. Think about $5 \cdot x = 3 \cdot x + 4$.

   a. What is the inverse of $3 \cdot x$?

   b. Use the addition principle. Add $^{-}3 \cdot x$. Copy and complete: $^{-}3 \cdot x + 5 \cdot x = {}^{-}3 \cdot x +$ ________

   c. Simplify the right side.

   d. Now all the variables are on one side. Simplify the left side and solve the equation.

2. Think about $3 + x = 2 \cdot x$.

   a. Use the addition principle. Add ^{-}x.

   b. Simplify the left side.

   c. Now all the variables are on one side. Simplify the right side and solve the equation.

3. Let's try $2 \cdot y + {}^{-}2 = {}^{-}3 \cdot y + 3$.

   a. What is the inverse of $^{-}3 \cdot y$?

   b. Use the addition principle. Add $3 \cdot y$.

   c. Simplify the right side.

   d. Now all the variables are on one side. Simplify the left side.

   e. Use the addition principle. Add 2.

   f. Solve the equation.

To solve an equation like $3 \cdot x = 2 \cdot x + 1$, we first use the addition principle. We add $^{-}2 \cdot x$.

4. Solve.

Example

$$\begin{aligned} 5 + {}^{-}2 \cdot y &= 3 \cdot y + {}^{-}5 \\ 5 + {}^{-}2 \cdot y + 2 \cdot y &= 3 \cdot y + 2 \cdot y + {}^{-}5 \\ 5 &= 5 \cdot y + {}^{-}5 \\ 5 + 5 &= 5 \cdot y + {}^{-}5 + 5 \\ 10 &= 5 \cdot y \\ 2 &= y \end{aligned}$$

Check

$$\begin{array}{r|l} \multicolumn{2}{c}{5 + {}^{-}2 \cdot y = 3 \cdot y + {}^{-}5} \\ \hline 5 + {}^{-}2 \cdot 2 & 3 \cdot 2 + {}^{-}5 \\ 5 + {}^{-}4 & 6 + {}^{-}5 \\ 1 & 1 \end{array}$$

a. $4 \cdot y = 7 \cdot y + {}^{-}3$ **b.** $6 \cdot x + 1 = 2 \cdot x + 9$

c. ${}^{-}9 \cdot x = {}^{-}3 \cdot x + {}^{-}12$ **d.** ${}^{-}4 \cdot x + {}^{-}3 = x + 2$

e. ${}^{-}7 \cdot y = 4 \cdot y + {}^{-}55$ **f.** ${}^{-}15 + 3 \cdot y = 23 + {}^{-}4 \cdot y$

EXERCISES

Solve.

1. $4 \cdot x = 3 \cdot x + 7$

2. $5 \cdot y + {}^{-}2 = 3 \cdot y$

3. ${}^{-}8 \cdot x = 4 \cdot x + {}^{-}24$

4. $3 \cdot y = 24 + {}^{-}5 \cdot y$

5. $9 \cdot x = 2 \cdot x + 14$

6. $2 \cdot x = 4 + {}^{-}2 \cdot x$

7. $3 \cdot y = 2 \cdot y + 5$

8. ${}^{-}2 \cdot x + 4 = 2 \cdot x$

9. $3 \cdot y = 7 \cdot y + 8$

10. $2 \cdot x = {}^{-}12 + {}^{-}2 \cdot x$

11. $5 \cdot y = 9 + 2 \cdot y$

12. $18 + 4 \cdot y = 10 \cdot y$

13. $5 + {}^{-}2 \cdot y = 3 \cdot y + {}^{-}5$

14. $5 \cdot x + {}^{-}1 = 13 + {}^{-}2 \cdot x$

15. $7 \cdot y + 5 = 2 \cdot y + 10$

16. $2 \cdot x + 2 = {}^{-}2 \cdot x + 18$

17. $6 \cdot y + 20 = y + 5$

18. ${}^{-}2 + 6 \cdot x = 4 \cdot x + 10$

19. $2 \cdot y + 2 = {}^{-}10 \cdot y + {}^{-}10$

20. ${}^{-}17 + 9 \cdot x = {}^{-}8 \cdot x + 17$

PROBLEM SOLVING

1. Mrs. Smith bought 3 dozen eggs. Nine of them were spoiled. What per cent of the eggs were spoiled?

2. Mr. Smart invested $1,200 in stocks. When the price went down, he sold them at a 12% loss. What was his loss?

3. Mr. Golin paid 20% of his salary in taxes. Last year his tax bill came to $3,750. What was his salary for the year?

4. A Boy Scout troop saved $26.25 to buy badges. Each badge cost $.75. How many can they buy?

5. Alan's father drove 306.9 miles on a trip. He used 18.6 gallons of gasoline. How many miles did he average per gallon?

6. The eighth grade class sold tickets for their variety show. Each ticket cost 75¢. How much money did they receive?

7. Cary carried the ball five times in a football game. He gained 8 yards, gained 4 yards, lost 6 yards, gained 7 yards, and then lost 9 yards.

   a. What was his total gain?

   b. What was his average gain per carry?

CHAPTER REVIEW

Solve.

1. $x + 1 = 7$
2. $y + 2 = 18$
3. $t - 4 = 12$
4. $w - 8 = 20$
5. $t + {}^{-}7 = {}^{-}16$
6. $u + 24 = {}^{-}20$
7. $w + {}^{-}15 = 24$
8. $x + 5 = 72$
9. $v + 15 = 40$
10. $u + 24 = 70$
11. $3 \cdot s = 27$
12. $2 \cdot t = 14$
13. $5 \cdot t = 100$
14. $12 \cdot y = {}^{-}84$
15. ${}^{-}6 \cdot t = 54$
16. $8 \cdot t = {}^{-}16$
17. $8 \cdot u = {}^{-}24$
18. $5 \cdot t = 30$
19. $4 = \frac{1}{8} \cdot t$
20. $\frac{1}{6} \cdot x = 6$
21. $2 \cdot y + 5 = 11$
22. $9 \cdot x + 1 = 82$
23. $3 \cdot x + 4 = 19$
24. $5 \cdot w + 2 = 37$
25. $2 \cdot t + 2 = 32$
26. $\frac{1}{4} \cdot t + 6 = 10$
27. $8 \cdot u + 42 = u$
28. $7 \cdot y + 12 = y$
29. $8 \cdot y + y = 45$
30. $9 \cdot t + t = 50$
31. $4 \cdot x + 7 \cdot x = 121$
32. $2 \cdot y + 9 \cdot y = 55$
33. $12 \cdot s + 2 \cdot s = {}^{-}56$
34. $5 \cdot y + 3 \cdot y = 48$
35. $6 \cdot t + t = 49$
36. $7 \cdot s + s = 88$

37. $x + 5 \cdot x = 24$

38. $s + 9 \cdot s = {}^-40$

39. $6 \cdot y + {}^-y = 25$

40. ${}^-9 \cdot w + w = 80$

41. $7 \cdot t - 3 = 25$

42. $6 \cdot y - 8 = 10$

43. $7 \cdot x - 5 = 30$

44. $9 \cdot u - 2 = 52$

45. $t - {}^-7 = 5$

46. $w - 8 = {}^-10$

47. $7 \cdot y - 2 = 61$

48. $12 \cdot u + 2 = {}^-58$

49. $10 \cdot x - 3 = 57$

50. $8 \cdot u - 5 = 59$

51. $58 - x = 40$

52. $6 \cdot w - 2 = 52$

53. $8 \cdot w - 14 = w$

54. $12 \cdot y = 22 + y$

55. $7 \cdot y + 12 = y$

56. $8 \cdot t + 42 = t$

57. $3 \cdot x + x = 12$

58. $2 \cdot y + y = 18$

59. $y + 2 \cdot y = 27$

60. $u + 15 \cdot u = 32$

61. $6 \cdot t - 2 \cdot t = 12$

62. ${}^-8 \cdot x - 2 \cdot x = 40$

Solve these problems.

63. What is the price of 5 gallons of paint on sale at 2 gallons for $13 or $7 per gallon?

64. Sue has $20. She bought a $4 pitcher. How many $3 vases can she also buy?

Terms You Should Know

equation	related sentence
solution	opposite operations
solve	addition principle
variable	multiplication principle

CHAPTER TEST

Solve.

1. $x + 2 = 24$
2. $y + 20 = 165$
3. $t - 2 = 15$
4. $n - 22 = {}^-12$
5. $t + 5 = 39$
6. $r + 20 = 58$
7. $2 \cdot y = 16$
8. $2 \cdot m = {}^-18$
9. $3 \cdot x = 27$
10. $\frac{1}{9} \cdot w = 9$
11. $3 \cdot x + 8 = {}^-4$
12. $5 \cdot x + 8 = {}^-22$
13. $2 \cdot y + 5 = 25$
14. $3 \cdot t + 24 = {}^-39$
15. $6 \cdot y + y = 70$
16. $8 \cdot v + v = {}^-18$
17. $3 \cdot t + 3 \cdot t = {}^-18$
18. $15 \cdot x + 5 \cdot x = {}^-60$
19. $r + 3 \cdot r = 44$
20. $t + 5 \cdot t = {}^-138$
21. $10 \cdot x + 27 = x$
22. $3 \cdot y + 8 = y$
23. $7 \cdot y + 30 = y$
24. $u = 14 + 8 \cdot u$
25. $2 \cdot t - 4 = {}^-16$
26. $2 \cdot y - 7 = 33$
27. $5 \cdot x + x = 36$
28. $y + 8 \cdot y = {}^-27$
29. $3 \cdot w - 2 \cdot w = 10$
30. ${}^-9 \cdot y - 6 \cdot y = 45$

Solve these problems.

31. Lyn bought 12 pencils for \$1.08. How much did each pencil cost?

32. Don bought 7 albums on sale at 2 for \$4.95 or 1 for \$3. How much change will he get from \$20?

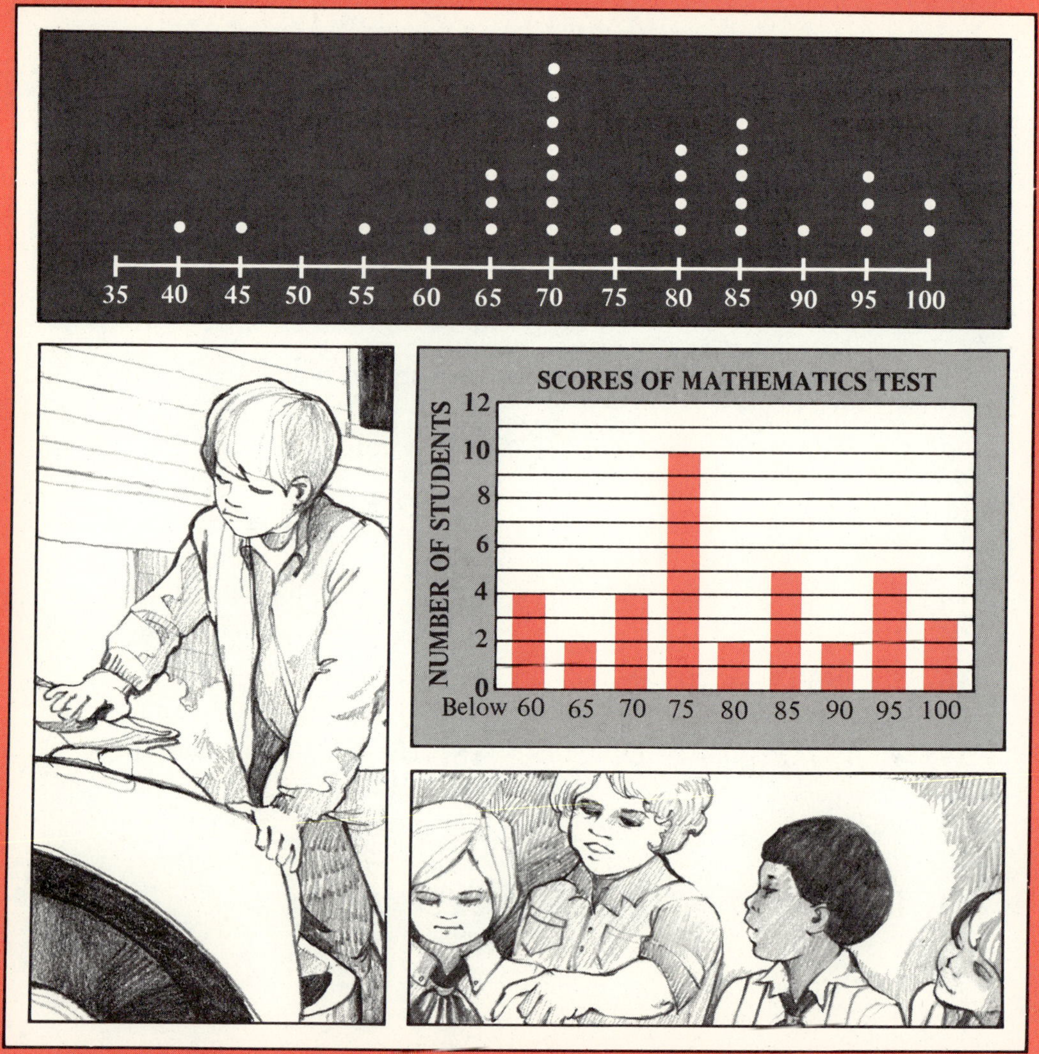

CHAPTER THIRTEEN

Probability and Statistics

PROBABILITY

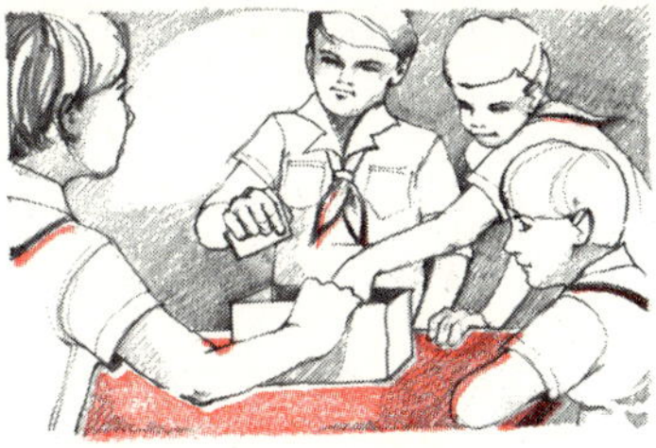

Only one boy could represent Troop 212 at a Scout Jamboree. Each boy wrote his name on a slip of paper and put it in a box. There are 28 boys in the troop.

1. How many chances has each boy for his name to be drawn?

2. How many possible choices of a representative are there?

The **probability** of an event is the ratio of the number of favorable possibilities to the total number of possibilities.

3. Douglas Carlson is a member of Boy Scout Troop 212. There is one chance out of 28 that his name will be drawn. Write a fractional numeral for the probability.

4. There are two boys named Stephen in the troop. What is the probability that a boy named Stephen will be selected?

5. **a.** How many chances are there that a girl's name will be chosen?

   b. What is the probability that a girl will be chosen?

The probability of an event that cannot occur is 0.

6. **a.** How many slips of paper have boys' names on them?

   b. What is the probability that a boy will be chosen?

The probability of an event that is certain to occur is 1.

7. The troop is divided into 4 patrols of 7 members each. The patrols are the Leopard Patrol, the Beaver Patrol, the Lion Patrol, and the Bear Patrol.

   a. What is the probability that a member of the Bear Patrol will be selected?

   b. What is the probability that a member of the Wildcat Patrol will be selected?

   c. What is the probability that a Boy Scout will be selected?

EXERCISES

Solve these problems.

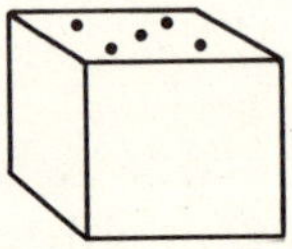

1. What is the probability that a die, when tossed, will land with a 5 up?

2. What is the probability of a die landing with an odd number up?

3. What is the probability of a die landing with a composite number up?

A jar contains 3 red marbles, 2 white marbles, and 7 blue marbles. A marble is to be chosen from a jar.

4. What is the probability that a blue marble will be chosen?

5. What is the probability that a red marble will be chosen?

6. What is the probability that a white marble will be chosen?

7. What is the probability that the marble chosen will be red, white, or blue?

8. What is the probability that a green marble will be chosen?

A special trick deck of cards contains 10 spades, 2 hearts, and 8 diamonds. A card is chosen from the deck.

9. What is the probability that a diamond will be chosen?

10. What is the probability that a heart will be chosen?

11. What is the probability that a spade, heart, or diamond will be chosen?

12. What is the probability that a club will be chosen?

UNEXPLORED TERRITORY

A wooden cube has all of its faces painted red. It is cut into smaller cubes by making six saw cuts as shown. These cubes are put into a bag and mixed. One cube is drawn from the bag.

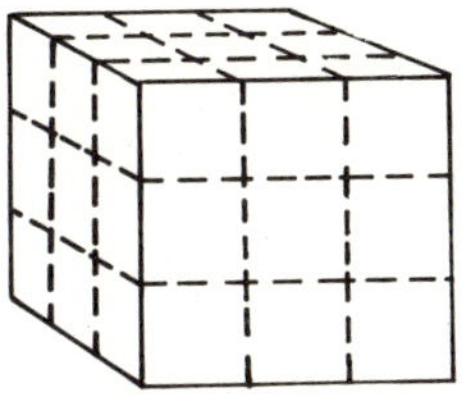

1. What is the probability that it will have three red faces?

2. What is the probability that it will have two red faces?

3. What is the probability that it will have one red face?

4. What is the probability that it will have no red faces?

5. What is the probability that it will have at least two red faces?

6. What is the probability that it will not have four unpainted faces?

7. What is the probability that it will not have less than two painted faces?

COMPOUND PROBABILITY

Let's consider a case of compound probability. We will examine two events which cannot occur at the same time.

1. There are 5 red marbles, 3 white marbles, and 5 black marbles in a bag. Melba will draw one marble from the bag without looking.

   a. What is the probability that she will draw a white marble?

   b. What is the probability that she will draw a black marble?

   c. Add your answers to parts a and b.

2. a. What is the total number of white and black marbles?

   b. Melba cannot draw both a white marble and a black marble at the same time. What is the probability that she will draw either a white marble or a black marble?

3. How do your answers to Items 1c and 2b compare?

Suppose two events cannot occur at the same time. Then the probability that one or the other of them will occur is the **sum** of their probabilities.

4. The probability that when a die is rolled, a 3 or a 4 will land up is $\frac{2}{6}$, or $\frac{1}{3}$. What is the probability that a 1, 2, or 3 will land up?

EXERCISES

1. What is the probability that, when a die is tossed, a 5 or a 6 will land up?

2. There are 4 yellow blocks, 6 green blocks, 8 blue blocks, and 2 red blocks in a bag. A block is to be drawn.

   a. What is the probability that a yellow or a blue block will be selected?

   b. What is the probability that a yellow or a red block will be selected?

   c. What is the probability that a yellow, a green, a blue, or a red block will be selected?

   d. What is the probability that a blue or a black block will be selected?

   e. What is the probability that a green block will not be selected?

   f. What is the probability that a pink block will be selected?

3. A circular board has 10 notches on it and a pointer hinged at the center. The notches are labeled *A, B, C, D, E, F, G, H, I,* and *J*. If the pointer is spun, what is the probability of it stopping on *A, C,* or *H*?

4. There are 225 eighth grade students at Hilltop Middle School. Seventy-five of them study French; 25 study Latin; 25 study Spanish; and the rest do not study a foreign language. No student is allowed to study two languages. A student's name is drawn from a box. What is the probability that the student studies a foreign language?

ANOTHER TYPE OF COMPOUND PROBABILITY

Now let's consider another type of compound probability.

LET'S EXPLORE

A bag contains a red, a black, and a yellow jellybean. Craig will draw a jellybean from the bag, replace it, mix them up, and then draw another jellybean.

1. What is the probability that he will draw a red jellybean the first time?

2. He replaces the first jellybean and draws again. What is the probability that he will draw a black jellybean the second time?

3. Did the result of the first drawing affect the result of the second drawing?

4. The diagram below shows all the possible results of two drawings.

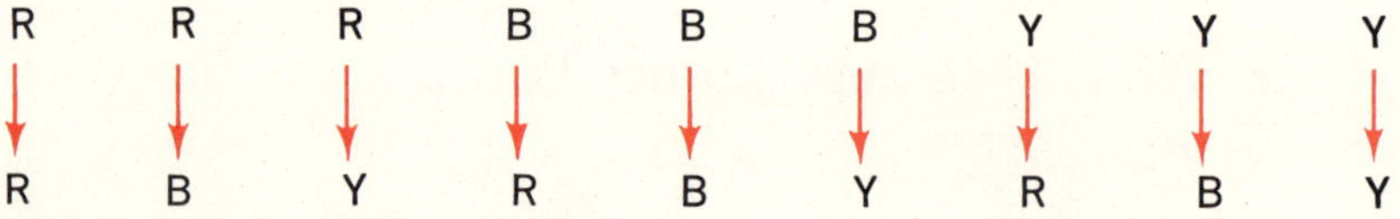

   What is the total number of possible results?

5. What is the probability that Craig will draw a red jellybean first and then a black jellybean?

6. Multiply your answers to Items 1 and 2.

7. Compare your answers to Items 5 and 6.

> Suppose we have two events which do not affect each other. The probability that both events will occur is the **product** of their probabilities.

8. A bag contains 3 yellow marbles and 2 black marbles. Kristin will draw a marble, replace it, and then draw another.

   a. What is the probability that she will draw a yellow marble the first time?

   b. She replaces the first marble and draws again. What is the probability that she will draw a black marble the second time?

   c. What is the probability that Kristin will draw a yellow marble and then a black one? (*Hint* Multiply your answers to parts a and b.)

EXERCISES

1. One block has letters *A, B, C, D, E,* and *F* on its faces. Another has 1, 2, 3, 4, 5, and 6 on its faces. The blocks are rolled together.

   a. What is the probability that 3 and *D* will land up?

   b. What is the probability that *B* and an even number will land up?

   c. What is the probability that a consonant and 2 will land up?

2. Three of the ten candidates for May Queen and two of the seven candidates for her escort are from Mrs. Kirby's class. The queen and her escort are to be selected by drawing names from a box. What is the probability that they will both be from Mrs. Kirby's class?

3. A nickel and a dime are tossed. What is the probability that they will both come up heads?

4. A bag contains 3 red blocks and 5 green blocks. One block is drawn and then replaced. A second drawing is made. What is the probability that a red one will be drawn both times?

DRAWING CONCLUSIONS

Mrs. Brooks took her class to the mathematics laboratory to perform an experiment. They had a large container of 1,500 marbles. Some were red and some were black. Without looking, each student drew out a marble, recorded the color, and then returned the marble to the container. Here are the results.

SAMPLINGS

Student	*Marble*	*Student*	*Marble*
1	R	16	B
2	B	17	R
3	R	18	B
4	R	19	R
5	B	20	B
6	B	21	B
7	R	22	B
8	B	23	B
9	B	24	B
10	R	25	B
11	B	26	R
12	B	27	B
13	R	28	B
14	B	29	B
15	B	30	R

1. a. How many marbles were drawn?

   b. How many red marbles were drawn?

   c. How many black marbles were drawn?

2. 10 out of 30, or $\frac{1}{3}$ of the marbles drawn were red. What part of the marbles drawn were black?

3. a. Find $\frac{1}{3}$ of 1,500.

   b. Find $\frac{2}{3}$ of 1,500.

4. Without counting them, the students estimated that there are probably about 500 red marbles in the container. How many black marbles are probably in the container?

EXERCISES

Solve these problems.

1. There are 900 marbles in a box. Some are blue, some yellow, and some red. A marble is drawn from the box, the color is recorded, and then the marble is returned. This procedure is repeated 20 times. Use the record below to estimate the number of marbles of each color in the box.

SAMPLINGS

Trial	*Color*	*Trial*	*Color*
1	B	11	B
2	Y	12	B
3	R	13	Y
4	R	14	Y
5	Y	15	R
6	Y	16	Y
7	Y	17	Y
8	B	18	R
9	R	19	Y
10	Y	20	B

2. Carl and Edward are candidates for president of the Student Council. The school newspaper decided to conduct a poll and try to predict the outcome of the election.

   There are 1,200 students in the school.
   80 students were interviewed.
   50 were for Carl.
   30 were for Edward.

   Based on their survey, how many votes should the newspaper predict for each boy in the election?

FREQUENCY DISTRIBUTION

Mrs. Eckstein's mathematics class recorded the temperature at noon every day during April. The class decided to make a frequency distribution table.

1. Here are the temperatures.

42°	45°	41°	40°	41°	44°	45°	47°	48°	45°
46°	48°	50°	49°	49°	46°	46°	47°	47°	46°
48°	51°	48°	49°	47°	47°	46°	46°	51°	49°

   a. What was the least temperature?

   b. What was the greatest temperature?

The **range** of a set of measurements goes from the least measurement to the greatest.

2. The range of the temperatures is 40° to 51°. Copy and complete the table.

TEMPERATURES

Temperature	*Tally*	*Frequency*
51°	||	2
50°	|	1
49°	||||	4
48°	||||	4
47°	~~||||~~	
46°	~~||||~~ |	
45°	|||	
44°	|	
43°		
42°	|	
41°	||	
40°	|	

A frequency distribution table arranges the data in either ascending or descending order and shows the number of times each item occurs.

3. A *dot frequency diagram* is another way to show data. It is a type of graph. A dot frequency diagram for scores of 12, 15, 17, 15, 14, 12, 16, 15, 14, 17, and 20 looks like this.

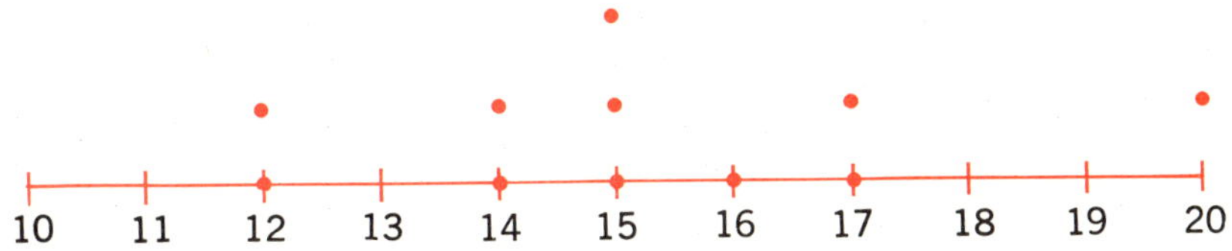

Make a dot frequency diagram for the temperatures recorded by Mrs. Eckstein's class.

EXERCISES

Solve these problems.

1. Take 10 coins and toss them. Count the number of tails and record that number. Do this 20 times.

   a. What is the range of the data?

   b. Make a frequency distribution table and a dot frequency diagram for the results.

2. Here are the yearly salaries for the employees of Tiny Tots Toy Store.

$4,500	$4,200	$5,100	$5,000
$4,000	$4,000	$4,500	$4,200
$4,800	$5,000	$4,500	$4,800
$4,000	$4,500	$4,800	$5,200
$4,800	$5,200	$4,000	$4,200

   a. What is the range?

   b. Make a dot frequency diagram to show the information.

THE MEAN

One of the most frequently used methods of analyzing data is to find the *mean*, or average.

1. The price of a dozen eggs varies throughout the year. Last year the Wise Egg Farm set their prices every two weeks. They sold one dozen large eggs at each of these prices.

50¢	50¢	50¢	52¢	54¢	55¢	55¢
55¢	60¢	60¢	65¢	65¢	70¢	70¢
70¢	70¢	80¢	80¢	75¢	75¢	70¢
66¢	66¢	59¢	58¢	58¢		

   a. Add the weekly prices.

   b. Divide by 26 to find the average weekly price of eggs last year.

The **mean** of any statistical data is found by adding the numbers and dividing the sum by the number of addends. The mean is also called the **average**.

2. Find the mean of these test scores.

   85 90 86 70 80 88 73 78 82 84

EXERCISES

Solve these problems.

1. Eric recorded the temperature at 4:00 pm every day for one week last July. The temperatures were 82°, 84°, 85°, 90°, 92°, 86°, and 88°. Find the mean 4:00 pm temperature for the week.

2. The Girl Scouts sold cookies to raise money. Mary sold \$21 worth of cookies; Sue sold \$18 worth; Kandi sold \$24 worth; Sherry sold \$32 worth; and Shirley sold \$25 worth. What was the mean of the sales of the cookies?

3. Sam took a survey and found that houses in his neighborhood had the following values.

VALUES OF HOUSES

Amount	*Tally*	*Frequency*
$32,000	|	1
31,000		0
30,000	||	2
29,000	|||	3
28,000	|	1
27,000		0
26,000	|	1
25,000	|||	3
24,000	卌 |	6
23,000	卌	5
22,000	|||	3
21,000	卌 |	6
20,000	卌 |	6
19,000		0
18,000	||	2
17,000	|||	3
16,000		0
15,000	|	1

a. What is the range?

b. What is the mean value of a house. (*Hint* Consider one $32,000 house, no $31,000 houses, two $30,000 houses, etc.)

UNEXPLORED TERRITORY

Two dice are rolled.

1. What is the probability that the sum of the numbers is 2?

2. What is the probability that the sum of the numbers is 3? (*Hint* There are two ways of getting a sum of 3— 1 + 2 and 2 + 1. So to find the answer, compute the probability of 1 and 2, or 2 and 1.)

3. What is the probability that the sum is 4?

ESTIMATING TO FIND THE MEAN

Mrs. Greenwald found these prices for six different brands of canned peaches.

35¢ 38¢ 41¢ 36¢ 40¢ 32¢

The cans were all the same size. She asked her class to find the mean price of a can of peaches.

1. Gary added the six prices and divided by 6. Use Gary's method to find the mean. *37¢*

2. Carolyn said she could find it another way. She estimated that the mean would be about 39¢.

   Since 35¢ is 4¢ less than 39¢, she wrote $^{-}4$.
   For 38¢ she wrote $^{-}1$.
   41¢ is 2¢ more than 39¢, so she wrote 2.

   a. What did she write for 36¢?

   b. What did she write for 40¢?

   c. What did she write for 32¢?

   d. She added $^{-}4$, $^{-}1$, 2, $^{-}3$, 1, and $^{-}7$. What was her answer?

   e. Then she divided the sum by 6. What was her quotient?

   f. Then she added the quotient to 39. What was her answer?

   g. Compare her final answer to Gary's.

3. Suppose Carolyn's original estimate had been 35¢. Use her method to compute the mean price.

To find a mean, we can do as follows:
(a) Estimate the mean.
(b) Find the difference between each number and our estimate.
(c) Find the average of these differences.
(d) Add this average to our estimate.

4. Estimate to find the mean.

168	150	165	185	140	200	175

EXERCISES

Estimate to find the mean of each.

1. 12, 16, 15, 14, 20, 13, 17
2. 3, 6, 4, 2, 8, 3, 6, 5
3. 200, 50, 150, 350, 100
4. 186, 190, 170, 198, 183, 181, 178
5. 728, 748, 622, 680, 720, 700
6. 100, 300, 500, 700, 800, 200, 400, 600

Solve these problems.

7. In chemistry class, water was added to 15 beakers of a concentrated solution of acid. The following changes in temperature were recorded.

40.0°	40.7°	40.9°	41.2°	39.8°
38.5°	39.2°	37.9°	41.0°	41.1°
40.2°	40.5°	39.2°	38.6°	38.0°

Find the mean change in temperature to the nearest tenth of a degree.

8. The average rise and fall of prices listed on the stock exchange is recorded daily. Here is a record of the average price changes during a 3-week period.

.5	.2	$^-1.0$	.3	$^-.3$
$^-.1$	1.3	.9	$^-.2$	$^-.5$
$^-.5$	.2	.3	1.3	$^-.1$

Find the mean stock price change for the 3-week period. Give your answer to the nearest thousandth of a point.

THE MODE

The *mode* is sometimes used to help interpret statistical data.

1. Make a frequency distribution chart for these temperatures.

   81° 61° 80° 81° 82° 79° 80° 78° 80° 79°

2. What temperature occurs most often?

In statistical data, the number which occurs most often is the **mode.**

3. What is the range of the temperatures in Item 1?

4. How many temperatures are below 78°?

5. What is the mean temperature?

6. Which represents the typical temperature better, the mean or the mode? Explain.

7. Look at this data.

   50 45 38 50 52 48 48 51

   a. What numbers occur most frequently?

   b. How many modes are there for this data?

Data may have more than one mode.

8. Look at this data.

   100 125 130 140 150

   a. What number occurs most frequently?

   b. Does any number occur more than once?

If no number occurs more than once in a set of data, there is no mode.

9. Find the mode of each.

a. 3, 5, 7, 4, 2, 1

b. 5, 8, 6, 4, 3, 5, 7, 4, 5, 3, 2

c. 2, 3, 5, 4, 4, 3, 2, 2, 4, 5

d. 40, 42, 41, 45, 41, 42, 40, 41, 48, 55

EXERCISES

Find the mode of each.

1. 100, 150, 125, 130, 125, 110

2. 12, 13, 17, 11, 9, 13, 14, 15

3. 30, 32, 28, 35, 40, 27, 41

4. 12, 14, 18, 16, 14, 16, 20, 11, 15

5. 78, 79, 40, 32, 80, 76, 78, 80

6. 55, 50, 42, 100, 62, 30

7. 25, 40, 48, 50, 40, 45, 70, 50, 55

8. 17, 20, 60, 30, 48, 60, 25, 12, 38, 60

9. 2, 4, 8, 4, 3, 7, 6, 3, 2, 1, 7, 4, 6, 3, 4, 2, 6

10. 20, 30, 40, 30, 16, 24, 32, 28, 30, 28, 32

Solve this problem.

11. These are the scores made on Mr. Preston's mathematics test.

82	85	88	80	82	82	82
63	80	82	90	97	89	81
81	45	82	83	71	86	84

Find the mean and the mode of the scores.

THE MEDIAN

Sometimes the range of measures is very great. In that case the *median* may be more useful than the mean or mode for interpreting the data.

1. a. Arrange these numbers in descending order.

   80 81 87 85 87 70 42 100 95

   b. How many numbers did you list?

   c. What is the middle one?

To find the **median** of any statistical data, we arrange the data in ascending or descending order. Then we pick the middle number.

2. a. Arrange these numbers in ascending order.

   20 50 32 26 60 10 30 40

   b. How many numbers did you list?

   c. Is there a middle one?

   d. What are the two middle scores?

   e. What is their average?

If a set of numbers has an even number of elements, we average the two middle ones to find the median.

3. Find the median of each.

   a. 3, 5, 8, 4, 2

   b. 2, 8, 7, 6, 5, 4, 3, 2, 5

   c. 2, 4, 8, 7, 6, 5, 10, 5, 5, 5

   d. 150, 142, 148, 100, 125, 130

   e. 50, 70, 58, 64, 72, 56, 60

   f. 100, 200, 500, 800, 300, 100, 200

   g. 125, 160, 180, 190, 150, 145, 140, 165, 175

EXERCISES

Find the median of each.

1. 100, 90, 82, 48, 76

2. 90, 92, 98, 76, 42, 81, 80, 87, 36, 78, 67

3. 2, 7, 9, 8, 6, 5, 8, 6, 5, 3, 2, 1, 10, 8, 7, 6, 5

4. 25, 12, 18, 30, 16, 40, 16, 18

5. 2, 6, 8, 3, 4, 7, 6, 3, 2, 10, 9, 12, 11, 15

Solve these problems.

6. Mrs. Jackson found these prices for six different brands of creamed corn.

   32¢ 36¢ 38¢ 35¢ 34¢ 35¢

   What is the median price?

7. Mr. Cooley decided to use the median score to help him scale the grades on a science test. He would use the median as a *C* grade. Find the median of these scores.

82	88	80	94	70
65	84	91	76	100
40	75	81	85	98
72	81	80	75	91
75	71	89	88	80
68	89	78	80	92

8. Find the median of these scores on a mathematics test.

80	84	85	92	94	81
78	88	98	96	86	85
84	80	74	95	93	89
76	88	88	86	70	100
98	92	90	94	88	91

PROBLEM SOLVING

1. The thirty employees of the Classy Car Sales made the following contributions to the United Fund.

CONTRIBUTIONS		
Amount	*Tally*	*Distribution*
$150.00	𝍷𝍷𝍷	3
50.00	𝍷	1
12.50	𝍷	1
10.00	𝍸 𝍸	10
7.50	𝍷𝍷𝍷	3
5.00	𝍸 𝍷	6
2.00	𝍷𝍷𝍷𝍷	4
0	𝍷𝍷	2

a. What is the median?

b. What is the mode?

c. What is the mean?

d. Does the mean, the median, or the mode best describe an individual's contribution? Explain.

2. These were the temperatures at 12 noon in Pearl River last November.

25°	20°	20°	22°	30°	38°
42°	40°	38°	39°	41°	40°
37°	48°	35°	39°	40°	39°
28°	20°	30°	18°	25°	30°
38°	39°	35°	29°	31°	15°

a. What is the median?

b. What is the mode?

c. What is the mean?

d. What would describe the daily temperature best, the mean, the median, or the mode? Explain.

3. The following prices for 1 pound of green beans were found in seven different stores.

27¢ 32¢ 29¢ 31¢ 29¢ 28¢

a. What is the median price?

b. What is the mode?

c. What is the mean?

4. Mark compared the prices of a certain model bicycle in five different stores. He found these prices.

$59.95 $62.95 $60.00 $59.95 $62.00

a. What was the range?

b. What was the median price?

c. What was the mode?

d. What was the mean?

5. The following scores were made on a mathematics test.

78	76	70	84	86	80
82	80	88	78	70	90
98	86	84	80	76	80
80	60	78	76	78	92
82	80	50	80	82	80

a. What was the range?

b. Make a frequency distribution table of the scores.

c. What was the mode?

d. What was the median?

e. What was the mean?

6. The six members of the Badger Patrol sold the following number of boxes of Girl Scout cookies.

30 42 28 36 40 36

Their goal as a patrol was to average 35 boxes each. Did they reach their goal?

BROKEN-LINE GRAPHS

A broken-line graph provides a clear picture of some statistical data.

1. Here is a graph of the temperatures recorded at 6:00 am for one week.

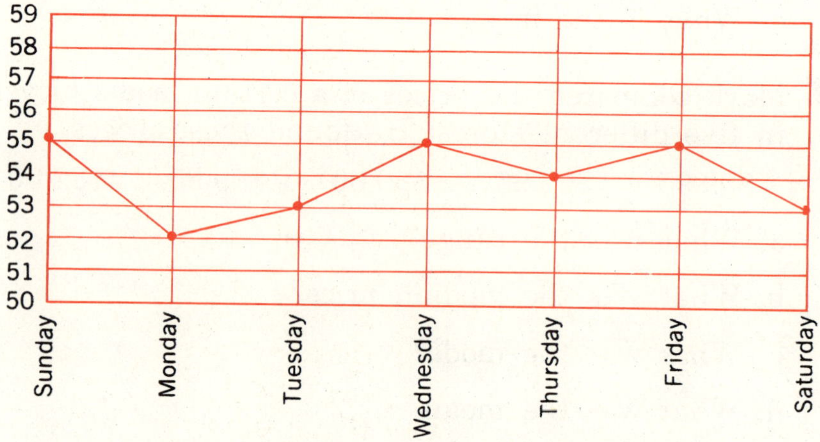

   a. What was the temperature on each day?

   b. On what days did the temperature fall?

   c. On what days did the temperature rise?

   d. What was the greatest temperature of the week?

   e. What was the least temperature of the week?

   f. What is the mode?

   g. What is the median?

   h. What is the mean?

2. a. Make a broken-line graph to show these attendances at ten basketball games.

520	485	550	600	675
650	680	700	650	920

   b. What is the greatest attendance?

   c. What is the least attendance?

d. What is the range of attendance?

e. What is the mode?

f. What is the median?

g. What is the mean?

EXERCISES

Solve these problems.

1. Here is a graph of the weekly sales at the Handy Appliance Store.

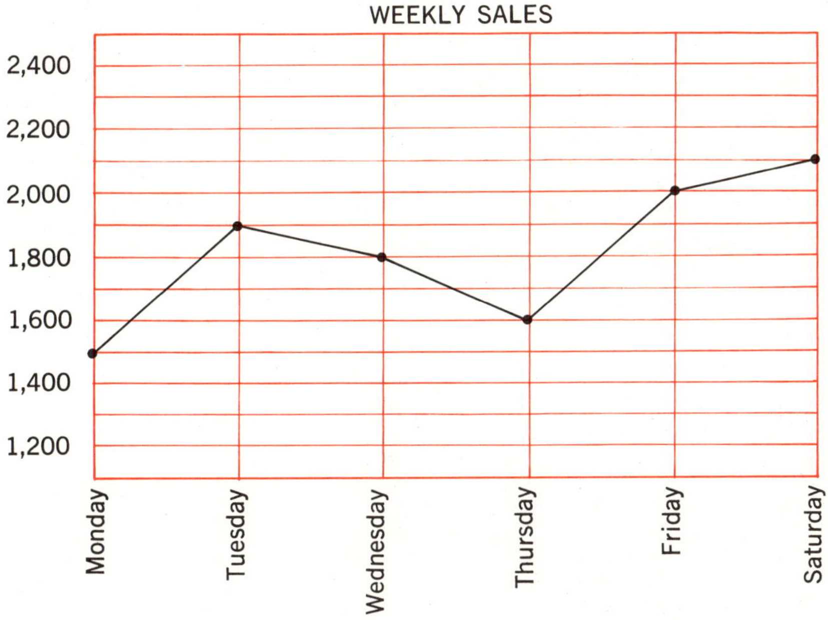

a. On what day were the sales greatest?

b. On what day were the sales least?

c. What is the range?

d. What is the mode?

e. What is the median?

f. What is the mean?

2. a. Make a broken-line graph to show the weekly attendance at Murray Middle School. The attendance was as follows:

Monday	1,212
Tuesday	1,200
Wednesday	1,250
Thursday	1,175
Friday	1,200

b. When was the attendance greatest?

c. When was the attendance least?

d. What is the range?

e. What is the mode?

f. What is the median?

g. What is the mean?

3. a. Make a broken-line graph to show the weekly contributions to the United Fund at the Busy Body Works.

1st week	$100
2nd week	$80
3rd week	$90
4th week	$150
5th week	$120
6th week	$200

b. What is the range?

c. What is the mean?

d. What is the median?

4. The following noon temperatures were recorded at Gauley Bridge for one week last July.

82° 86° 85° 85° 80° 84° 88°

a. What was the range?

b. Make a broken-line graph to show temperatures.

DO YOU REMEMBER?

Add.

1. 28.6 + 43.9
2. 76.4 + 8.3
3. 34.75 + 29.8
4. 8.63 + 24.656
5. 14.9 + .6
6. .18 + 29.4 + 1.578

7. $\begin{array}{r} 19.42 \\ 6.734 \\ \underline{246.9} \end{array}$
8. $\begin{array}{r} 84.63 \\ 6.456 \\ \underline{380.005} \end{array}$
9. $\begin{array}{r} 124.06 \\ 8.642 \\ \underline{75.09} \end{array}$
10. $\begin{array}{r} 138.07 \\ 81.391 \\ \underline{72.2} \end{array}$

Subtract.

11. .8 − .5
12. 29.6 − 5.8
13. 29.4 − 18.65
14. 142.08 − 2.563
15. 42 − 2.07
16. 9.884 − .0643

17. $\begin{array}{r} 4.005 \\ \underline{-\ 1.26} \end{array}$
18. $\begin{array}{r} 19.2 \\ \underline{-\ 12.476} \end{array}$
19. $\begin{array}{r} 75.00 \\ \underline{-\ 47.654} \end{array}$
20. $\begin{array}{r} 2.7 \\ \underline{-\ .009} \end{array}$

Multiply.

21. 1.8 × .43
22. 56.4 × 8.3
23. 82.56 × 2.04
24. 1.064 × .054
25. .042 × 9.5
26. 8.32 × 264.8
27. 85.6 × 3.5
28. .064 × 45.832

Divide.

29. $17\overline{)7.14}$
30. $.06\overline{).054}$
31. $1.25\overline{)25.0}$
32. $7.3\overline{)2.117}$
33. $.15\overline{).1230}$
34. $1.3\overline{).0624}$
35. $.008\overline{).001888}$
36. $2.8\overline{)1.484}$
37. $.042\overline{).000126}$

BAR GRAPHS

Sometimes a dot frequency diagram is useful in making a bar graph. Here is a dot frequency diagram for the scores on a mathematics test.

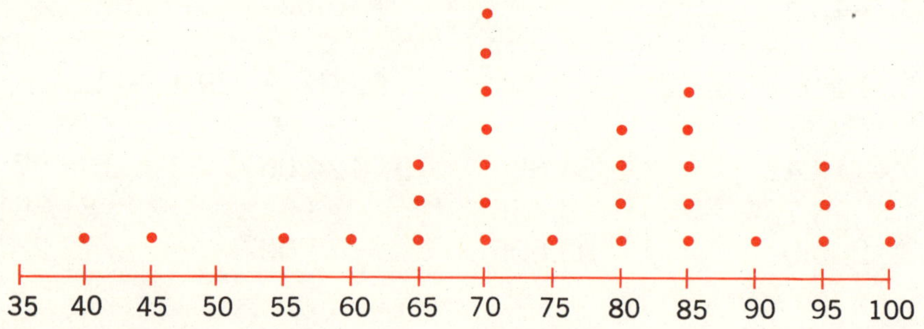

1. a. A graph of the information must show what range of scores?

   b. A bar graph must allow for how many students for any one score?

2. Shown below is part of the bar graph for the information given above. Copy and complete it.

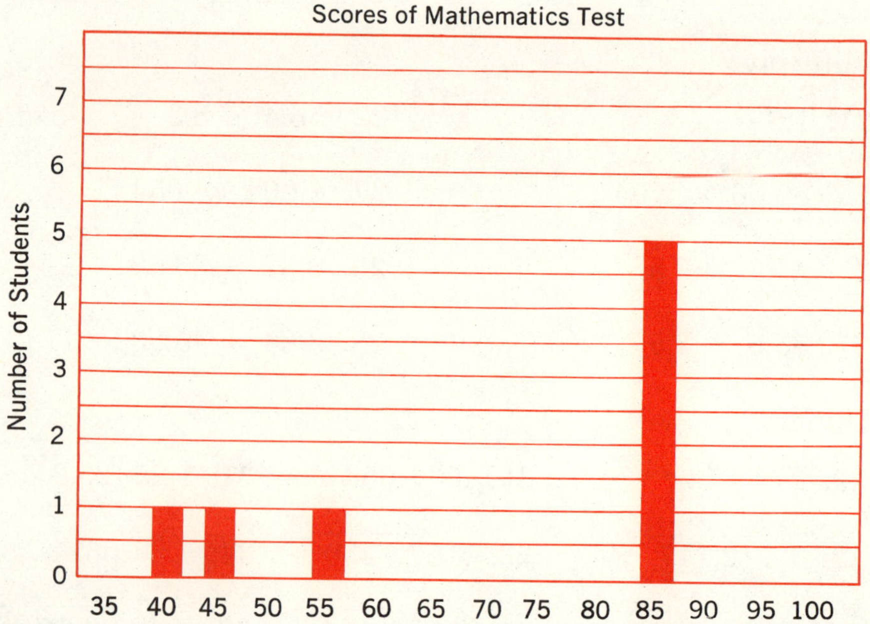

EXERCISES

Solve these problems.

1. Make a bar graph to show the distribution of the final grades in Mr. Dean's eighth grade mathematics class.

A	B	B	C	B	F	D
C	B	A	C	C	C	D
F	C	C	B	C	A	C
C	A	B	C	D	C	C
B	C	D	C	F	B	B

2. A car dealer decided to keep a record of the colors of cars ordered by customers. He wanted to know color preferences to make out future orders. Every time a salesman sold a car, he wrote the color on a sheet posted on the bulletin board. The following were recorded.

Dk. Blue	Dk. Green	Lt. Green	Lt. Green
Red	Black	Yellow	Dk. Blue
Black	Lt. Green	White	Red
White	Lt. Blue	Dk. Green	White
Lt. Blue	Yellow	Lt. Green	Yellow
Dk. Green	Yellow	Yellow	Lt. Green
Red	Red	Dk. Blue	Lt. Green
Lt. Blue	White	Lt. Blue	White
Yellow	Dk. Blue	Lt. Blue	Yellow
White	Lt. Blue	Yellow	Lt. Blue

Make a bar graph to show the car color preference.

3. Make a bar graph to show the weekly sales of the ten salesmen of the Handy Gadget Appliance Store.

Mr. Green	$1,200	Mr. Hayes	$1,500
Miss Williams	1,500	Mr. Smythers	1,600
Mrs. Brent	1,000	Miss Keyo	1,000
Mr. Grant	1,200	Mrs. Matthews	1,300
Mr. Barron	1,400	Mrs. Fox	1,200

CIRCLE GRAPHS

When data represent parts of a whole, a circle graph is often helpful. The interior of a circle is separated into parts by central angles.

1. **a.** How many degrees are there in a circle?

   b. What size angle would show $\frac{1}{3}$ of the circle?

   c. What size angle would show $\frac{1}{12}$ of the circle?

   d. What size angle would show $\frac{1}{4}$ of the circle?

2. The Jamison family had the following expenses for the month of February.

Food	\$200	Recreation	\$ 15
House payment	\$165	Doctors	\$ 25
Utilities	\$ 40	Car payment	\$110
Clothing	\$ 30	Miscellaneous	\$ 15

Copy and complete.

EXPENSES FOR FEBRUARY

Item	*Amount*	*Fractional Part*	*Measure of Central Angle*
Food	200	$\frac{200}{600} = \frac{1}{3}$	$\frac{1}{3} \times 360° = 120°$
House Payment	165	$\frac{165}{600} = \frac{11}{40}$	
Utilities	40		
Clothing	30		
Recreation	15		
Doctor	25		
Car Payment	110		
Miscellaneous	15		
Totals	600	1	360°

3. The information in Item 2 is shown on the circle graph below. Tell how many degrees are in each central angle.

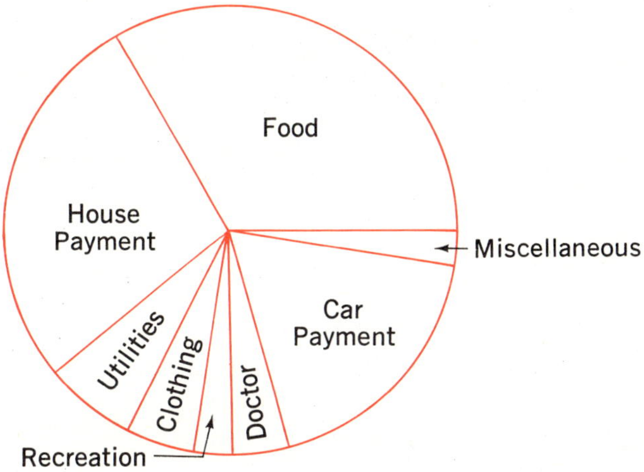

EXERCISES

Solve these problems.

1. A Boy Scout troop sold 600 tickets to a jamboree. The Bear Patrol sold 125 tickets; the Lion Patrol sold 100 tickets; the Wolf Patrol sold 150; the Alligator Patrol sold 100; and the Cougar Patrol sold 125. Make a circle graph to show the total ticket sales for the troop.

2. Scott, Bill, Sam, and Kirk are candidates for president of the Student Government. They each received the following number of votes.

   Scott – 250　　Sam – 200
   Bill – 300　　Kirk – 350

   Make a circle graph to show the votes for Student Government President.

3. The weight of sugar is about 42% carbon, 52% oxygen, and 6% hydrogen. Make a circle graph to show the weight of sugar.

PROBLEM SOLVING

1. Mr. Spanos is building a new city parking lot which is to be 220 feet by 140 feet. He allows about 100 square feet per car, including aisle space. About how many cars will his lot hold?

2. Mr. Roth estimated that it would cost $15 per cubic yard to pour a concrete wall. Find the cost of pouring a wall 30 feet long, 9 feet high, and 2 feet thick.

3. Concrete weighs about 150 pounds per cubic foot. Find the weight of the wall in Exercise 2.

4. Mr. Kelly, the electrician, bought 98 feet of BX cable at $.225 per foot. Find the total cost.

5. Mr. Kelly estimates that 3% of the wire used in Exercise 4 was wasted. Find the amount of wire wasted.

6. To complete a rush job, a machinist worked five 8-hour days at his regular wage of $5.90 per hour. He then worked 8 hours on Saturday at time and one-half. Finally, he worked 3 hours on Sunday at double pay. Find his total wages.

7. A certain type of solder contains 65% tin. The rest is lead. Find the amount of tin and lead in 25 pounds of solder.

CHAPTER REVIEW

Solve these problems.

1. A die is tossed.

   a. What is the probability that it will land with a 3 up?

   b. What is the probability that it will land with a prime number up?

   c. What is the probability that it will land with a 5 or 6 up?

2. A jar contains 5 yellow disks, 2 black disks, and 4 green disks. A disk is chosen from the box.

   a. What is the probability that a green, black, or yellow disk will be chosen?

   b. What is the probability that a white disk will be chosen?

3. A box contains 5 red and 7 white crayons. Jane will draw a crayon, replace it and then draw another.

   a. What is the probability that she will draw a red crayon the second time?

   b. What is the probability that she will draw a red crayon both the first and second time?

4. Robyn recorded the following noon temperatures for two weeks.

80°	84°	82°	82°	81°	80°	83°
85°	82°	81°	84°	84°	84°	82°

   a. What was the range of temperatures?

   b. Make a frequency distribution table of the temperatures.

   c. What were the median, mode, and mean?

   d. Make a broken-line graph of the temperatures.

5. The following were test grades from Mr. Thompson's mathematics class.

A	B	A	B	B	D
C	F	C	A	C	D
F	C	A	C	F	C
B	A	D	A	D	C
C	C	C	B	B	D

a. Make a dot frequency diagram of the scores.

b. Make a bar graph of the scores.

c. Make a circle graph of the scores.

d. Make a broken-line graph of the scores.

6. There are 1,000 marbles in a box. Some are red, some white, and some blue. A marble is drawn, the color recorded, and the marble is returned. Another marble is drawn. This procedure is repeated 16 times. Use the record below to estimate the number of marbles of each color in the box.

SAMPLINGS

Trial	*Color*	*Trial*	*Color*
1	R	8	W
2	R	90	W
3	W	11	W
4	W	12	R
5	B	13	W
6	R	14	W
7	W	15	B
8	W	16	W

Terms You Should Know

probability
compound probability
mean
median
broken-line graph
frequency distribution chart
dot frequency diagram
mode
bar graph
circle graph

CHAPTER TEST

Solve these problems.

1. A die is tossed.

   a. What is the probability that the die will land with a 2 up?

   b. What is the probability that the die will land with a 1 or a 4 up?

2. A box contains 10 black, 6 yellow, and 4 blue marbles.

   a. What is the probability of selecting a blue marble?

   b. What is the probability of selecting a red marble?

   c. What is the probability of selecting a black, yellow, or blue marble?

   d. A marble is selected, replaced, and then another is selected. What is the probability of selecting a black one followed by a yellow one?

3. A box contains 500 marbles. Some are blue and some are white. A marble is selected, replaced, and another is drawn. This is repeated 20 times. Twelve blue and eight white marbles are drawn. Estimate the number of marbles of each color in the box.

4. The following is the 1970 sales record of Ace Company.

   a. Make a broken-line graph of this data.

Jan.	\$130,000	May	\$150,000	Sept.	\$170,000
Feb.	\$120,000	June	\$160,000	Oct.	\$160,000
March	\$100,000	July	\$180,000	Nov.	\$120,000
April	\$150,000	Aug.	\$200,000	Dec.	\$130,000

   b. What are the range, median, mode, and mean?

5. Last semester 8 students made A's, 10 made B's, 12 made C's, 8 made D's, and 2 made F's in Algebra I. Make a bar graph of these scores.

Symbol List

$\ldots$	continues forever	1
L.C.D.	least common denominator	28
$\cong$	is congruent to	52
$\triangle ABC$	triangle *ABC*	52
$\overline{AB}$	line segment from *A* to *B*	58
$\angle ABC$	angle *ABC*	60
145°	145 degrees	61
$\overrightarrow{AB}$	ray from *A* through *B*	62
$\overleftrightarrow{AB}$	line through *A* and *B*	64
$\parallel$	is parallel to	64
$\perp$	is perpendicular to	65
$<$	is less than	85
$>$	is greater than	85
$\approx$	is approximately equal to	92
35%	thirty-five per cent	113
$^{-}6$	negative six	147
$\lvert 9 \rvert$	the absolute value of nine	164
7^3	the third power of 7	177
π	the number approx. $\frac{22}{7}$, or 3.14	209
$.53\overline{53}$	53 repeats forever	254
$m\ \overline{AB}$	the measure of $\overline{AB}$	274
$\sim$	is similar to	275
$m \angle A$	the measure of angle *A*	277
$\sqrt{\ }$	square root	289

PRACTICE EXERCISES

Add.

4	2	3	2	1	2	4	6	5	3
5	2	4	5	6	4	2	2	3	3
4	5	6	7	2	3	4	5	1	2
4	2	3	2	7	6	3	4	7	3
1	0	3	1	0	4	3	2	0	8
1	2	2	7	6	1	5	6	0	0
6	6	7	6	8	9	8	9	7	8
8	9	7	8	9	9	7	8	9	8
6	5	7	8	9	5	3	2	7	4
6	9	6	2	4	7	7	9	4	8
1	2	3	4	5	9	8	7	6	5
9	8	9	7	8	2	3	5	4	5

Subtract.

5	6	8	9	4	5	6	7	5	4
3	4	2	7	1	1	5	2	2	2
8	6	9	8	7	7	9	8	6	5
6	3	4	5	3	5	6	3	2	4
16	34	28	19	38	27	19	26	38	25
4	2	3	6	4	5	7	5	6	3
10	15	10	18	17	14	12	16	11	13
3	7	6	9	9	6	5	8	6	7
33	45	21	73	35	56	28	52	46	31
4	8	2	8	6	7	9	7	7	9
42	51	56	35	63	71	62	77	34	33
4	2	9	9	6	3	4	8	9	5

Multiply.

5	1	1	3	3	1	2	1	4	2
1	3	4	3	2	1	4	5	2	3
4	2	4	0	4	3	4	5	2	2
6	5	3	8	4	7	0	6	9	7
5	0	1	4	1	2	3	1	5	3
5	6	7	8	2	0	9	8	3	5
8	1	5	7	3	8	5	6	6	5
1	6	7	4	8	3	9	6	7	4
6	6	6	3	5	7	6	4	0	8
3	4	5	1	2	3	8	9	7	8
9	9	0	7	8	5	1	3	6	2
4	2	3	6	7	8	9	4	2	2
9	0	9	1	0	3	2	4	2	4
0	9	9	0	0	6	6	5	8	7
8	7	0	9	7	7	0	6	7	5
9	7	5	6	1	9	2	1	5	0
9	8	7	6	4	0	9	8	9	8
8	5	8	0	1	4	1	6	7	2
0	6	8	9	7	9	8	7	3	2
1	9	0	5	0	3	4	2	0	1
23	12	31	22	33	34	11	23	12	11
3	3	2	3	2	2	4	2	4	3
29	18	15	26	12	14	26	13	25	24
2	4	3	2	5	7	3	6	3	4
12	25	14	19	28	31	15	21	17	20
2	4	3	2	3	2	5	4	5	5

Divide.

$6 \div 3$	$8 \div 2$	$5 \div 5$	$8 \div 4$	$1 \div 1$
$6 \div 2$	$0 \div 3$	$4 \div 4$	$3 \div 3$	$5 \div 1$
$7 \div 7$	$25 \div 5$	$36 \div 6$	$0 \div 1$	$12 \div 3$
$16 \div 4$	$21 \div 7$	$30 \div 6$	$18 \div 9$	$14 \div 7$
$28 \div 4$	$35 \div 7$	$6 \div 6$	$8 \div 1$	$15 \div 5$
$15 \div 3$	$8 \div 8$	$27 \div 9$	$2 \div 2$	$32 \div 8$
$3 \div 1$	$30 \div 5$	$24 \div 4$	$18 \div 3$	$20 \div 4$
$42 \div 7$	$36 \div 6$	$45 \div 9$	$24 \div 3$	$0 \div 8$
$0 \div 5$	$56 \div 7$	$42 \div 6$	$18 \div 2$	$36 \div 4$
$64 \div 8$	$36 \div 9$	$0 \div 2$	$21 \div 3$	$10 \div 2$
$35 \div 5$	$16 \div 8$	$20 \div 5$	$0 \div 9$	$18 \div 6$
$81 \div 9$	$4 \div 2$	$12 \div 2$	$12 \div 4$	$9 \div 9$
$4 \div 1$	$56 \div 8$	$40 \div 5$	$72 \div 8$	$0 \div 7$
$63 \div 9$	$7 \div 1$	$54 \div 6$	$0 \div 4$	$72 \div 9$
$54 \div 9$	$16 \div 2$	$63 \div 7$	$48 \div 6$	$9 \div 1$
$2 \div 1$	$14 \div 2$	$32 \div 4$	$27 \div 3$	$45 \div 5$
$28 \div 7$	$12 \div 6$	$24 \div 8$	$6 \div 1$	$0 \div 6$
$40 \div 8$	$49 \div 7$	$3 \div 3$	$48 \div 8$	$10 \div 5$
$99 \div 9$	$0 \div 10$	$15 \div 15$	$16 \div 2$	$12 \div 1$

(3) Add.

1. 31 + 58 2. 52 + 46 3. 25 + 43 4. 312 + 515 5. 243 + 156

1. 31 58	2. 52 46	3. 25 43	4. 312 515	5. 243 156
6. 38 49	7. 27 36	8. 98 83	9. 178 291	10. 479 542

Subtract.

11. 32 11	12. 48 26	13. 37 15	14. 276 145	15. 403 295
16. 31 18	17. 52 37	18. 63 49	19. 498 299	20. 506 387

(8) Multiply.

1. 36 34	2. 18 25	3. 27 32	4. 213 46	5. 502 58
6. 721 38	7. 450 27	8. 912 87	9. 362 143	10. 525 248

(18) Divide.

1. $8\overline{)568}$ 2. $7\overline{)392}$ 3. $18\overline{)450}$ 4. $26\overline{)1{,}274}$

5. $53\overline{)4{,}982}$ 6. $17\overline{)7{,}752}$ 7. $52\overline{)25{,}116}$ 8. $32\overline{)9{,}250}$

9. $23\overline{)10{,}495}$ 10. $93\overline{)76{,}742}$ 11. $69\overline{)50{,}692}$ 12. $71\overline{)70{,}963}$

(25) Find the simplest fractional numeral.

1. $\frac{12}{16}$ 2. $\frac{8}{32}$ 3. $\frac{25}{50}$ 4. $\frac{9}{63}$ 5. $\frac{16}{72}$

6. $\frac{32}{40}$ 7. $\frac{25}{45}$ 8. $\frac{10}{55}$ 9. $\frac{5}{7}$ 10. $\frac{21}{24}$

11. $\frac{2}{32}$ 12. $\frac{15}{21}$ 13. $\frac{28}{42}$ 14. $\frac{44}{66}$ 15. $\frac{17}{51}$

(28) Find the L.C.D.

1. $\frac{1}{2}, \frac{3}{5}$ 2. $\frac{2}{3}, \frac{5}{7}$ 3. $\frac{1}{4}, \frac{5}{8}$

4. $\frac{3}{22}, \frac{5}{33}$ 5. $\frac{3}{10}, \frac{2}{25}$ 6. $\frac{7}{8}, \frac{5}{12}$

7. $\frac{1}{3}, \frac{3}{4}, \frac{5}{8}$ 8. $\frac{2}{7}, \frac{1}{2}, \frac{3}{10}$ 9. $\frac{9}{10}, \frac{5}{6}, \frac{4}{15}$

(30) Add. Write simplest fractional numerals for the sums.

1. $\frac{1}{2}$, $\frac{2}{3}$ 2. $\frac{5}{8}$, $\frac{3}{5}$ 3. $\frac{1}{2}$, $\frac{3}{4}$ 4. $\frac{1}{3}$, $\frac{1}{6}$ 5. $\frac{1}{10}$, $\frac{3}{5}$

6. $\frac{1}{4}$, $\frac{2}{3}$, $\frac{5}{6}$ 7. $\frac{1}{2}$, $\frac{5}{8}$, $\frac{1}{5}$ 8. $\frac{3}{8}$, $\frac{1}{7}$, $\frac{3}{10}$ 9. $\frac{5}{12}$, $\frac{3}{8}$, $\frac{1}{10}$ 10. $\frac{2}{5}$, $\frac{1}{4}$, $\frac{1}{6}$

(33) Subtract. Write simplest fractional numerals for the differences.

1. $\frac{3}{4}$, $\frac{1}{4}$ 2. $\frac{5}{8}$, $\frac{1}{4}$ 3. $\frac{9}{16}$, $\frac{1}{4}$ 4. $\frac{1}{2}$, $\frac{2}{5}$ 5. $\frac{20}{21}$, $\frac{3}{7}$

6. $\frac{7}{12}$, $\frac{1}{3}$ 7. $\frac{9}{14}$, $\frac{1}{2}$ 8. $\frac{10}{21}$, $\frac{3}{14}$ 9. $\frac{7}{15}$, $\frac{3}{10}$ 10. $\frac{5}{8}$, $\frac{5}{12}$

(34) Add and simplify.

1. $8\frac{1}{2}$, $2\frac{2}{3}$ 2. $1\frac{3}{5}$, $2\frac{7}{10}$ 3. $25\frac{3}{4}$, $6\frac{5}{8}$ 4. $7\frac{5}{9}$, $12\frac{2}{3}$ 5. $18\frac{3}{25}$, $10\frac{5}{9}$

6. $13\frac{4}{5}$, $25\frac{5}{8}$ 7. $6\frac{5}{12}$, $22\frac{5}{9}$ 8. $13\frac{4}{11}$, $12\frac{3}{5}$ 9. $51\frac{4}{7}$, $23\frac{2}{3}$ 10. $48\frac{5}{9}$, $27\frac{5}{6}$

(34) Subtract and simplify.

1. $28\frac{5}{8} - 12\frac{3}{8}$
2. $15\frac{1}{2} - 7\frac{1}{3}$
3. $22\frac{5}{6} - 15\frac{2}{3}$
4. $23 - 4\frac{5}{8}$
5. $25\frac{1}{3} - 2\frac{5}{6}$
6. $28 - 4\frac{3}{7}$
7. $21\frac{1}{5} - 9\frac{3}{4}$
8. $17\frac{2}{3} - 15\frac{7}{9}$
9. $25\frac{3}{10} - 21\frac{3}{8}$
10. $29\frac{1}{4} - 12\frac{3}{5}$

(37) Multiply and simplify.

1. $\frac{1}{2} \times \frac{1}{3}$
2. $\frac{2}{5} \times \frac{3}{7}$
3. $\frac{5}{9} \times \frac{3}{4}$
4. $\frac{3}{5} \times \frac{1}{6}$
5. $\frac{9}{10} \times \frac{5}{12}$
6. $\frac{2}{5} \times \frac{15}{16}$
7. $\frac{7}{8} \times \frac{4}{21}$
8. $\frac{5}{6} \times \frac{9}{10}$
9. $\frac{20}{21} \times \frac{6}{25}$
10. $\frac{8}{9} \times \frac{15}{16}$
11. $\frac{7}{25} \times \frac{10}{21}$
12. $\frac{3}{5} \times \frac{20}{21}$

(41) Divide and simplify.

1. $\frac{1}{2} \div \frac{1}{3}$
2. $\frac{1}{6} \div \frac{2}{3}$
3. $\frac{3}{4} \div \frac{7}{8}$
4. $12 \div \frac{3}{4}$
5. $\frac{8}{9} \div \frac{2}{3}$
6. $\frac{10}{11} \div \frac{5}{6}$
7. $\frac{1}{4} \div \frac{9}{10}$
8. $\frac{5}{8} \div \frac{11}{12}$
9. $\frac{13}{25} \div \frac{7}{10}$
10. $25 \div \frac{1}{2}$
11. $\frac{3}{11} \div \frac{6}{55}$
12. $\frac{2}{5} \div \frac{8}{15}$

(45) Multiply and simplify.

1. $4\frac{1}{4} \times \frac{7}{8}$
2. $5\frac{3}{7} \cdot 6\frac{3}{8}$
3. $12\frac{5}{7} \cdot 2\frac{2}{9}$
4. $13\frac{1}{2} \times 6\frac{3}{11}$
5. $3\frac{2}{7} \cdot 5\frac{2}{7}$
6. $7\frac{1}{3} \times 3\frac{8}{9}$

Divide and simplify.

7. $3\frac{1}{9} \div \frac{4}{7}$
8. $2\frac{4}{9} \div 1\frac{2}{3}$
9. $11\frac{5}{7} \div 12\frac{1}{2}$
10. $6\frac{2}{7} \div \frac{3}{13}$
11. $12\frac{8}{9} \div 4\frac{2}{5}$
12. $8\frac{3}{4} \div 3\frac{3}{8}$

(83) Write decimal numerals.

1. $\frac{7}{100}$ 2. $8\frac{3}{10}$ 3. $12\frac{3}{1,000}$ 4. $\frac{32}{10}$

Write word names.

5. .8 6. 2.3 7. .006 8. 10.004

Write simplest fractional numerals or mixed numerals.

9. .3 10. 2.4 11. .005 12. 30.01

Write decimals.

13. Twenty-two and seven thousandths

14. One hundred twelve and one hundred three ten-thousandths

(85) Which are true and which are false?

1. $.15 > .2$ 2. $2.15 < .1234$ 3. $0 < .24$

4. $2.01 > .492$ 5. $12.3 < 4.27$ 6. $.0001 < .01$

List in order from greater to lesser.

7. .023, .4, .05 8. .13, 2.5, .0048

9. 1.6204, 1.396, 1.9 10. .165, .1653, 1.6501

(88) Add.

1. $\begin{array}{r} .5 \\ \underline{.9} \end{array}$ 2. $\begin{array}{r} .8 \\ \underline{.7} \end{array}$ 3. $\begin{array}{l} .3 \\ \underline{.407} \end{array}$ 4. $\begin{array}{r} .349 \\ \underline{1.2} \end{array}$

5. $\begin{array}{r} .492 \\ \underline{8.42} \end{array}$ 6. $\begin{array}{r} 1.3 \\ \underline{24.87} \end{array}$ 7. $\begin{array}{r} 23.4 \\ 1.871 \\ \underline{12.46} \end{array}$ 8. $\begin{array}{r} 1.0021 \\ .102 \\ \underline{31.4} \end{array}$

9. $.38 + .102 + 2.7 + 1.5$

10. $5.2 + 12.004 + .00036 + 1.23 + 14$

(90) Subtract.

1. 2.58 − 1.24
2. 7.01 − 3.482
3. 1.236 − .58
4. 24.3 − 9.807
5. 12.1 − 8.479
6. 3.02 − .5861
7. 2.134 − 1.2
8. 17. − 2.589
9. 3.142 − .00123
10. 12.2 − 3.147

(92) Copy and complete.

	Number	Nearest Ten	Nearest One	Nearest Tenth	Nearest Hundredth	Nearest Thousandth
1.	123.4781					
2.	281.0364					
3.	782.0056					
4.	100.0007					
5.	1,234.7002					

(95) Multiply.

1. 25.2 × 1.3
2. 12.04 × 3.5
3. .103 × 7.1
4. .0048 × .0027
5. 1.82 × .0034
6. .635 × .00261
7. .00481 × .0293
8. .2613 × .00127

(98) Divide.

1. 9) 8.1
2. 6) .0036
3. 25) .0125
4. 15) 2.25
5. 1,300) 2.6
6. 16) .96
7. 12) .096
8. 8) 36

(100) Divide.

1. .5) 23
2. .15) 30
3. .4) 9.8
4. .05) .7
5. .001) .34
6. .06) 8.4
7. 1.1) .0099
8. .025) .00625

(102) Find decimal numerals.

1. $\frac{5}{8}$ 2. $\frac{5}{9}$ 3. $\frac{4}{3}$ 4. $\frac{13}{12}$ 5. $\frac{9}{20}$

6. $\frac{4}{15}$ 7. $\frac{7}{6}$ 8. $\frac{3}{25}$ 9. $\frac{3}{7}$ 10. $\frac{9}{5}$

(106) Write the decimal equivalent.

1. $\frac{5}{6}$ 2. $\frac{1}{2}$ 3. $\frac{3}{4}$ 4. $\frac{1}{3}$ 5. $\frac{9}{20}$

6. $\frac{1}{6}$ 7. $\frac{2}{5}$ 8. $\frac{3}{8}$ 9. $\frac{2}{3}$ 10. $\frac{1}{4}$

Write the fractional equivalent.

11. $.3\overline{3}$ 12. $.83\overline{3}$ 13. .25 14. .6 15. .625

16. .875 17. .8 18. .5 19. $.16\overline{6}$ 20. .2

(113) Write simplest fractional numerals.

1. 40% 2. 26% 3. 412% 4. 300% 5. $\frac{1}{2}$%

Write decimal numerals.

6. 21% 7. 1.2% 8. .05% 9. 120% 10. 500%

(116) Write per cent notation.

1. $\frac{3}{100}$ 2. $\frac{75}{100}$ 3. $\frac{1}{4}$ 4. $\frac{7}{10}$ 5. .5

6. .001 7. $120\frac{1}{2}$ 8. $\frac{5}{16}$ 9. $\frac{7}{9}$ 10. 3.2

(120) Write a number sentence and solve.

1. What is 5% of 400? 2. What is 20% of 10?

3. What is .1% of 1,000? 4. What is 62% of 35?

5. What is 1.5% of 80? 6. What is 75% of 4?

7. What is 60% of 1,200? 8. What is 90% of 300?

(122) Write simplest fractional numerals for these ratios.

1. 8 to 12
2. 5:7
3. 24:60
4. 30 out of 40
5. $\frac{12}{18}$
6. 20 to 10

(124) Write per cent notations.

1. 12 is what per cent of 36?
2. 5 is what per cent of 50?
3. 20 is what per cent of 15?
4. 25 is what per cent of 60?
5. 80 is what per cent of 200?
6. 200 is what per cent of 80?

(126) Write a number sentence and solve.

1. 32 is 50% of what number?
2. 12 is 200% of what number?
3. 1.5 is 25% of what number?
4. 60 is .25% of what number?
5. 20 is 125% of what number?
6. 5 is 10% of what number?

(128) Write number sentences and solve.

1. What is 6% of 50?
2. What is 15% of 34?
3. 16 is what per cent of 24?
4. 25 is 20% of what number?
5. 20 is what per cent of 10?
6. 50 is what per cent of 400?
7. 80 is $133\frac{1}{3}$% of what number?
8. .5 is 10% of what number?
9. 2 is 25% of what number?
10. 100 is what per cent of 20?

(134) Copy and complete.

	Principal	*Interest Rate Per Year*	*Time*	*Interest*
1.	$300	8%	1 yr.	
2.	$100	10%	6 mos.	
3.	$50	9%	3 mos.	
4.	$1,000	8%	9 mos.	
5.	$2,500	10%	6 mos.	

(138) Copy and complete.

	Marked Price	*Rate of Discount*	*Discount*	*Sale Price*
1.	$100	10%		
2.	$50		$10	
3.	$9	$33\frac{1}{3}\%$		
4.	$25			$10
5.	$500		$50	
6.	$1,000	5%		

(140) Copy and complete.

	Sales	*Rate of Commission*	*Commission*
1.	$500	10%	
2.	$1,000		$150
3.	$220	12%	
4.	$1,500		$300
5.	$28,000	7%	
6.	$35,000		700

(152) Add.

1. $^{-}5 + {}^{-}8$ 2. $12 + {}^{-}3$ 3. $7 + 10$

4. $^{-}24 + 13$ 5. $^{-}60 + {}^{-}82$ 6. $^{-}24 + 36$

7. $25 + {}^{-}18$ 8. $25 + 30$ 9. $^{-}8 + 17$

10. $200 + {}^{-}134$ 11. $^{-}6 + {}^{-}38$ 12. $^{-}52 + 70$

(156) Name the inverse of each.

1. $^{-}18$ 2. 25 3. 0 4. 27 5. $^{-}34$

Subtract by adding the inverse.

6. $^{-}7 - {}^{-}5$ 7. $^{-}3 - 7$ 8. $25 - {}^{-}8$

9. $^{-}24 - 26$ 10. $82 - {}^{-}30$ 11. $^{-}14 - {}^{-}36$

12. $^{-}6 - {}^{-}22$ 13. $^{-}30 - 50$ 14. $25 - 62$

15. $30 - {}^{-}60$ 16. $100 - 126$ 17. $^{-}17 - {}^{-}23$

(160) Multiply.

1. $^{-}4 \times 8$ 2. $^{-}7 \times {}^{-}3$ 3. $5 \times {}^{-}2$

4. 6×20 5. $0 \times {}^{-}18$ 6. $^{-}7 \times {}^{-}25$

7. $^{-}3 \times 14$ 8. $27 \times {}^{-}12$ 9. $^{-}34 \times {}^{-}12$

10. $^{-}27 \times 10$ 11. $35 \times {}^{-}25$ 12. 16×58

(162) Divide.

1. $25 \div {}^{-}5$ 2. $^{-}100 \div {}^{-}4$ 3. $75 \div 15$

4. $30 \div {}^{-}6$ 5. $^{-}125 \div {}^{-}25$ 6. $^{-}169 \div 13$

7. $^{-}39 \div {}^{-}13$ 8. $42 \div {}^{-}14$ 9. $^{-}64 \div {}^{-}4$

10. $^{-}88 \div 22$ 11. $51 \div {}^{-}17$ 12. $48 \div 16$

(164) Write simplest names.

1. $|{}^{-}8|$ 2. $|0|$ 3. $|17|$ 4. $|{}^{-}48|$ 5. $|100|$

Which are true and which are false?

6. $|{}^{-}25| > |5|$ 7. $|0| < |{}^{-}18|$ 8. $|{}^{-}20| > |16|$

(168) Graph these ordered pairs.

1. $(2, {}^{-}1)$ 2. $(1, {}^{-}2)$ 3. $(0, 0)$ 4. $({}^{-}30)$ 5. $(0, 5)$

6. $(3, 3)$ 7. $({}^{-}3, {}^{-}3)$ 8. $({}^{-}3, 3)$ 9. $(3, {}^{-}3)$ 10. $(6, 0)$

(170) Graph these sentences.

1. $x + y = 4$ 2. $x = {}^{-}y$ 3. $x - y = 0$

4. $2 \cdot x = y$ 5. $x - y = 3$ 6. $x + y = {}^{-}1$

(177) Write exponential notation.

1. $3 \cdot 3 \cdot 3 \cdot 3 \cdot 3 \cdot 3$ 2. $7 \cdot 7 \cdot 7 \cdot 7 \cdot 7 \cdot 7 \cdot 7$

3. $5 \cdot 5 \cdot 5$ 4. $11 \cdot 11 \cdot 11 \cdot 11 \cdot 11$

5. 243 6. 27

Write standard numerals.

7. 2^4 8. 3^6 9. 4^3 10. 5^4

(180) Write exponential notation.

1. $2^3 \cdot 2^5$ 2. $5^4 \cdot 5^0$

3. $3^4 \cdot 3^2 \cdot 3^1$ 4. $2^6 \cdot 2^4 \cdot 2^0 \cdot 2^5$

5. $2^2 \cdot 3^4 \cdot 2^5$ 6. $3^4 \cdot 5^0 \cdot 5^3 \cdot 3^1$

7. $2^3 \cdot 2^4 \cdot 3^4 \cdot 3^2 \cdot 5^7$ 8. $5^7 \cdot 5^3 \cdot 2^3 \cdot 2^2$

9. $2^5 \cdot 5^3 \cdot 3^4 \cdot 3^2 \cdot 5^0$ 10. $2^0 \cdot 3^0 \cdot 7^3 \cdot 7^2 \cdot 7^1 \cdot 3^4 \cdot 2^1$

(182) Write exponential notation.

1. $\frac{2^3}{2^1}$ 2. $\frac{2^5}{2^3}$ 3. $\frac{3^{14}}{3^9}$

4. $\frac{2^3 \cdot 3^4 \cdot 2^5}{2^2 \cdot 3^2}$ 5. $\frac{3^0 \cdot 3^2 \cdot 5^4 \cdot 5^1}{3^1 \cdot 5^3}$

(184) Write exponential notation.

1. $\frac{1}{5^4}$ 2. $\frac{1}{3^2}$ 3. $\frac{1}{2^5}$ 4. $\frac{1}{3^7}$

Write fractional numerals.

5. 3^{-12} 6. 5^{-6} 7. 2^{-8} 8. 5^{-7}

(188) Simplify.

1. $(3^2)^4$ 2. $(2^5)^3$ 3. $(3^7)^5$ 4. $(7^2)^1$

5. $(3^5)^0$ 6. $(1^6)^7$ 7. $(2^3)^6$ 8. $(2^8)^7$

(189) Write expanded numerals.

1. 25.123 2. 1.4825 3. 63.004

4. 125.1273 5. 43.0275 6. 1,237.48902

(190) Simplify.

1. $(3^2)^{-4}$ 2. $(5^{-2})^{-3}$ 3. $(11^{-5})^6$ 4. $(7^{-3})^{-4}$

5. $(2^6)^{-3}$ 6. $(2^{-5})^4$ 7. $(5^{-2})^{-3}$ 8. $(7^{-2})^0$

(193) Write standard numerals.

1. 2.489×10^2 2. 38×10^3 3. 12.402×10^4

4. $.13 \times 10^{-2}$ 5. 843×10^{-1} 6. 14.6×10^{-4}

(195) Write scientific notation.

1. 12,000,000 2. 6,342,000,000

3. .013 4. .0000007

(195) Write scientific notation.

1. 1,271.5
2. .00000000023
3. 1,400,000,000,000,000
4. 38,000,000,000

(206) Copy and complete.

1. 3 yd^2 = ____ ft^2
2. 2 ft^2 = ____ $in.^2$
3. 5 cm^2 = ____ mm^2
4. 5 m^2 = ____ cm^2
5. 54 ft^2 = ____ yd^2
6. 1,000 cm^2 = ____ m^2
7. 108 ft^2 = ____ yd^2
8. 1,000,000 mm^2 = ____ cm^2
9. 10 cm^2 = ____ m^2
10. 288 $in.^2$ = ____ ft^2

(235) Show why each is a rational number.

1. 0
2. $^-3$
3. 12
4. 2.5
5. $3\frac{1}{4}$
6. $^-21$
7. .002
8. $1\frac{1}{2}$

(236) Which are true and which are false?

1. $\frac{2}{3} < 0$
2. $\frac{^-1}{2} > \frac{^-3}{4}$
3. $\frac{1}{5} > \frac{^-1}{2}$
4. $\frac{^-2}{3} < 0$
5. $0 > \frac{5}{8}$
6. $0 < \frac{^-1}{8}$
7. $\frac{^-7}{6} > \frac{5}{6}$
8. $1\frac{1}{2} > \frac{5}{3}$
9. $^-2\frac{1}{4} < 1\frac{1}{8}$

(238) Name the inverse of each.

1. $\frac{^-1}{2}$
2. $\frac{5}{8}$
3. $^-2.3$
4. $\frac{^-10}{7}$

Write the simplest name for each.

5. $\left|\frac{^-1}{2}\right|$
6. $\left|\frac{^-7}{5}\right|$
7. $\left|\frac{9}{10}\right|$
8. $|0|$

(241) Multiply and simplify.

1. $\frac{^-1}{5} \times \frac{^-1}{3}$ 2. $\frac{1}{7} \times \frac{^-2}{5}$ 3. $\frac{^-1}{2} \times \frac{7}{10}$

4. $\frac{0}{3} \times \frac{5}{8}$ 5. $\frac{^-6}{7} \times \frac{14}{15}$ 6. $\frac{^-7}{8} \cdot \frac{^-4}{5}$

7. $\frac{3}{8} \cdot \frac{^-2}{9}$ 8. $\frac{^-7}{8} \cdot \frac{^-4}{3} \cdot \frac{6}{7}$ 9. $\frac{^-1}{3} \cdot \frac{^-25}{8} \cdot \frac{^-4}{15}$

(242) Name the reciprocal of each.

1. $\frac{3}{4}$ 2. $\frac{^-10}{7}$ 3. $\frac{0}{3}$ 4. $\frac{3}{15}$ 5. $\frac{^-21}{20}$

(243) Divide and simplify.

1. $\frac{2}{3} \div \frac{^-1}{3}$ 2. $\frac{^-1}{2} \div \frac{^-1}{4}$ 3. $\frac{1}{4} \div \frac{^-1}{2}$

4. $\frac{^-7}{8} \div \frac{^-3}{4}$ 5. $\frac{^-2}{5} \div \frac{4}{3}$ 6. $\frac{^-3}{7} \div \frac{^-9}{14}$

7. $\frac{^-5}{27} \div \frac{2}{3}$ 8. $\frac{7}{15} \div \frac{^-4}{5}$ 9. $\frac{3}{4} \div \frac{7}{8}$

(244) Add and simplify.

1. $\frac{1}{2} + \frac{^-3}{4}$ 2. $\frac{^-8}{9} + \frac{^-1}{3}$ 3. $\frac{^-5}{6} + \frac{1}{9}$

4. $\frac{^-2}{5} + \frac{^-7}{10}$ 5. $\frac{3}{8} + \frac{^-5}{6}$ 6. $\frac{7}{12} + \frac{1}{3}$

7. $\frac{^-9}{10} + \frac{4}{5}$ 8. $\frac{1}{3} + \frac{^-1}{4}$ 9. $\frac{^-1}{3} + \frac{^-1}{4}$

(246) Add and simplify.

1. $\frac{7}{8} + \frac{^-7}{8}$ 2. $\frac{3}{16} + \frac{^-8}{5}$ 3. $\frac{^-3}{4} + \frac{^-1}{2}$

4. $\frac{1}{3} + \frac{^-3}{5}$ 5. $\frac{^-5}{9} + \frac{^-1}{6}$ 6. $\frac{7}{8} + \frac{3}{5}$

7. $\frac{9}{10} + \frac{^-5}{8}$ 8. $\frac{^-2}{21} + \frac{5}{14}$ 9. $\frac{7}{8} + \frac{^-5}{12}$

(248) Subtract and simplify.

1. $\frac{^-9}{10} - \frac{^-1}{5}$ 2. $\frac{^-1}{3} - \frac{^-5}{6}$ 3. $\frac{7}{8} - \frac{^-1}{2}$

4. $\frac{1}{2} - {}^-\frac{7}{8}$ 5. $\frac{^-3}{10} - \frac{5}{3}$ 6. $\frac{^-7}{8} - \frac{1}{4}$

7. $\frac{3}{20} - \frac{^-1}{5}$ 8. $\frac{^-5}{11} - \frac{2}{7}$ 9. $\frac{1}{3} - \frac{^-1}{2}$

(250) Find two square roots of each. Use the table on page 406.

1. $\frac{49}{25}$ *$\frac{7}{5}$,* 2. $\frac{100}{81}$ 3. $\frac{9}{64}$ 4. $\frac{25}{49}$

5. $\frac{400}{9}$ *$\frac{20}{3}$,* 6. $\frac{289}{196}$ 7. $\frac{^-25}{9}$ 8. $\frac{169}{121}$

(254) Find decimal numerals.

1. $\frac{7}{9}$ 2. $\frac{^-2}{5}$ 3. $\frac{^-1}{3}$ 4. $\frac{5}{12}$

5. $\frac{^-3}{7}$ 6. $\frac{^-13}{6}$ 7. $\frac{5}{8}$ 8. $\frac{^-9}{20}$

(256) Find simplest fractional numerals.

1. $.25\overline{5}$ 2. $.32\overline{2}$ 3. $.87\overline{7}$

4. $.63\overline{3}$ 5. $.29\overline{9}$ 6. $.14\overline{4}$

(258) Find simplest fractional numerals.

1. $.24\overline{24}$ 2. $.4\overline{4}$ 3. $.12\overline{12}$

4. $.123\overline{123}$ 5. $.80\overline{80}$ 6. $.35\overline{35}$

(289) Which of these real numbers are irrational?

1. $\sqrt{149}$ 2. $\sqrt{1{,}600}$ 3. $\sqrt{169}$ 4. $\sqrt{81}$

7. $\sqrt{100}$ 5. $\sqrt{92}$ 6. $\sqrt{500}$ 8. $\sqrt{10{,}000}$

(298) Approximate these square roots to the nearest tenth.

1. $\sqrt{7}$ 2. $\sqrt{50}$ 3. $\sqrt{3}$ 4. $\sqrt{200}$ 5. $\sqrt{120}$

6. $\sqrt{1,200}$ 7. $\sqrt{230}$ 8. $\sqrt{156}$ 9. $\sqrt{180}$ 10. $\sqrt{135}$

(302) Graph these equations.

1. $x + y = {}^{-}4$ 2. $x - y = 3$ 3. $4 \cdot x = y$

4. $\frac{1}{3} \cdot x = y$ 5. $x + y = 2$ 6. $x - y = {}^{-}1$

(313) Solve these equations.

1. $x + 3 = 7$ 2. $y + {}^{-}2 = 12$ 3. $x + {}^{-}4 = {}^{-}8$

4. $t + 7 = 25$ 5. $t + {}^{-}3 = {}^{-}24$ 6. $s + {}^{-}7 = 0$

7. $25 = s + {}^{-}8$ 8. ${}^{-}15 = 3 + u$ 9. $u + {}^{-}14 = 70$

10. $36 = y + {}^{-}2$ 11. ${}^{-}18 = w + {}^{-}5$ 12. $y + 17 = 0$

(316) Solve these equations.

1. $2 \cdot y = 16$ 2. $5 \cdot u = 45$ 3. $3 \cdot t = 18$

4. ${}^{-}2 \cdot x = 22$ 5. $w \cdot 4 = {}^{-}20$ 6. $12 \cdot y = {}^{-}60$

7. ${}^{-}5 \cdot t = 35$ 8. $10 \cdot v = 100$ 9. ${}^{-}5 \cdot z = {}^{-}40$

10. $12 \cdot s = {}^{-}48$ 11. ${}^{-}15 \cdot x = {}^{-}30$ 12. $7 \cdot t = 21$

(320) Solve. Use the addition principle.

1. $x + 5 = 27$ 2. $y + {}^{-}3 = 6$

3. $t + {}^{-}18 = {}^{-}3$ 4. $x + {}^{-}2 = {}^{-}20$

5. $y + {}^{-}3 = 0$ 6. $12 = w + 7$

7. $w + {}^{-}2 = 25$ 8. $u + 17 = {}^{-}12$

(324) Solve. Use the multiplication principle.

1. $2 \cdot y = 24$ 2. $^{-}3 \cdot x = 15$ 3. $^{-}7 \cdot y = {}^{-}35$

4. $\frac{1}{10} \cdot x = 100$ 5. $\frac{1}{2} \cdot t = {}^{-}16$ 6. $^{-}8 = \frac{1}{3} \cdot z$

7. $25 \cdot w = 75$ 8. $^{-}50 \cdot t = 250$ 9. $^{-}4 = \frac{1}{2} \cdot w$

10. $3 \cdot x = 48$ 11. $^{-}2 \cdot y = {}^{-}16$ 12. $7 \cdot t = {}^{-}49$

(326) Solve. Use the addition and multiplication principles.

1. $2 \cdot y + 3 = 11$ 2. $4 \cdot x + 2 = 18$

3. $2 \cdot s + 7 = 1$ 4. $\frac{1}{2} \cdot z + 5 = 10$

5. $20 = \frac{1}{3} \cdot t + 9$ 6. $25 = \frac{1}{2} \cdot t + 25$

7. $4 \cdot x + {}^{-}20 = 0$ 8. $^{-}3 \cdot s + {}^{-}6 = {}^{-}24$

9. $7 \cdot y + 7 = {}^{-}14$ 10. $\frac{1}{3} \cdot v + 4 = 11$

(330) Solve these equations.

1. $2 \cdot x + 5 \cdot x = 14$ 2. $3 \cdot y + 7 \cdot y = 0$

3. $\frac{1}{3} \cdot t + \frac{2}{3} \cdot t = 6$ 4. $8 \cdot s + {}^{-}2 \cdot s = 30$

5. $5 \cdot y + {}^{-}8 \cdot y = {}^{-}36$ 6. $^{-}2 \cdot u + {}^{-}6 \cdot u = 64$

7. $7 \cdot v + 3 \cdot v = 30$ 8. $6 \cdot w + {}^{-}4 \cdot w = 12$

9. $32 = 4 \cdot t + 4 \cdot t$ 10. $^{-}50 = 2 \cdot x + 8 \cdot x$

(332) Solve these equations.

1. $2 \cdot x + x = 9$ 2. $7 \cdot t + t = 24$

3. $3 \cdot y + y = 20$ 4. $^{-}14 = 6 \cdot z + z$

(332) Solve these equations.

1. $8 \cdot x + x = {}^{-}45$
2. ${}^{-}60 = 9 \cdot z + z$
3. $25 = 4 \cdot t + t$
4. $3 \cdot x + x = {}^{-}16$
5. $20 = {}^{-}5 \cdot s + s$
6. $6 \cdot s + s = 42$

(334) Solve these equations.

1. ${}^{-}x = 8$
2. ${}^{-}t = 0$
3. $2 \cdot y + {}^{-}y = 14$
4. $6 \cdot w + {}^{-}w = 30$
5. $5 \cdot x + {}^{-}x = 0$
6. $12 \cdot x + {}^{-}x = 121$
7. $24 = {}^{-}y + {}^{-}7 \cdot y$
8. $40 = {}^{-}3 \cdot y + {}^{-}y$
9. $8 \cdot t + {}^{-}t = 35$
10. ${}^{-}6 \cdot u + {}^{-}u = {}^{-}14$

(336) Solve these equations.

1. $2 \cdot x - 5 = 5$
2. $5 \cdot y - 8 = 22$
3. ${}^{-}3 \cdot v - 10 = {}^{-}25$
4. ${}^{-}t - 5 \cdot t = 18$
5. ${}^{-}12 \cdot t - t = 39$
6. $25 \cdot t - 16 = 34$
7. $2 \cdot t - 5 \cdot t = 12$
8. $15 = 4 \cdot w - w$
9. $100 = {}^{-}5 \cdot s - {}^{-}s$
10. $10 \cdot x - x = 36$

(338) Solve these equations.

1. $2 \cdot x = x + 8$
2. $3 \cdot y = y + {}^{-}4$
3. $7 \cdot y + {}^{-}15 = 2 \cdot y$
4. $3 \cdot x = {}^{-}3 \cdot x + 18$
5. $4 \cdot w = 2 \cdot w + 24$
6. $w = 5 \cdot w + 100$
7. $7 \cdot u = 6 \cdot u + {}^{-}3$
8. $8 \cdot x = 2 \cdot x - 24$
9. $3 \cdot v + 24 = {}^{-}5 \cdot v$
10. $7 \cdot t + 1 = 6 \cdot t$

(1) Solve these problems.

1. Jan bought a glass of milk for 25¢, a tuna-salad sandwich for 45¢, and a cup of soup for 25¢. How much change did she receive from a dollar?

2. Bill made 8 baskets and 5 free throws in a game. Each basket is worth two points and each free throw is worth one. How many points did Bill score?

3. The Best Bakery Company is packaging cupcakes. There are 6 cupcakes to a box. How many boxes are necessary to package 1,848 cupcakes?

4. Mrs. Green has 150 sheets of paper. She wishes to distribute them evenly among her 30 students. How many sheets of paper will each student get?

5. Karen earned $2 babysitting Monday night, $4 Thursday, and $5 Saturday. How much did she earn?

(25) Solve these problems.

1. Mr. Reynolds went fishing one week-end. He caught a $4\frac{1}{2}$-pound bass, a 5-pound bass, a $3\frac{1}{2}$-pound bass, and a 5-pound pike. How many pounds of fish did he catch?

2. Mr. Johnson needs 5 shelves, each $2\frac{3}{4}$ feet long. How many 8-foot boards will he need to cut all five shelves?

3. Mrs. Kelly's baby weighed $8\frac{1}{2}$ pounds when he was born. When he was six weeks old, he weighed $9\frac{1}{4}$ pounds. How much weight did the baby gain?

4. On his paper route, David walks $1\frac{3}{4}$ miles each day except Sunday. On Sunday, he walks $2\frac{1}{2}$ miles. How many miles does he walk per week?

(83) Solve these problems.

1. Mr. Wright ordered 250,000 campaign buttons at \$.01 each. How much will they cost?

2. Joan wants to buy new ice skates which cost \$24.00. She has \$15.75. How much more does she need?

3. Robert sold 15 boxes of candy at \$1.25 per box. Gary and Steve each sold 20 boxes. How much money did they receive for their sales?

4. José bought six table tennis balls for \$1.98. How much did each ball cost?

5. Mrs. Matthews bought the following items at a grocery store.

   2 lb green beans at \$.29 per lb
   6 cans dog food at 2 cans for \$.45
   3 lb tomatoes at \$.39 per lb
   1 head lettuce at \$.25 per head
   1 doz oranges at 6 for \$.59
   2 lb ground beef at \$.89 per lb
   1 gal ice cream at \$1.19 per gal

   Find Mrs. Matthews total grocery bill.

(113) Solve these problems.

1. Bert estimated that on school days he spends 25% of his time in school. How many hours a day does he spend in school?

2. Myron answered 18 questions correctly on a test of 20 questions. What per cent did he answer correctly?

3. On Monday, 20% of the 350 students at Green Junior High were absent. How many students were absent?

4. Susan received a grade of 80% on a test of 40 questions. How many questions did she miss?

(113) Solve these problems.

1. Ken won the election for president of the student body. He received 55% of the votes. There are 720 students. How many votes did Ken receive?

2. Find the interest on $50 at 8% for 6 months.

3. How much is $1,000 worth at the end of 6 months, invested at 12% compounded quarterly?

4. Jane bought a dress on sale at a 15% discount. The dress was marked $20. What was the sale price?

5. Linda is an apprentice beautician. She earns a salary of $40 per week plus 20% of her fees over $100. One week her fees totaled $300. Find her week's pay.

(147) Solve these problems.

1. The temperature of a solution changed from $^{-}5^{\circ}$ to 20°. What was the difference in temperature?

2. Holli bought two 8-bottle cartons of soda and three 6-bottle cartons. How many bottles did she buy in all?

3. The temperature on January 11 was $^{-}12^{\circ}$ in Butte, Montana and 78° in Tampa, Florida. What was the difference in the temperature?

4. A scuba diver made a dive to $^{-}21$ ft. The next dive was to $^{-}38$ ft. How much deeper was the second dive?

5. The following temperatures for one week were recorded at 1:00 P.M. daily: 5°, $^{-}2^{\circ}$, 6°, 3°, $^{-}10^{\circ}$, $^{-}1^{\circ}$, $^{-}2^{\circ}$. Find the average temperature for the week.

6. Mr. Robinson had a bank balance of $150.13. He made deposits of $100 and $225. He wrote checks for $75, $25.50, $150, $12.75, and $22.80. What was his new balance?

(177) Solve these problems.

1. The national budget one year was 6.2×10^9 dollars. Write standard notation for this.

2. A blueprint called for a measurement to be 1.5×10^{-2} inches. Mr. Brown's measurement for it was 1.5×10^{-1}. How much was the measurement off?

3. The budget for Tops Company last year was 1.2×10^5 dollars. This year the budget is 2×10^5 dollars.

   a. How much did the budget increase?

   b. Write standard notation for your answer.

4. The North Star is approximately 2,400,000,000,000,000 miles from the earth. Write scientific notation for this distance.

(203) Solve these problems.

1. Find the area of a square whose perimeter is 48 ft.

2. Donald wishes to put a fence around his small garden. The garden is rectangular and measures 30 feet by 20 feet. How many feet of fencing will he need?

3. Mr. Smith, a contractor, estimates that he uses 5 split blocks per square foot. He wishes to build a wall 25 feet long and 6 feet high. How many split blocks will he need?

4. Mr. Preston has a lawn sprinkler which throws water a maximum of 15 feet. The sprinkler rotates in a complete circle. How many square feet of lawn can be sprinkled without moving the sprinkler? Use 3.14 for π.

5. How many cubic yards of concrete are needed to pour a sidewalk that is 81 feet long and 4 feet wide? The concrete is to be 4 inches thick.

(235) Solve these problems.

1. A photographer schedules one-half hour for each portrait. He works an eight-hour day, six days a week. How many portraits can be scheduled each week?

2. How many 3-pound boxes of cookies can be packaged from 237 pounds of cookies?

3. Jim drove 291.1 miles. His car used 20.5 gallons of gasoline. How many miles per gallon did it average?

4. Mrs. Scott wants her new clothesline pole to be six feet above the ground. The part below the ground must be at least $\frac{1}{4}$ as long as that above the ground. What length pole is needed?

5. The change in the stock market index was $^{-}.2$ for five consecutive days. How much did the index change in all during that time?

(287) Solve these problems.

1. The courtyard at Wood Junior High is rectangular and measures 40 feet by 30 feet. Find the length of a walk which lies diagonally across the courtyard.

2. A guy wire is attached to a pole 8 feet from the ground. The wire is attached to a stake that is 5 feet from the base of the pole. How long is the wire? Give your answer to the nearest tenth.

3. The base of a 12 foot ladder is placed 7 feet from a building. The top of the ladder rests on the building. How far up the building does the ladder reach? Give your answer to the nearest tenth.

4. Find the perimeter of a triangular flower garden shaped like a right triangle. The legs of the triangle each measure 6 feet. Give your answer to the nearest tenth.

(313) Solve these problems.

1. Maureen bought a sweater marked \$16 on sale for \$12. What was the rate of discount?

2. Mr. Kirby receives a commission of 12% of his sales. How much is his commission on a sale of \$1,000?

3. The price of bananas increased from 15¢ per pound to 18¢. What was the per cent of increase?

4. The bus fare was recently raised from 32¢ to 40¢. What was the per cent of increase?

5. Mrs. Brady paid an interest rate of 10% on \$1,000 for 6 months. How much interest did she pay?

6. John paid \$2.70 for a baseball at a 10% discount sale. What was the original price of the baseball?

(345) Solve these problems.

1. What is the probability that a die, when tossed, will land with an odd number up?

2. Two die are tossed. What is the probability that they will both land with a 5 up?

3. There are 16 red and 20 blue marbles in a box. What is the probability of drawing a white marble?

4. There are 25 blue and 30 yellow pencils in a box. What is the probability of drawing a blue one?

5. The following temperatures were recorded on Tuesday: $^-8°$, $^-6°$, $^-4°$, 0°, 2°, 8°, 9°, 13°, 13°, 14°, 14°, 14°. Find the mean temperature recorded.

6. Sue took five arithmetic tests during one marking period. Her scores were 85, 88, 90, 88, and 87. What were the mean, median, and mode of her scores?

Tables of Measure

LENGTH

12 inches (in.) = 1 foot (ft)
3 feet = 1 yard (yd)
1 rod = $16\frac{1}{2}$ feet
5,280 feet = 1 mile (mi)
1 kilometer (km) = 1,000 meters (m)
1 hectometer (hm) = 100 meters
1 dekameter (dkm) = 10 meters
10 decimeters (dm) = 1 meter
100 centimeters (cm) = 1 meter
1,000 millimeters (mm) = 1 meter
1 inch ≈ 2.5 centimeters
1 centimeter ≈ $\frac{3}{8}$ inch

WEIGHT OR MASS

1 ton (T) = 2,000 pounds (lb)
1 pound = 16 ounces (oz)
1 ounce = 437.5 grains (gr)
1 kilogram = 1,000 grams (gm)
1 pound ≈ 454 grams

CAPACITY

1 gallon (gal) = 4 quarts (qt)
1 quart = 2 pints (pt)
1 pint = 2 cups (c)
1 pint = 16 fluid ounces (fl oz)
1 liter (l) = 1,000 milliliters (ml)
1 gallon ≈ $3\frac{3}{4}$ liters (l)

TIME

1 day = 24 hours (hr)
1 hour = 60 minutes (min)
1 minute = 60 seconds (sec)
1 week (wk) = 7 days
1 year (yr) = $365\frac{1}{4}$ days
1 decade = 10 years
1 century = 100 years

SQUARES AND SQUARE ROOTS

n	n^2	$\sqrt{n}$	n	n^2	$\sqrt{n}$
1	1	1.000	**51**	2,601	7.141
2	4	1.414	**52**	2,704	7.211
3	9	1.732	**53**	2,809	7.280
4	16	2.000	**54**	2,916	7.348
5	25	2.236	**55**	3,025	7.416
6	36	2.449	**56**	3,136	7.483
7	49	2.646	**57**	3,249	7.550
8	64	2.828	**58**	3,364	7.616
9	81	3.000	**59**	3,481	7.681
10	100	3.162	**60**	3,600	7.746
11	121	3.317	**61**	3,721	7.810
12	144	3.464	**62**	3,844	7.874
13	169	3.606	**63**	3,969	7.937
14	196	3.742	**64**	4,096	8.000
15	225	3.873	**65**	4,225	8.062
16	256	4.000	**66**	4,356	8.124
17	289	4.123	**67**	4,489	8.185
18	324	4.243	**68**	4,624	8.246
19	361	4.359	**69**	4,761	8.307
20	400	4.472	**70**	4,900	8.367
21	441	4.583	**71**	5,041	8.426
22	484	4.690	**72**	5,184	8.485
23	529	4.796	**73**	5,329	8.544
24	576	4.899	**74**	5,476	8.602
25	625	5.000	**75**	5,625	8.660
26	676	5.099	**76**	5,776	8.718
27	729	5.196	**77**	5,929	8.775
28	784	5.292	**78**	6,084	8.832
29	841	5.385	**79**	6,241	8.888
30	900	5.477	**80**	6,400	8.944
31	961	5.568	**81**	6,561	9.000
32	1,024	5.657	**82**	6,724	9.055
33	1,089	5.745	**83**	6,889	9.110
34	1,156	5.831	**84**	7,056	9.165
35	1,225	5.916	**85**	7,225	9.220
36	1,296	6.000	**86**	7,396	9.274
37	1,369	6.083	**87**	7,569	9.327
38	1,444	6.164	**88**	7,744	9.381
39	1,521	6.245	**89**	7,921	9.434
40	1,600	6.325	**90**	8,100	9.487
41	1,681	6.403	**91**	8,281	9.539
42	1,764	6.481	**92**	8,464	9.592
43	1,849	6.557	**93**	8,649	9.644
44	1,936	6.633	**94**	8,836	9.695
45	2,025	6.708	**95**	9,025	9.747
46	2,116	6.782	**96**	9,216	9.798
47	2,209	6.856	**97**	9,409	9.849
48	2,304	6.928	**98**	9,604	9.899
49	2,401	7.000	**99**	9,801	9.950
50	2,500	7.071	**100**	10,000	10.000

Glossary

This glossary contains an illustration or a brief description of important terms used in this book.

Absolute value On a number line, the distance from a number to 0.

Acute angle An angle whose measure is between 0° and 90°.

Addition principle If $a = b$ is true, then $a + c = b + c$ is also true for any number c.

Adjacent angles Two angles which have a common vertex and a common side between them.

Algorithm A procedure for doing a computation.

Altitude of a triangle A segment from the vertex of a triangle perpendicular to the line containing the opposite side.

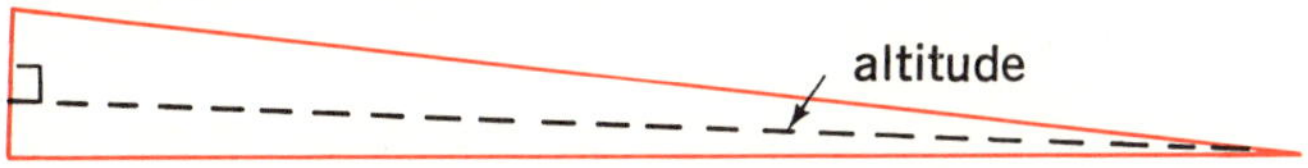

Area The measure of a two-dimensional region.

Associative property (addition) For any numbers a, b, and c, the following is true.

$$(a + b) + c = a + (b + c)$$

Associative property (multiplication) For any numbers a, b, and c, the following is true.

$$(a \cdot b) \cdot c = a \cdot (b \cdot c)$$

Average The number halfway between two numbers is their average.

Base A side or face of a geometric figure.

Bisect To cut in half.

Center of a circle The point in the interior of the circle which is the same distance from every point of the circle.

Central angle An angle whose vertex is the center of a circle.

Circle A set of all points on a plane that are the same distance from a given point.

Circumference The distance around a circle.

Closed Curve A curve that can be traced more than once (in the same direction) without lifting the pencil.

Commission Money paid a salesman. The commission is part of the amount of his sales.

Commutative property (addition) For any numbers a and b, the following is true.

$$a + b = b + a$$

Commutative property (multiplication) For any numbers a and b, the following is true.

$$a \cdot b = b \cdot a$$

Compass An instrument for marking distances. It is also used to draw circles.

Complementary angles Two angles whose measures total 90°.

Composite number A natural number with more than two different factors.

Examples 4, 6, 8, 9, 10

Compound interest Interest earned on previously earned interest as well as on the original principal.

Compound probability Probability calculated on the basis of two or more events.

Computer A device for doing calculations.

Cone A three-dimensional figure with a circle for its base and a curved lateral surface.

Congruent Figures are congruent if they can be made to look exactly alike by reflection or rotation or both.

Coordinates An ordered pair of numbers which tells the location of a point with respect to two perpendicular number lines.

Curve A set of points that can be drawn without lifting the pencil.

Cylinder A three-dimensional figure with two parallel congruent bases. If the bases are circles, it is a circular cylinder.

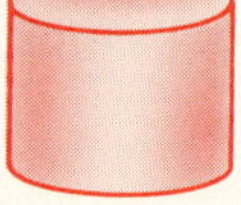

Decagon A polygon with ten sides.

Decimal numeral A base-ten numeral.

Decimal system The base-ten system of numeration.

Degree A unit of measure for angles.

Denominator The number named by the numeral below the fraction bar in a fractional numeral.

Diagonal A segment, other than a side, joining two vertices of a geometric figure. $\overline{AC}$ is a diagonal.

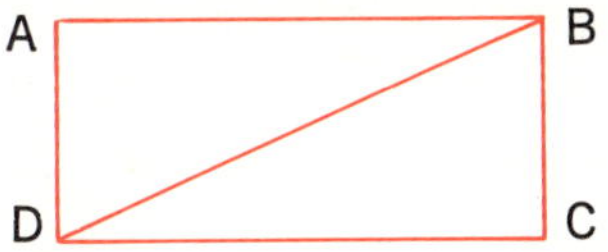

Diameter A segment which contains the center of the circle, and whose endpoints are on the circle.

Discount The amount of money that is subtracted from the marked price of an article to obtain the sale price.

Distributive property For any numbers a, b, and c, the following is true.

$$a \cdot (b + c) = (a \cdot b) + (a \cdot c)$$

Dividend The number being divided. In $10 \div 2 = 5$, the dividend is 10.

Divisible A number is divisible by another number if the latter number is a factor of the first.

Divisor The number by which another is being divided. In $10 \div 2 = 5$, the divisor is 2.

Dodecagon A polygon with 12 sides.

Equation A number sentence with = for its verb.

Equivalent fractional numerals Fractional numerals that name the same number.

Equilateral triangle A triangle whose three sides have the same length.

Estimate To find an approximate answer.

Even number A number that has the number 2 as a factor.

Expanded numeral A numeral which shows the meaning of a decimal numeral or another base numeral.
Example $257 = (2 \cdot 100) + (5 \cdot 10) + (7 \cdot 1)$

Exponent A number which tells how many times a number is used as a factor. In 5^3 the exponent is 3.

Exponential notation A kind of numeral in which exponent symbols are used.
Examples 5^6, 9^8, 7^{-3}, 3^0

Factor To factor a number is to find a factorization of the number.

Factor (of a number) A divisor of that number.

Factorization An expression that shows factors of a number. A factorization of 28 is $4 \cdot 7$.

Flow chart A chart which outlines the steps in a procedure.

Fraction Any number of arithmetic that is not a whole number.

Fractional numeral A numeral made up of two numerals with a horizontal bar between them, such as $\frac{5}{8}$ or $\frac{8.32}{.07}$.

Geometric figure Any set of points.

Graph A way to picture number information.

Greatest common factor The greatest factor common to two or more numbers. The greatest common factor of 20 and 25 is 5.

Heptagon A polygon having seven sides.

Hexagon A polygon having six sides.

Hypotenuse The longest side of a right triangle.

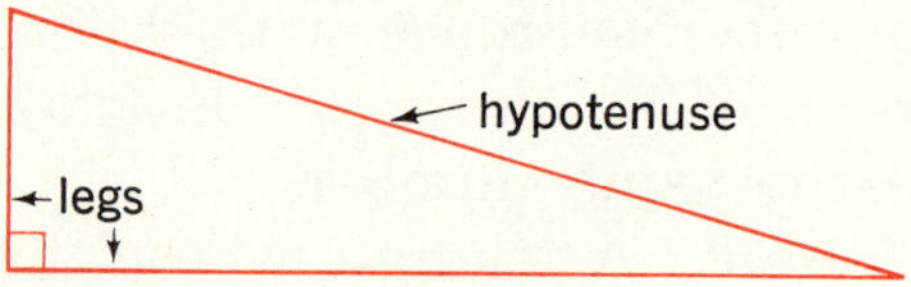

Inequality Any number sentence which uses $\neq$, $<$, or $>$.

Input Information that goes into a computer.

Integers This is the set of integers:

$$\{\ldots\ ^-6,\ ^-5,\ ^-4,\ ^-3,\ ^-2,\ ^-1,\ 0,\ 1,\ 2,\ 3,\ 4,\ 5,\ 6,\ \ldots\}$$

Interest Money paid for the use of money.

Inverse of an integer The integer which when added to the original integer will give 0.

Irrational number A number which can be named by a non-terminating, non-repeating decimal.

Isosceles triangle A triangle having two or more sides the same length.

Lateral area All of the area of a cylinder, prism, or cone except its base(s).

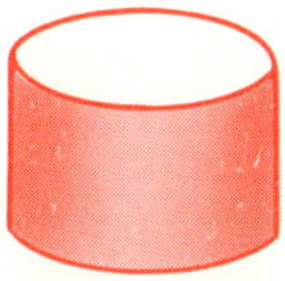

Least common denominator The least common multiple of two or more denominators.

Least common multiple The smallest natural number that is a multiple common to two or more numbers.

Line A straight path that goes on without end in both directions. A line has no endpoints.

Marked price The original price before a discount.

Mean A number found by adding statistical data and dividing the sum by the number of addends. The mean is also called the average.

Measure The number assigned to something so that its size may be compared with others of the same kind.

Median Arrange statistical data in ascending or descending order. Then pick the middle number.

Example 53 72 86 95 98

↑ median (86)

Mixed numeral A numeral which contains a standard numeral for a whole number and a fractional numeral for a number less than one. $1\frac{3}{4}$ is a mixed numeral.

Mode The number which occurs most often in statistical data.

Multiple (of a number) A number that has that number as a factor.

Multiplication principle If $a = b$ is true, then $a \cdot c = b \cdot c$ is also true for any number c.

Natural numbers The set of numbers $\{1, 2, 3, \ldots\}$. They are used for counting.

Negative integer An integer less than zero.

Nonagon A polygon having nine sides.

Nomograph A graph used for computing.

Number line A line, used in graphing, in which points are associated with numbers.

Numbers of arithmetic The set of numbers that can be named by fractional numerals with whole number numerators and natural number denominators. Whole numbers and fractions are numbers of arithmetic.

Numeral A name of a number.

Numerator The number named by the numeral above the fraction bar of a fractional numeral.

Obtuse angle An angle whose measure is between 90° and 180°.

Octagon A polygon having eight sides.

Odd Number A number that does not have 2 as a factor.

One-dimensional Having length only.

Operation For two members of a set, an operation is a way of finding a third member of the set, associated with the first two. There must be at most one answer.

Ordered pair A pair of numbers for which one is first and the other second. In the ordered pair of numbers (3,6) the 3 is first and the 6 is second.

$$(3,6) \neq (6,3)$$

Origin The point of intersection of the two axes of a graph. Its coordinates are (0,0).

Output Information given by a computer.

Proportion A statement (usually an equation) that says two ratios are equivalent.

Parallel lines Lines on a plane which do not intersect.

Parallelogram A quadrilateral with two pairs of parallel sides.

Pentagon A polygon having five sides.

Per cent May be thought of as the ratio of some number to 100.

'8 per cent,' or '8%,' means '$8 \times \frac{1}{100}$,' or '$8 \times .01$.'

Perimeter The sum of the measures of the sides of a polygon.

Perpendicular lines Lines which form right angles.

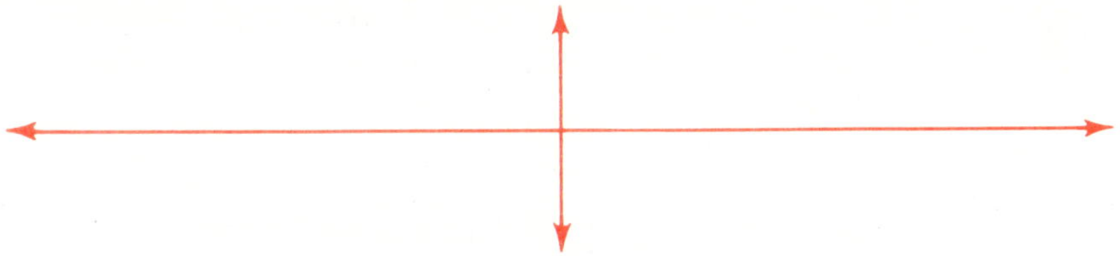

Plane A flat surface with no end.

Polygon A closed curve made up of segments.

Positive integer An integer greater than zero.

Power A number which can be named using exponential notation. For example, 8 is the third power of 2 because 8 can be named 2^3.

Prime factorization A factorization which shows only prime numbers.

Prime number A natural number with exactly two different factors.

Examples 2, 3, 5, 7, 11, 13

Principal The amount of money on which interest is to be paid.

Prism A three-dimensional figure with two parallel congruent bases which are polygons.

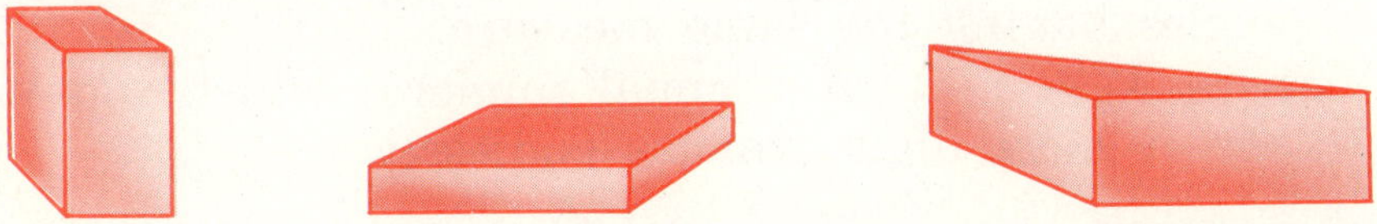

Probability The ratio of the number of favorable possibilities to the total number of possibilities.

Protractor An instrument for measuring angles.

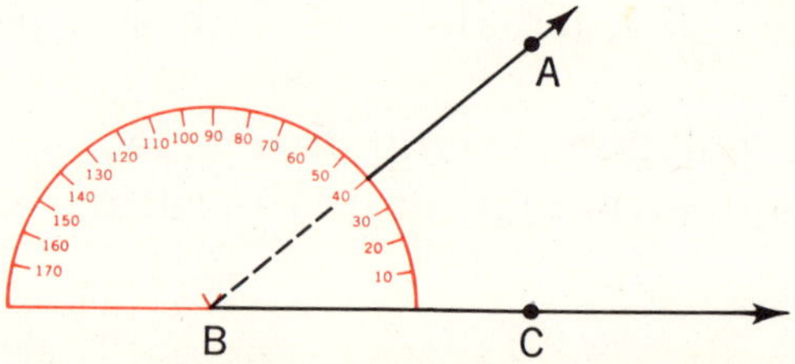

Pythagorean property of right triangles If a and b are measures of the legs and c is the measure of the hypotenuse of a right triangle, then $a^2 + b^2 = c^2$.

Quadrilateral A polygon having four sides.

Quarterly Occurring every 3 months.

Quotient The result of dividing.

Radius A segment having the center of a circle as one endpoint and a point of the circle as the other endpoint.

Rational numbers The set of numbers of arithmetic and their additive inverses.

Ratio A comparison of two numbers by division. The quotient of two numbers.

Ray A straight path that goes on without end in one direction. A ray has one endpoint.

Real numbers The rational and irrational numbers.

Reciprocal If the product of two numbers is 1, the numbers are reciprocals of each other. The reciprocal of $\frac{2}{3}$ is $\frac{3}{2}$.

Rectangle A parallelogram with four right angles.

Reflection A mirror image of a figure.

Region A closed figure and its inside.

Regular polygon A polygon with all sides and all angles having the same measure.

Repeating decimal A decimal numeral which repeats one or more digit symbols endlessly.

Example $\frac{1}{3} = .333\ldots$ or $.3\overline{3}$

Rhombus A parallelogram with all sides the same length.

Right angle An angle whose measure is 90°.

Right triangle A triangle with one right angle.

Rotation The turning of a figure.

Sale price The difference between the marked price and the discount.

Scalene triangle A triangle with all three sides of different length.

Scientific notation A kind of numeral using exponential notation.

Examples 10^4 $\quad 2 \cdot 3 \times 10^{-3}$

Segment A set of points made up of two given points and all the points between them.

Similar figures Figures which have the same shape. Corresponding angles are congruent and corresponding sides have the same ratio.

Simplest fractional numeral A fractional numeral with 1 as the greatest common factor of the numerator and denominator.

Simplify To write the simplest numeral.

Solve To find all of the solutions of a sentence.

Solution A number which makes a number sentence true.

Sphere A set of all points in space the same distance from a given point.

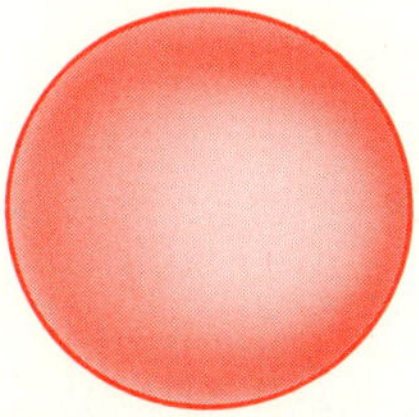

Square A rectangle with all sides the same length.

Square root of a number A number which when multiplied by itself gives that number.

Example $\sqrt{9} = 3$, because $3 \cdot 3 = 9$

Straight angle An angle whose measure is 180°.

Straightedge An instrument for drawing segments.

Subunit A unit which is made by dividing a larger unit into parts.

Example $\frac{1}{4}$-inch is a subunit of the inch.

Surface area The total area of the surface of a three-dimensional figure.

Supplementary angles Two angles whose measures total 180°.

Three-dimensional Having length, width, and height.

Transversal A line which intersects two other lines in the same plane.

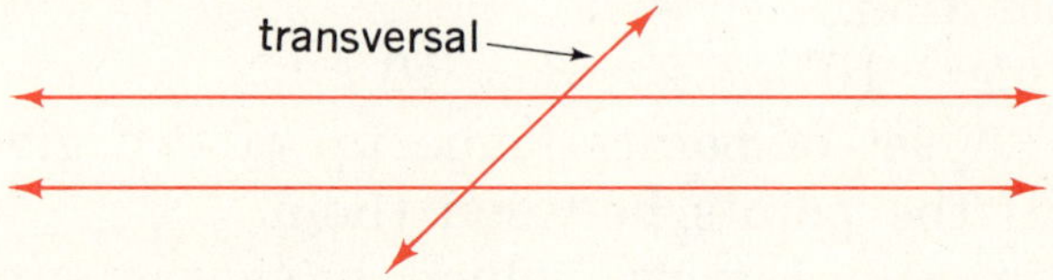

Trapezoid A quadrilateral with at least one pair of sides parallel.

Triangle A polygon having three sides.

Two-dimensional Have length and width only.

Variable A symbol which holds a place for a numeral in a number sentence.

Vertex An endpoint which is common to two sides of a polygon or angle.

Vertical angles Opposite angles formed by two lines that meet.

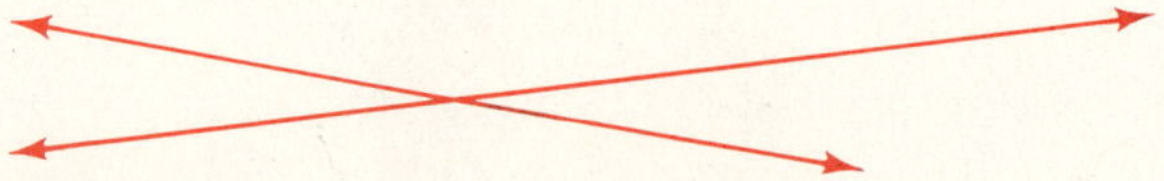

Volume The measure of a three-dimensional figure and its inside.

Weight The amount of force on an object due to the earth's gravity.

Whole number A natural number or zero: {0, 1, 2, 3, . . .}.

Index